РОСТ КРИСТАЛЛОВ

ROST KRISTALLOV

GROWTH OF CRYSTALS

VOLUME 16

Growth of Crystals

Volume 16

Edited by

Kh. S. Bagdasarov and É. L. Lube
Institute of Crystallography
Academy of Sciences of the USSR, Moscow

Translated by

Dennis W. Wester

CONSULTANTS BUREAU • NEW YORK AND LONDON

The Library of Congress cataloged the first volume of this title as follows:

Growth of crystals. v. [1]
   New York, Consultants Bureau, 1958 –

    v. illus., diagrs. 28 cm.
   Vols. 1, 3 –   constitute reports of 1st –   Conference on Crystal
Growth, 1956 –   v. 2 contains interim reports between the 1st and
2nd Conference on Crystal Growth, Institute of Crystallography,
Academy of Sciences, USSR.

   "Authorized translation from the Russian" (varies slightly)
   Editors: 1958-   A. V. Shubnikov and N. N. Sheftal.

   1. Crystal – Growth. I. Shubnikov, Aleksei Vasil'evich, ed. II.
Sheftal', N. N., ed. III. Consultants Bureau Enterprises, inc., New
York, IV. Soveshchanie po rostu kristallov. V. Akademiia nauk
SSSR. Institut kristallografii.
QD921.R633   548.5   58-1212

ISBN 0-306-18116-9

The original Russian text was published for the Institute of Crystallography of the
Academy of Sciences of the USSR by Nauka Press in Moscow in 1986

PREFACE

Papers from the Sixth All-Union Conference on Growth of Crystals comprise Volume 16 of this series. The articles were chosen with a view to more fully elucidate the basic problems of crystal growth as reflected in domestic and foreign reviews and in original studies. This volume consists of six parts.

Part I is devoted to mechanisms of crystal growth that are important for production of materials with given properties. This part examines the temporal evolution of an inhomogeneous state and the array of semicellular and eutectic structures during microstructure formation, the effect of impurity on the nonequilibrium vacancy concentration in a growing crystal, and the role of soluble and insoluble impurities in the birth and growth of crystals.

Part II deals with the synthesis and electrophysical properties of novel solid electrolytes that are promising for practical use, analysis and correlation of the large amount of data on growth by the Bridgman—Stockbarger method of single crystals of fluorite phases far from stoichiometry, and the hydrothermal chemistry and growth of hexagonal germanium dioxide.

Crystallization of films is of great importance in microelectronics. Part III gives a review of production methods of layers by electroliquid epitaxy and a discussion of questions on the epitaxy of semiconductors and metallic and dielectric films from molecular beams, the analytical methods used, and the properties of the films that are formed. Relations between growth conditions, structure, and cathodoluminescence of ZnSe on GaAs are studied. Kinetic relationships during gallium arsenide surface doping using gas-phase epitaxy are examined. Finally, results on photostimulated growth of doped PbTe are analyzed.

Crystallization of biological substances is treated in Part IV. Efforts in this direction have recently been accelerated. Conversion of biological molecules into the crystalline state is necessary for structural studies. The results obtained are then used in genetic engineering, medicine, pharmacology, biosynthesis, and other areas. Purification methods of the starting compounds can be applied to the crystallization of other organic and inorganic materials from solution. We hope to draw the attention of theoreticians to the mechanisms of formation of crystals composed of very large molecules.

The development of experimental methods for studying crystallization processes is dealt with in Part V. These include an investigation of melt structure and crystallization processes of various substances at high temperature using combined light scattering and direct observation of changes in acoustic emission of the actual crystal structure during growth from the melt. A review of contemporary electron-microscopic methods as applied to defect structure of semiconductors is also given.

The difficulties of direct scrutiny of crystallization processes are well known.   Therefore, mathematical modeling methods are very effective for simulation of these processes. Part VI includes a review of mathematical modeling of crystallization from the melt, liquid- and gas-phase epitaxy, and simulation of melt hydrodynamics during Bridgman crystallization with accelerated crucible rotation. The influence of electromagnetic methods on hydrodynamics and heat and mass transfer in crystals grown from the melt is discussed.

We are indebted to coworkers S. N. Ryadnov, D. D. Perlov, and M. Yu. Muromskii of the Institute of Crystallography, Academy of Sciences of the USSR, for their help in preparing this volume.

Kh. S. Bagdasarov

É. L. Lube

CONTENTS

Part I

MECHANISMS OF CRYSTAL GROWTH
FROM THE MELT

EVOLUTION AND SELECTION DURING GROWTH OF SEMICELLULAR AND EUTECTIC PATTERNS

E. A. Brener, M. B. Geilikman, and E. D. Temkin

Formation of microstructure during production remains a central theme of materials science. The problem of pattern selection is characteristic of a whole range of nonequilibrium systems. Growth of dendrites and snowflakes, as well as eutectic and cellular crystallization [1, 2], besides such well-known examples as Rayleigh–Benard thermal convection, are currently being investigated widely. These nonlinear systems, despite their differences, have a common feature. The initial homogeneous state becomes unstable at a certain critical value of a definite controlling parameter and a continuous array of heterogeneous states becomes possible. Undercooling of the initial liquid phase or a temperature gradient (at a fixed crystallization rate) is the controlling parameter in the case at hand. The question of the temporal evolution of the heterogeneous state and, ultimately, the selection of that pattern which is in fact adopted arises. This work examines that problem relative to enlargement of the grain structure and selection of the eutectic structure during crystallization from the melt.

1. GRAIN STRUCTURE ENLARGEMENT

We now examine enlargement of the cellular structure during crystallization of a new growing phase. We will examine such enlargement only with respect to processes that occur at the crystallization front. Cellular interfaces in the bulk of the new phase are considered to be frozen. First, the cellular pattern can arise during growth of an ordered crystal in which the cells are distinguished by their orientation. Second, cells can be single crystalline regions, or grains, which are disoriented relative to each other due to growth initiated at various centers or dislocations which join into a small-angled interface. In either of these cases, the cellular interfaces have additional surface energy. Therefore, an enlargement of the pattern which lowers the system energy should occur with time. Qualitatively, this mechanism leads to the following. Since the cellular or block interfaces have surface energy, a local distortion of the crystallization front (Fig. 1) is required at the ternary points for maintenance of equilibrium conditions. The local temperature at the front and the orientation of the block interfaces relative to the growth direction will be different for blocks of various sizes. For this reason, the block sizes change with time. Blocks that have sizes larger than some average value are increasing while those with a smaller size are decreasing. The latter eventually collapse and the scale of the characteristic pattern enlarges with time.

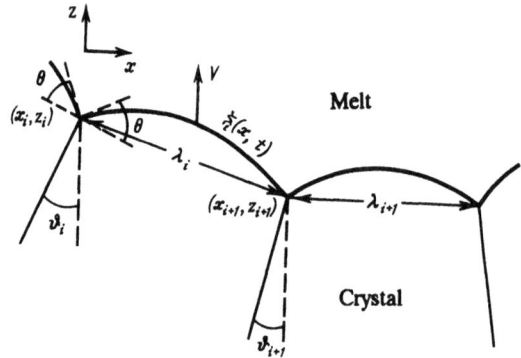

Fig. 1. Irregular semicellular pattern.

1.1. Crystallization of an Irregular Grain Pattern

We now examine in more detail crystal growth from a thin melt during directed crystallization at a rate V with a constant temperature gradient G. A certain block pattern (Fig. 1) exists near the crystallization front. If the kinetic processes at the interface occur rapidly enough, local equilibrium is achieved at the advancing crystallization front

$$G\zeta + \gamma K = 0. \tag{1}$$

Here ζ is calculated from the point where the temperature is equal to the melting point,

$$K = \frac{\partial}{\partial x}\left[\partial\zeta/\partial x \Big/ \sqrt{1 + (\partial\zeta/\partial x)^2}\,\right],$$

where K is the local curvature of the front and γ is a constant proportional to the interface energy. Calculation of the final kinetic rate at the crystallization front as well as examination of isothermal growth were carried out in [3]. A continuous crystallization front and local thermodynamic equilibrium should be preserved at the ternary points

$$z_i = \zeta_i(x_i) = \zeta_{i-1}(x_i). \tag{2}$$

This means that the angle θ between the tangent to the front and the perpendicular to the block interface is fixed at this point (Fig. 1). Neglecting surface energy anisotropy, we will consider this angle to be constant for all ternary points. Then, the equilibrium condition at the ternary points can be written as (Fig. 1)

$$\begin{aligned}\partial\zeta_i/\partial x &= \tan(\theta - \vartheta_i),\\ \partial\zeta_{i-1}/\partial x &= -\tan(\theta + \vartheta_i),\end{aligned} \tag{3}$$

where ϑ_i is the angle between the growth direction and the tangent to the block interface at the ith ternary point. We integrate equilibrium equation (1) for the ith block. The interfaces of this block with the neighboring $(i - 1)$ and $(i + 1)$ blocks are inclined at angles of ϑ_i and ϑ_{i+1}, respectively, while the ternary points have the ordinates z_i and z_{i+1}. At small inclination angles between the crystallization front and the block interfaces (i.e., at θ and $\vartheta_i \ll 1$), integrating and taking into account (3) leads to the following relations between ordinates of the ternary points z_i and z_{i+1} and the angles ϑ_i and ϑ_{i+1}:

$$\begin{aligned}z_i &= L_G\left\{\frac{\{2(\theta + \vartheta_{i+1}) + (\theta - \vartheta_i)[\exp(\lambda_i/L_G) + \exp(-\lambda_i/L_G)]\}}{[\exp(-\lambda_i/L_G) - \exp(\lambda_i/L_G)]}\right\},\\[2mm] z_{i+1} &= L_G\left\{\frac{\{2(\theta - \vartheta_i) + (\theta + \vartheta_{i+1})[\exp(\lambda_i/L_G) + \exp(-\lambda_i/L_G)]\}}{[\exp(-\lambda_i/L_G) - \exp(\lambda_i/L_G)]}\right\},\end{aligned} \tag{4}$$

where $L_G = (\gamma/G)^{1/2}$ and λ_i is the size of the ith block (Fig. 1). Equations (4), which can also be written for other blocks, allows the difference between angles $(\vartheta_{i+1} - \vartheta_i)$ to be found. This, in turn, determines the rate of change of block width λ_i: at small angles ϑ_i we have

$$\dot{\lambda}_i = V(\vartheta_{i+1} - \vartheta_i),$$

where V is the rate of advance of the crystallization front. We now examine two limiting cases.

   a. Small Block Sizes, $\lambda_i \ll L_G$ From Eqs. (4) in this case we have

$$z_{i+1} = z_i = -L_G^2 \frac{2\theta + (\vartheta_{i+1} - \vartheta_i)}{\lambda_i} = \text{const} = -L_G^2 \frac{2\theta}{\lambda_0}. \tag{5}$$

The ordinate of the ternary point z_i in this case does not depend on the number i and is constant along the front. The characteristic pattern scale λ_0 is included in the right side of (5). This has the following physical meaning. In a regular pattern with block width λ_0 (in such a pattern, all angles $\vartheta_i = 0$), the ternary points are located at the same level as in the examined irregular pattern. Solving (5), we find

$$\dot{\lambda}_i \simeq V(\vartheta_{i+1} - \vartheta_i) = 2V\theta(\lambda_i/\lambda_0 - 1). \tag{6}$$

   b. Large Block Sizes, $\lambda_i \gg L_G$ From Eqs. (4) and the continuity condition (2) we obtain

$$\vartheta_i \simeq \theta[\exp(-\lambda_i/L_G) - \exp(-\lambda_{i-1}/L_G)].$$

Therefore,

$$\dot{\lambda}_i = V(\vartheta_{i+1} - \vartheta_i) = V\theta[\exp(-\lambda_{i+1}/L_G) - 2\exp(-\lambda_i/L_G) + \exp(-\lambda_{i-1}/L_G)].$$

In contrast to the previous case, the rate $\dot{\lambda}_i$ depends not only on the local value of λ_i but also on the sizes of neighboring blocks. To find the law that governs enlargement of the block pattern with time, we examine the equation of growth $\dot{\lambda}_i$ at average field where the quantities $\exp(-\lambda_{i+1}/L_G)$ and $\exp(-\lambda_{i-1}/L_G)$ are replaced by their average values defined as

$$\exp(-\lambda_0/L_G) = \langle \exp(-\lambda_i/L_G) \rangle;$$

then

$$\dot{\lambda}_i = 2V\theta[\exp(-\lambda_0/L_G) - \exp(-\lambda_i/L_G)]. \tag{7}$$

Equations (6) and (7) show that blocks that have a size larger than λ_0 continue to grow while blocks with a size smaller than λ_0 are diminished. The latter eventually disappear. Precisely due to this mechanism the number of blocks in the system decreases while the pattern scale λ_0 increases with time. A problem analogous to that of coalescence, examined in [4], arises.

1.2. Law of Pattern Enlargement with Time

   We now introduce the time-dependent distribution function of blocks by size $f(\lambda, t)$. This is defined such that $f(\lambda, t)d\lambda$ is the number of blocks per unit length of front with sizes in the range λ to $\lambda + d\lambda$. This distribution function satisfies the continuity equation

$$\frac{\partial f}{\partial t} + \frac{\partial}{\partial \lambda}(f\dot{\lambda}) = 0, \tag{8}$$

where λ is determined by (6) and (7). The function f is normalized by the condition

$$\int_0^\infty \lambda f(\lambda, t)\, d\lambda = 1. \tag{9}$$

Equation (8) is written for the distribution function f, which is homogeneous along the planar front. (With macroscopic curvature of the front K, an additional factor $\partial(f\dot{S})/\partial S$ appears in (8), where S is the coordinate along the arc of the macrofront and

$$\dot{S} = \int_0^S V_n K\, dS$$

is the rate of change of block coordinates whose interfaces are situated perpendicular to the macrofront which is advancing relative to the normal at a rate V_n.) We now examine (8) for the two limiting cases indicated above.

a. *Small Block Sizes*, $\lambda_i \ll L_G$. In order to detemine the asymptote of the distribution function $f(\lambda, t)$ and the function $\lambda_0(t)$ at large times, we find a self-consistent solution for (6), (8), and (9) of the form

$$f(\lambda, t) = A\,\lambda_0^{-2}(t)\,\varphi(u), \tag{10}$$

where $u = \lambda/\lambda_0(t)$ and A is a constant determined by the normalization condition of (9). Substituting (10) into (8), we find the asymptotic law for the change of $\lambda_0(t)$ and the form of the self-consistent function $\varphi(u)$

$$\lambda_0(t) = 2\theta\, Vt/\nu, \tag{11}$$

$$\varphi(u) = \begin{cases} \exp(-u) & \text{at}\quad \nu = 1, \\[2mm] [u + \nu/(1-\nu)]^{(\nu-2)/(\nu-1)} & \text{at}\quad 0 < \nu < 1. \end{cases} \tag{12}$$

If ν was exponentially small in the initial size distribution of large blocks, then the self-consistent parameter $\nu = 1$. In the case where an integrated exponential "tail" with index δ occurs in the initial distribution, the value of ν is determined according to (12) by the relation $\delta = (\nu - 2)/(\nu - 1)$. The average value of the pattern distribution parameter $\langle\lambda\rangle = \lambda_0$. Thus, with small block sizes $\langle\lambda\rangle \ll L_G$, enlargement occurs linearly with time.

b. *Large Block Sizes*, $\lambda_i \gg L_G$. We introduce the dimensionless variables $\tau = 2V\theta t/L_G$ and $\tilde{\lambda} = \lambda/L_G$. The distribution function f satisfies the continuity equation (8)

$$\frac{\partial f}{\partial \tau} + \frac{\partial}{\partial \lambda}\{f[\exp(-\tilde{\lambda}_0) - \exp(-\tilde{\lambda})]\} = 0 \tag{13}$$

and is normalized by the condition of (9). In the region $\tilde{\lambda} > \tilde{\lambda}_0$ of (13), it is possible to neglect $\exp(-\tilde{\lambda})$ by comparison with $\exp(-\tilde{\lambda}_0)$ (recall that $\tilde{\lambda}_0 \gg 1$). Thus, the general solution of (13) has the form

$$f(\tilde{\lambda}, \tau) = \psi(\tilde{\lambda} - \int \exp[-\tilde{\lambda}_0(\tau)])\, d\tau,\quad \tilde{\lambda} > \tilde{\lambda}_0, \tag{14}$$

and the function ψ is uniquely determined by the form of the initial distribution function at large $\tilde{\lambda}$.

In the region $\tilde{\lambda} < \tilde{\lambda}_0$, neglecting $\exp(-\tilde{\lambda}_0)$ by comparison with $\exp(-\tilde{\lambda})$, we find

$$f(\tilde{\lambda}, \tau) = \exp(\tilde{\lambda})\, g[\tau + \exp(\tilde{\lambda})],\quad \tilde{\lambda} < \tilde{\lambda}_0, \tag{15}$$

where g is an arbitrary function. We determine this function from the intersection of (14) and (15) at $\tilde{\lambda} = \tilde{\lambda}_0$ and obtain an approximate solution by this.

From the normalization condition of (9) it is apparent that the distribution function at large $\tilde{\lambda}$ should fall rather quickly: either exponentially or in an exponential manner, $\tilde{\lambda}^{-\delta}$, $\delta > 2$. It can be shown that since the integral in (9) is independent of time in both cases, the argument of the ψ function in (14) at $\widetilde{\lambda} = \widetilde{\lambda}_0$ is dependent on time at the main approximate asymptote at large times τ. This condition gives the law

$$\widetilde{\lambda}_0 (\tau) = \ln \tau. \qquad (16)$$

Actually, $\psi(x) \sim e^{-\alpha x}$ for exponential decay of the distribution function at large $\widetilde{\lambda}$. From the intersection of (14) and (15) at $\widetilde{\lambda} = \widetilde{\lambda}_0$ we find

$$g(\tau) \exp (\widetilde{\lambda}_0) = \exp \{ -\alpha [\widetilde{\lambda}_0 - \int \exp (-\widetilde{\lambda}_0) \, d\tau]\} \; ,$$

and for the normalized integral we obtain

$$\int_0^\infty \widetilde{\lambda} f(\widetilde{\lambda}, \tau) \, d\widetilde{\lambda} \sim \widetilde{\lambda}_0 (\tau) \exp \{ -\alpha [\widetilde{\lambda}_0 - \int \exp (-\widetilde{\lambda}_0) \, d\tau]\} \; . \qquad (17)$$

The law (16) makes the exponential index in (17) independent of time. In the following approximation, it is necessary to accept, for both contraction and time independence of the preexponential, that

$$\widetilde{\lambda}_0 - \int \exp (-\widetilde{\lambda}_0) \, d\tau \simeq \frac{1}{\alpha} \ln \ln \tau \ll \ln \tau, \qquad (18)$$

i.e.,

$$\widetilde{\lambda}_0 \simeq \ln \tau + (a \ln \tau)^{-1} \; .$$

For exponential decay of the distribution function at large $\widetilde{\lambda}$ we have $\psi(x) \sim x^{-\delta}$ ($\delta > 2$) and instead of (18) we obtain

$$\widetilde{\lambda}_0 - \int \exp (-\widetilde{\lambda}_0) \, d\tau \simeq (\ln \tau)^{1/(\delta - 1)} \ll \ln \tau.$$

Thus, at $\lambda_0 \gg L_G$, the pattern scale at the asymptote of large times is enlarged logarithmically and slowly according to the law

$$\lambda_0(t) = L_G \ln (2 \, V \theta \, t / L_G). \qquad (19)$$

In conclusion of this section, we provide several numerical estimates. The following parameter values are taken as characteristic: $\gamma \approx 10^{-4}$ deg·cm, $G \approx 10^2$ deg·cm, $V \approx 10^{-3}$ cm/sec, and $\theta \approx 10^{-1}$. Then the characteristic value $L_G \approx 10^{-3}$ cm. While the characteristic pattern scale $\lambda_0 < L_G$, enlargement of the blocks is described by the law (11) with $\lambda_0 \sim V \theta t$. For time $t \sim L_G/V\theta \sim 10$ sec, the average block size reaches the value L_G (the crystal grows to a length $L \sim Vt \sim 10^{-2}$ cm with this). Afterwards, the enlargement law changes logarithmically according to (19). With this law, the characteristic block size is $\lambda_0 \sim L_G \ln(L\theta/L_G) \sim 10 L_G \sim 10^{-2}$ cm in a crystal which has grown to a length $L \approx 10^2$ cm. Clearly, from the estimates given, the enlargement of the pattern occurs mainly by a weakly logarithmic law. The macroscopic bulging of the crystallization front toward the melt (this could be due to an assortment of thermal conditions) can significantly accelerate block enlargement. Actually, the curvature of the macrofront K contributes additionally to the rate of change of block sizes λ. According to (7), $\lambda \sim V\theta \exp(-\lambda/L_G)$ with logarithmic growth. Therefore, with $\lambda K \gtrsim \theta \exp(-\lambda/L_G)$, i.e., with $\lambda \gtrsim L_G \ln(\theta/L_G K)$, pattern enlargement will occur not by a logarithmic but by an exponential law

$$\lambda_0 \sim L_G \ln (\theta/L_G K) \exp (V K t). \qquad (20)$$

The length of the crystal which grows upon transition to the law (20) is $L \sim K^{-1}$. We assume that upon growth of a crystal with a cross section of ~ 10 cm the curvature is of the order of the inverse cross-sectional size of the crystal, $K \approx 10^{-1}$ cm. Then, the block pattern disappears in effect at $\lambda_0 K \approx 1$. From (20) we find that this occurs for a crystal length

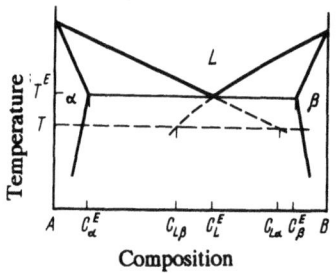

Fig. 2. Eutectic phase diagram.

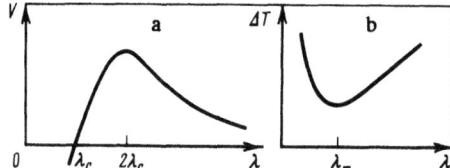

Fig. 3. Growth rate and undercooling at the crystallization front as functions of interlamellar distance upon crystallization of a regular eutectic. Isothermal (a) and directed (b) crystallization.

$$L \sim K^{-1} \ln (1/L_G K) \sim 10^2 \text{ cm}.$$

This mechanism of pattern enlargement should eventually lead to complete disappearance of all blocks. However, some final remaining block density is observed experimentally. This can be related to block generation processes which are not considered in this work as well as to the existence of special low-energy interfaces which are located along the growth axis. Besides this, the final block density can arise from a concave crystallization front. A stationary state arises due to the fact that the growth of the average size blocks examined above compensates for the decrease of this dimension due to the concave front. At small curvatures $L_G K \ll \theta$

$$\dot{\lambda}_0 \sim -\lambda_0 VK + V \theta \exp (-\lambda_0/L_G) = 0,$$

whence

$$\lambda_0 \sim L_G \ln (\theta/L_G K),$$

i.e., the stationary block size is on the order of L_G.

2. EUTECTIC PATTERN CRYSTALLIZATION

The pattern selection problem also arises for eutectic crystallization in a system with the phase diagram depicted in Fig. 2. The theory of stationary eutectic growth [5, 6] gives only the relation between the growth rate and the interlamellar distance λ of a regular pattern

$$V \sim \frac{\Delta T}{\lambda} (1 - \lambda_c (\Delta T)/\lambda) \tag{21}$$

or, if concerned with crystallization in a temperature gradient at a fixed rate V, between the λ parameter and undercooling at the crystallization front ΔT (Fig. 3). Therefore, additional concepts are needed for finding the λ at which the eutectic grows in reality.

As was noted by Cahn and discussed in [5], stationary growth is unstable at $\lambda < 2\lambda_c$, where $\lambda = 2\lambda_c$ is the maximum on the curve of (21) for $V(\lambda)$. On the other hand, as was shown by Jackson and Hunt [5], instability of a second type appears at sufficiently large λ. The crystallization front of the lamella collapses,

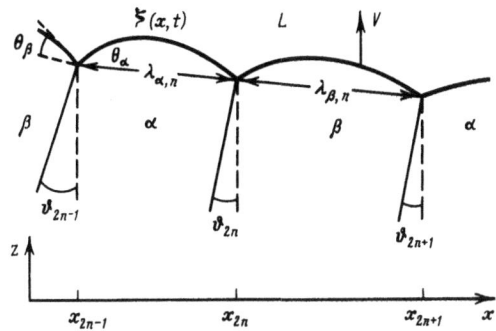

Fig. 4. Irregular eutectic pattern.

forming deep pockets as a result of which generation of lamellae of the second solid phase becomes possible. The hypothesis of marginal growth was then formulated, confirming that growth occurs under conditions corresponding to borderline stability. In [7] it was confirmed that $\lambda = 2\lambda_c$. This corresponds to the first type of instability and agrees with the earlier assumption of Zener [6] about the rate maximum (or the undercooling minimum).

It was proposed in [8] that λ corresponds to the instability limit relative to generation of new lamellae, i.e., instability of the second type. The criterion of maximal entropy production as well as some additional concepts [9] give a different answer: $\lambda = 3\lambda_c$, which corresponds to an inflection point on the $V(\lambda)$ curve. Finally, a sequential approach to the selection problem was formulated in [10] and consists of the following.

Fluctuations always exist in the real system. These will lead to decreased or increased thickness of some of the lamellae at the front. Instability at small λ arises for an eventual decreased lamellar thickness, which is followed by collapse of the lamellae. The pattern enlarges as a result of this. Conversely, instability of the second type sets in with increased lamellar thickness. This corresponds to collapse of the front with formation of pockets. This causes generation of new lamellae and, consequently, a division of the initial lamella. This process decreases the average interlamellar distance. The stationary pattern that arises as a result of selection corresponds to a dynamic equilibrium between these lamellar generation and collapse processes. Thus, fluctuations in the system and events of lamellar division and coalescence are responsible for the presence at the crystallization front of lamellae of different sizes, i.e., the eutectic pattern becomes irregular. Theories of stationary growth [5, 6] are insufficient in this situation. A dynamic equation that describes the evolution of an irregular structure is required. The question of the evolution and selection of a eutectic structure is multifaceted. Effective interaction between lamellae is achieved through a diffusional field in the melt. Construction of simple but physically meaningful models is expedient due to the complexity of this problem. Therefore, a simplified model was used for description of the dynamics of the eutectic pattern in [10]. Evolution of the pattern in this model occurs as a result of bending of the crystallization macrofront, which takes place at the ternary points of the lamellae with the melt. Perpendicularity of the lamellae to the macrofront is postulated in this case. This assertion has not yet been verified.

We now examine the situation in which the additional assumption of perpendicularity of the lamellar macrofront is not required, namely: If the surface energy of the melt—solid interface is large by comparison with the surface energy between solid phases, then, as will be evident below, bending of the macrofront need not be considered. Evolution and, consequently, pattern selection occurs at the macroscopically planar front. This is physically justified, for example, in a eutectic system with isomorphous solid phases. If the compositions of these phases are mutually similar, then the surface energy of their interface, which depends on the difference in compositions, is small.

2.1. Irregular Eutectic Growth Equation at Long Wavelengths

We examine isothermal crystallization of a thin film in a eutectic system with the phase diagram shown in Fig. 2. The structure of the solid phase, which consists of alternating α- and β-lamellae, is shown in Fig. 4. The growth equations should be described for lamellae of each phase:

$$\frac{d\lambda_{\alpha,n}}{dt} = V(\vartheta_{2n} - \vartheta_{2n-1}), \quad \frac{d\lambda_{\beta,n}}{dt} = V(\vartheta_{2n+1} - \vartheta_{2n}), \tag{22}$$

where V is the rate of the crystallization front and we consider the lamellar turn angles, defined in Fig. 4, to be small: $\vartheta_i \ll 1$.

In order to find the ϑ_i angles, we write the local equilibrium condition for each of the lamella with the melt:

$$C_L - C_{L\alpha} - \Gamma_\alpha K = 0 \quad \text{for } \alpha\text{-lamella}, \tag{23a}$$

$$C_L - C_{L\beta} + \Gamma_\beta K = 0 \quad \text{for } \beta\text{-lamellae}, \tag{23b}$$

where C_L is the melt composition at the front, K is the local front curvature, and $C_{L\alpha}$ and $C_{L\beta}$ are equilibrium compositions at extended liquidus lines (Fig. 2). The capillary lengths Γ_α and Γ_β are proportional to the surface energies at the solid–melt interfaces.

Integrating the local equilibrium condition (23) along the front of each lamella, at small θ and ϑ_i we find

$$\vartheta_{2n} - \vartheta_{2n-1} = -2\theta_\alpha - \frac{1}{\Gamma_\alpha} \int_{\lambda_{\alpha,n}} [C_L(x) - C_{L\alpha}] \, dx, \tag{24}$$

$$\vartheta_{2n+1} - \vartheta_{2n} = -2\theta_\beta + \frac{1}{\Gamma_\beta} \int_{\lambda_{\beta,n}} [C_L(x) - C_{L\beta}] \, dx,$$

where θ_α and θ_β are the edge angles at the ternary point (Fig. 4).

Solution of the melt diffusion problem is required for exact integration over the lamellae using (24). The concentration field in the melt $C_L(x, z)$ can be treated in a quasistationary approximation, i.e., it is assumed to satisfy the stationary equation

$$\left(\frac{\partial^2}{\partial x^2} + \frac{\partial^2}{\partial z^2} + \frac{V}{D} \frac{\partial}{\partial z} \right) C_L = 0, \quad C_L(x, \infty) = C_\infty, \tag{25}$$

where D is the melt diffusion coefficient and C_∞ is the initial melt composition. Linear analysis of eutectic pattern disturbances, which was carried out in [11], can serve as a basis for the quasistationary approximation. The macroscopic crystallization front (the front which occurs at the ternary points) in this approximation is in fact planar and advances with a constant velocity V. As it turns out, with lamellar sizes which are in fact constant, a stationary macrofront shape can be established at small θ_α and θ_β values at the portion of the front which includes a large number of lamellae $n \sim 1/\theta \gg 1$. This macrofront deviates little from planarity. We will write the mass balance condition at the crystallization front so that it is analogous to that which was found for a regular eutectic [5], considering the front to be planar:

$$\frac{\partial C_L}{\partial z} \doteq \frac{V}{D}(C_s(x) - C_L(x)) \quad \text{at} \quad z = 0, \tag{26}$$

where C_s is the crystal composition, equal to C_α^E in the α-lamellae and C_β^E in the β-lamellae.

Solution of the diffusional problem (25) and (26) can be represented in the form

$$C_L = (\bar{C} - C_\infty) \exp(-Vz/D) + \tilde{C}_L, \tag{27}$$

where $\bar{C}_L$ is the average composition along the crystallization front. At small undercoolings, correspondingly small growth rates V, and not excessively large heterogeneous pattern scales $l(Vl/D << 1)$, the stationary diffusion equation of $\Delta \tilde{C}_L$ can be used for finding $\tilde{C}_L$

$$\tilde{C}_L (x, z = 0) = \frac{V}{\pi D} (C_\beta^E - C_\alpha^E) \int_{-\infty}^{\infty} g(x') \ln |x - x| dx'. \tag{28}$$

The function $g(x)$ depends on the actual succession of α- and β-lamellae along the front:

$$g(x) = \begin{cases} g_\alpha \equiv -(1 - \eta_0) & \text{for } \alpha\text{-lamellae,} \\ g_\beta \equiv \eta_0 & \text{for } \beta\text{-lamellae.} \end{cases} \tag{29}$$

Here η_0 is the average fraction of the α-phase which is found from the obvious relation

$$\eta_0 C_\alpha^E + (1 - \eta_0) C_\beta^E = C_\infty , \tag{30}$$

which reflects the coincidence of the solid and melt compositions under a stationary growth regime.

We will carry out further calculations considering that distortion of the eutectic pattern is long-waved, i.e., the characteristic hetergeneous scale $l >> \lambda$. Obviously, distortion of the pattern at the first moment after division or coalescence of lamellae is not long-waved. On the other hand, generation and collapse of lamellae results in a decrease or increase of the eutectic pattern period, i.e., finally to long-waved pattern changes. In the case where $\lambda_{S,n}$ ($S = \alpha, \beta$) is a slowly changing function of n (or the spatial x coordinate), we obtain $g(x)$ from (29) as the sum $g(x) = \bar{g}(x) + \tilde{g}(x)$, where

$$\bar{g}(x) = (g_a \lambda_{\alpha,n} + g_\beta \lambda_{\beta,n})/\lambda_n \equiv - [\eta(x) - \eta_0], \tag{31a}$$

$$\tilde{g}(x) = \begin{cases} -[1 - \eta(x)] & \text{for } \alpha\text{-lamellae,} \\ \eta(x) & \text{for } \beta\text{-lamellae} \end{cases} \tag{31b}$$

and $\eta(x) = \lambda_a(x)/\lambda(x)$. The function $\bar{g}(x)$ is the $g(x)$ value averaged over the λ_n period and slowly changes with the x coordinate. The function $\tilde{g}(x)$ changes sharply upon transition from α- to β-lamellae, but its average over the period λ_n is zero. The concentration $\tilde{C}_L(x)$ from (28) accounting for (31) takes the form

$$\tilde{C}_L(x) = \frac{V(C_\beta^E - C_\alpha^E)}{\pi D} \{ \int_{\infty}^{\infty} dx' \tilde{g}(x') \ln |x-x'| - \int_{\infty}^{\infty} dx' [\eta(x') - \eta_0] \ln |x-x'| \}. \tag{32}$$

If the small terms that contain derivatives of $\lambda_S(x)$ are neglected, then the first term of (32) upon integration of (24) will give the same result as in the regular eutectic but with local values of $\lambda_S(x)$. With regard to the second term in (32), it changes substantially at distances on the order of the heterogeneity scale $l >> \lambda$, and upon integration over λ_S it can be considered constant. Calculating the integral in (24) and substituting it into (22), we obtain

$$\dot{\lambda}_{S,n} = V \left\{ -2\theta_S \pm \frac{1}{\Gamma_S} \left[\lambda_{S,n} (\bar{C}_L - C_{LS}) \pm \frac{V(C_\beta^E - C_\alpha^E)}{D} \lambda^2(x) P(\eta) \right. \right.$$

$$\left. \left. - \lambda_S(x) \frac{V(C_\beta^E - C_\alpha^E)}{\pi D} \int_{-\infty}^{\infty} dx' [\eta(x') - \eta_0] \ln |x-x'| \right] \right\}, S = \alpha, \beta, \tag{33}$$

where

$$P(\eta) = \sum_{k=1}^{\infty} \frac{\sin^2 \pi k \eta}{(\pi k)^3} ;$$

the upper index relates to $S = \alpha$, the lower to $S = \beta$. Equation (33) shows that significant delocalization arises with a deviation of the fraction of the $\eta(x)$ phases from the average value η_0. The presence of such delocalization causes long-waved disturbance of $\eta(x)$ to relax quickly to its average value η_0. It is possible to set $\eta = \eta_0$ for description of the changes of λ in (33). This can be confirmed by examination of (33) with linear deviations from the average λ_0 and η_0 values, where λ_0 is the average of the $\lambda_n = \lambda_{\alpha,n} + \lambda_{\beta,n}$ value at the front. Substituting into (33) the variables $\Delta\eta = \eta - \eta_0$ and $\Delta\lambda = \lambda - \lambda_0$ in the form

$$\Delta\eta(x,t) = e^{-\omega t} \int_{-\infty}^{\infty} \frac{dk}{2\pi} e^{ikx} \Delta\eta_k, \quad \Delta\lambda(xt) = e^{-\omega t} \int_{\infty}^{\infty} \frac{dk}{2\pi} e^{ikx} \Delta\lambda_k,$$

we obtain

$$\omega_\lambda \simeq \frac{\lambda_0 V^2 (C_\beta^E - C_\alpha^E) P(\eta_0)}{D(\Gamma_\alpha + \Gamma_\beta)\eta_0(1-\eta_0)} \frac{\lambda_0 - 2\lambda_c}{\lambda_0 - \lambda_c}, \tag{34a}$$

$$\omega_\eta \simeq \frac{1}{|k|} \frac{V^2(C_\beta^E - C_\alpha^E)\eta_0(1-\eta_0)(\Gamma_\alpha + \Gamma_\beta)}{D\Gamma_\alpha\Gamma_\beta} \tag{34b}$$

From (34a), the eutectic pattern is seen to be unstable at $\lambda < 2\lambda_c$, where $\lambda = 2\lambda_c$ is the maximum on the curve of $V(\lambda)$ (Fig. 3a) which describes the growth rate of the regular eutectic [5]:

$$V(\lambda) = \frac{D}{\lambda} \frac{(C_{L\alpha} - C_{L\beta})(1 - \lambda_c/\lambda)\eta_0(1-\eta_0)}{(C_\beta^E - C_\alpha^E)P(\eta_0)}, \tag{35a}$$

$$\lambda_c = \left(\frac{2\theta_\alpha\Gamma_\alpha}{\eta_0} + \frac{2\theta_\beta\Gamma_\beta}{1-\eta_0}\right)/(C_{L\alpha} - C_{L\beta}). \tag{35b}$$

Considering that the characteristic time for relaxation of η to its average value is significantly smaller than the characteristic time for a change of λ ($\omega_\lambda/\omega_\eta \sim \lambda|k| \ll 1$), it is possible to set $\eta = \eta_0$ at the long-wave limit in (33). Thus, one independent equation remains which describes the relaxation $\lambda(x) = \lambda_\alpha(x) + \lambda_\beta(x)$. Using the condition that the whole macrofront length must be conserved

$$\frac{d}{dt} \sum_n (\lambda_{\alpha,n} + \lambda_{\beta,n}) = 0$$

and the condition that the average composition upon transition from the liquid to solid phase must be maintained

$$\sum_n [C_2^E \lambda_{\alpha,n} + C_\beta^E \lambda_{\beta,n} - C_\infty(\lambda_{\alpha,n} + \lambda_{\beta,n})] = 0,$$

we can eliminate $\bar{C}_L$ and transform (33) can be transformed into

$$\frac{d\Lambda}{dt} = \frac{V(C_{L\alpha} - C_{L\beta})}{\Gamma_\alpha + \Gamma_\beta} \left[-\frac{\Lambda_0 - 1}{\langle\Lambda^2\rangle} \Lambda^2 + \Lambda - 1\right], \tag{36a}$$

where

$$V = V(\Lambda_0)\Lambda_0^2/\langle\Lambda^2\rangle; \quad \Lambda = \lambda/\lambda_c; \tag{36b}$$

the symbol $\langle...\rangle$ denotes averaging along the front and $V(\Lambda_0)$ is determined by (35a). When the dispersion Λ is small and $\langle\Lambda^2\rangle \simeq \Lambda_0^2$, we obtain

$$\frac{d\Lambda}{dt} = \frac{V(C_{L\alpha} - C_{L\beta})}{\Gamma_\alpha + \Gamma_\beta} \frac{\Lambda_0 - 1}{\Lambda_0^2} (\Lambda - \Lambda_0)\left(\frac{\Lambda_0}{\Lambda_0 - 1} - \Lambda\right) \tag{36c}$$

and $V = V(\Lambda_0)$ is that for a regular eutectic. Equation (36c) has two stationary points at $\Lambda = \Lambda_0$ and at $\Lambda = \Lambda_0/(\Lambda_0 - 1)$. At $\Lambda_0 > 2$, the stationary point is stable while the second stationary point is unstable.

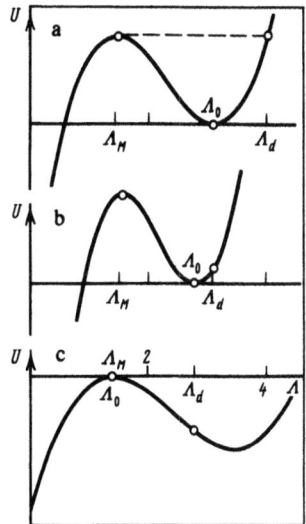

Fig. 5. Potential $U(\Lambda)$ calculated from the equation $\dot{\Lambda} = -dU/d\Lambda$. $\Lambda_d/2 > \Lambda_M$ (a), $\Lambda_d/2 = \Lambda_M$ (b), and $\Lambda_d/2 = \Lambda_M = \Lambda_0$ (c). For a and b, the average pattern period Λ_0 coincides with the potential minimum, and for c, with the maximum.

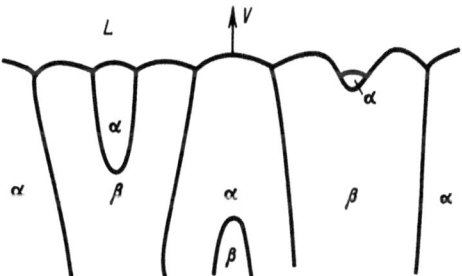

Fig. 6. Portion of the crystallization front showing collapse of β-lamellae and generation of α-lamellae.

2.2. Pattern Selection

To describe fluctuations and generation and collapse processes of lamellae, we introduce the lamellar size distribution function $f(\lambda, t)$, which obeys the Fokker–Planck equation with the source

$$\frac{\partial f}{\partial t} + \frac{\partial}{\partial \lambda}\left[\dot{\lambda}f - B\,\frac{\partial f}{\partial \lambda}\right] = I(\lambda), \tag{37}$$

where it is necessary to substitute (36) obtained above for $\dot{\lambda}$. The distribution function is defined such that $fd\lambda$ is the number of lamellae between λ to $\lambda + d\lambda$ per unit length of front. It is normalized by the natural condition

$$\int_0^\infty \lambda f d\lambda = 1. \tag{38}$$

In (37), B is the lamellar diffusion coefficient by size, which is assumed to be dependent on λ. The source $I(\lambda)$ arises as a result of lamellar division. Let $W(\lambda', \lambda'')$ be the probability of a lamella with size λ' dividing into two lamellae of sizes λ'' and $\lambda' - \lambda''$ in unit time. Then

$$I(\lambda) = \int_{\lambda}^{\infty} f(\lambda') \, W(\lambda', \lambda) \, d\lambda' - f(\lambda) \int_{0}^{\lambda/2} W(\lambda, \lambda'') \, d\lambda''. \tag{39}$$

In [10], a model in which the lamella, attaining a certain critical size λ_d, divides into halves was used. Then

$$I(\lambda) = 2j_d \delta \, (\lambda - \lambda_d/2), \tag{40}$$

where j_d is the flux of lamellae at $\lambda = \lambda_d$. Such critical division can be related to instability of the crystallization front at large lamellar sizes.

We now examine the stationary solution of (37) with the source $I(\lambda)$ from (40) at limiting conditions $f(0) = f(\lambda_d) = 0$:

$$f(\lambda) = \frac{j_d}{B} e^{-U(\lambda)/B} \int_{0}^{\lambda} e^{U(\lambda')/B} d\lambda' \quad \text{at} \quad \lambda < \lambda_d/2,$$

$$f(\lambda) = \frac{j_d}{B} e^{-U(\lambda)/B} \int_{\lambda}^{\lambda_d} e^{U(\lambda')/B} d\lambda' \quad \text{at} \quad \lambda > \lambda_d/2, \tag{41}$$

where $U(\lambda)$, defined as $U(\lambda) = -\int \dot{\lambda} d\lambda$, is the potential, which has extrema coinciding with the stationary points of (36) (Fig. 5). The constant j_d is determined from the normalization condition of (38), and the continuity condition $f(\lambda)$ at $\lambda = \lambda_d/2$ allows the value λ_0 to be determined. The potential $U(\lambda)$ depends parametrically on the λ_0 value.

2.2.1. Low Noise Level

At the limit $B \to 0$, the continuity condition $f(\lambda)$ at $\lambda = \lambda_d/2$ can be written as

$$U(\lambda_-) = U(\lambda_+), \tag{42}$$

where λ_- and λ_+ correspond to the largest potential values in the regions $\lambda < \lambda_d/2$ and $\lambda > \lambda_d/2$. Two possibilities exist:

1. $\lambda_d/2 > \lambda_M$, where λ_M is the potential coordinate maximum (Fig. 5a). Then $\lambda_+ = \lambda_d$, $\lambda_- = \lambda_M$, and (42) gives the solution

$$\Lambda_0 = \frac{1}{3} \, [\Lambda_d + 2 + \sqrt{\Lambda_d^2 - 2\Lambda_d + 4}], \tag{43}$$

which exists at $\Lambda_d \geq 16/5$.

2. $\lambda_d/2 = \lambda_M$. Then $\lambda_- = \lambda_d/2$, $U(\lambda_d) < U(\lambda_d/2)$, and therefore $\lambda_+ = \lambda_d/2$. This situation is realized at $\Lambda_d \leq 16/5$. The quantity Λ_0 is determined from the equation $\Lambda_M = \Lambda_d/2$. Two cases are possible upon solution of this equation:

$$\Lambda_0 = \Lambda_d/(\Lambda_d - 2) \quad \text{at} \quad 3 \leqslant \Lambda_d \leqslant 16/5, \tag{44}$$

$$\Lambda_0 = \Lambda_d/2 \quad \text{at} \quad 2 \leqslant \Lambda_d < 3. \tag{45}$$

Solution (44) corresponds to that where Λ_0 coincides with the potential minimum U (Fig. 5b), while for solution (45), Λ_0 coincides with the potential maximum U (Fig. 5c).

Thus, the Λ_0 value found at $\Lambda_d \geq 3$ corresponds to the minimum U (Fig. 5a, b). This means that an actually regular stable structure with period Λ_0 (with a sharp Gaussian distribution function near Λ_0) occurs at

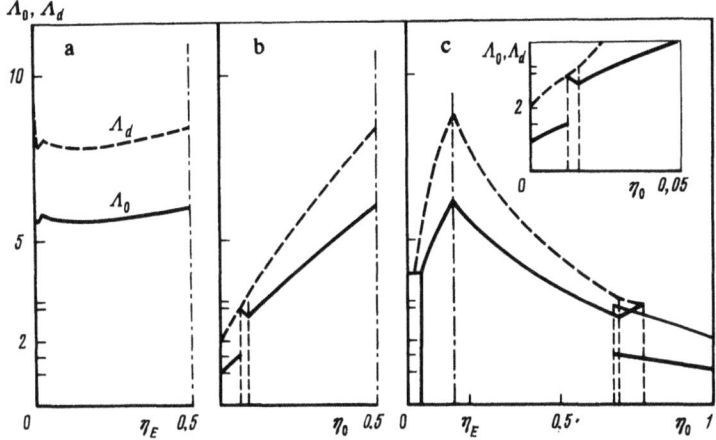

Fig. 7. Average interlamellar distance Λ_0 and distance Λ_d, at which division of lamellae occurs, as a function of fraction of α-phase η_0 in the crystallizing pattern [12]. Results are obtained at $\theta_\alpha = \theta_\beta = \pi/6$. (a) $\eta_0 = \eta_E$ (eutectic melts), α-lamellae are divided over the whole region $0 < \eta_E < 0.5$ and the case shown in Fig. 5a is realized; (b) $\eta_E = 0.5$ (symmetrical eutectic diagram), β-lamellae are divided over the whole region $0 < \eta_0 < 0.5$ and the cases shown in Fig. 5 are realized in the order c, b, a as η_0 is increased; and (c) $\eta_E = 0.2$, β-lamellae are divided over the range $0 < \eta_0 < 0.15$ and α-lamellae at $0.15 < \eta_0 < 1$. Two regions exist in the right part in each of which crystallization of two different patterns is possible.

a low noise level. Division and collapse of lamellae, which maintain the given stationary state, occur exponentially rarely in such a pattern.

At $2 \le \Lambda_d < 3$, the average Λ_0 value corresponds to the potential maximum and according to (45) $1 \le \Lambda_0 < 1.5$. Thus, the value Λ_0 falls in the region $\Lambda < 2$ (or $\lambda < 2\lambda_c$), where stationary growth of a regular structure with this period cannot exist as a result of its instability. However, solution (45) describes crystallization of an irregular chaotic pattern in which dynamic equilibrium between collapse of lamellae which have sizes smaller than average and division of lamellae which have sizes larger than average is maintained when they attain the values of Λ_d. In contrast to the preceding case, processes of division and collapse do not require fluctuational attainment of certain sizes. Frequencies of these processes are not exponentially trivial and the distribution function has an exponential character.

At $B \to 0$, the distribution function $f(\Lambda)$ for solutions (43) and (44) is concentrated near Λ_0. If even a small probability for division of these lamellae exists, they will divide with sizes near Λ_0. This means that $\Lambda_d = \Lambda_0$ should be observed. Thus, from (44), we find

$$\Lambda_0 = 3. \tag{46}$$

This universal solution does not depend on parameters which characterize fluctuation and division of lamellae. It is possible to arrive at this result [12] by solving (37) with the source of general form (39), i.e., by disregarding the model assumption relative to division exactly in halves with size Λ_d [10]. According to (43)-(45), Λ_d, the width into which the lamellae are divided, must be known for a final determination of the period Λ_0. Bulging portions, which facilitate generation of another solid phase (Fig. 6), arise with an increase of the lamellar size at the crystallization front. An estimate of Λ_d was given with the assumption [12] that a point with vertical inclination appears first with this size at the crystallization front. With this assumption, an equation that relates Λ_d to the average period Λ_0 is obtained. Solution of this equation together with (43)-(45) defines Λ_0 and Λ_d as functions of the melt composition (or η_0). Some results of calculations [12] from these formulas are given in Fig. 7. Plots of Λ_0 and Λ_d as functions of η in Fig. 7a and b are symmetric relative to the com-

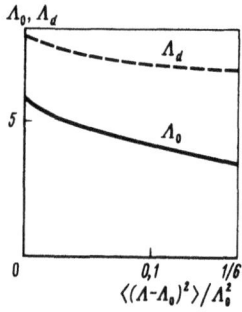

Fig. 8. Plot of Λ_0 and Λ_d as functions of the pattern period dispersion for a eutectic melt with $\eta_0 = \eta_E = 0.38$ [12]. Results were obtained at $\theta_\alpha = \theta_\beta = \pi/6$.

position $\eta = 0.5$. The α-lamellae are divided for eutectic melts with $\eta = \eta_E$ (Fig. 7a) at $0 < \eta \le 0.5$ and the β-lamellae are divided at $0.5 < \eta < 1$. For the symmetrical eutectic $\eta_E = 0.5$ (Fig. 7b), β-lamellae are divided at $0 < \eta \le 0.5$, α-lamellae at $0.5 < \eta < 1$. For the eutectic with $\eta_E = 0.2$ (Fig. 7c), β-lamellae are divided at $0 < \eta \le 0.15$, α-lamellae at $0.15 < \eta < 1$. Figure 7a shows that $\Lambda_0 \approx 6$ for eutectics and depends weakly on the position of the eutectic point.

2.2.2. Effect of Noise Level

Two substantial effects arise at the final noise level. First, all sharp periods (breaks and jumps in Fig. 7) from one growth regime to another are smoothed. Second, a decrease of the average pattern period Λ_0 occurs with an increase of noise or, which is the same, with an increase of the lamellar thickness dispersion. Figure 8 shows results of a calculation of Λ_0 and Λ_d as functions of the pattern dispersion for the eutectic $\eta_0 = \eta_E = 0.38$ (corresponding to the Pb–Sn system). Results of λ_d estimation [12] were also used in this calculation. The term containing λ can be neglected in (37) at high temperatures ($B \gg U$). Then it is easily shown analytically that the dispersion tends toward the final limit $\langle(\Lambda - \Lambda_0)^2\rangle/\Lambda_0^2 = 1/6$, and $\Lambda_0 = \Lambda_d/2$. This is the limiting value, $\Lambda_0 = 3.4$, for the system examined in Fig. 8.

Growth of a eutectic pattern under isothermal conditions was examined in this work. However, all results can in fact be applied to growth in a temperature gradient with the condition that the temperature drop on the λ_0 scales be much smaller than the undercooling required for growth with a given rate V and lamellar size λ_0. This is fulfilled at $\lambda_0 \ll L_G$, where L_G is the characteristic thermal length introduced in (4). In this case, (35a) determines the supersaturation $\Delta C = C_{L\alpha} - C_{L\beta}$ (and together with it the undercooling). The average size of lamellae, which arise as a result of selection processes, during growth in a temperature gradient are more conveniently expressed not through the size λ_c but through the characteristic size, which corresponds to the supersaturation minimum $\Delta C(\lambda)$, from (35a) at a given growth rate V. This transition is derived from the formula

$$\lambda_0/\lambda_m = \sqrt{\lambda_0/\lambda_c - 1.}$$ (47)

Thus, the universal low-temperature solution $\lambda_0 = 3\lambda_c$ in the isothermal case during growth in a temperature gradient has the form

$$\lambda_0 = \lambda_m \sqrt{2.}$$

The completed examination shows that the average Λ_0 value arising as a result of selection, depending on the division model, can lie in a certain range of values. The upper limit of this range corresponds to division of lamellae with $\Lambda = \Lambda_d$, i.e., at the point of stability loss of the lamellar front. The value Λ_0, which corresponds to this upper limit, is given in Fig. 7. In principle, division can occur even at small Λ values as a result of generation processes due to thermal fluctuations. This leads to a decrease of Λ_0. The minimal Λ value is observed in the case where lamellar division occurs not near the point of stability loss, but near the average Λ_0 value. Solution of (46) with $\Lambda_0 = 3$ corresponds to this case. Thus, for example, for the Pb–Sn eutectic at $\eta = \eta_E = 0.38$ we obtain $3 \leq \lambda_0/\lambda_c \leq 5.84$, which gives $2^{1/2} \leq \lambda_0/\lambda_m \leq 2.20$, in agreement with (47). We used data on eutectic crystallization in thin films [13] for comparison of this theoretical result with experiment. Experimental values of λ_0/λ_m have a wide scatter in various experiments and span the range 1.2-2.0. It should be noted that patterns which occur in these experiments are practically without defects. This corresponds to a low level of fluctuations. The selection process in this case can be rather lengthy and the observed patterns do not correspond to the final stationary state. If the selection process does not occur, then the patterns which arise can have periods which lie between two stability limits $1 < \lambda_0/\lambda_m \lesssim 4$ [5, 13].

3. CONCLUSIONS

A new mechanism of grain enlargement during solidification, related to processes at the crystallization front, is proposed and studied. The enlargement process, which is analogous to coalescence, occurs according to the law $\lambda_0 \sim t$ at small grain sizes ($\lambda_0 \ll L_G$) and according to the law $\lambda_0 \sim \ln t$ at $\lambda_0 \gg L_G$. Enlargement may be accelerated, creating a front which protrudes into the melt. In this case, $\lambda_0 \sim e^{at}$. A stationary density of blocks with average size $\lambda_0 \sim L_G$ is established at the protruding front.

A theory for selection of a eutectic pattern, which sequentially considers collapse and generation of new lamellae during crystallization, is developed. Pattern selection leads to the average interlamellar distance λ_0 being found in the range $2^{1/2} < \lambda_0/\lambda_m \lesssim b$, the lower limit of which is realized in systems with facile lamellar generation, the upper limit, in systems with difficult generation. For eutectics, $b \approx 2.2$ and depends weakly on the position of the eutectic point.

REFERENCES

1.   J. S. Langer, "Instabilities and pattern formation in crystal growth," *Rev. Mod. Phys.*, **52**, No. 1, 1-28 (1980).
2.   "Symposium on establishment of microstructural spacing during dendritic and cooperative growth," *Metall. Trans. A*, **15**, No. 6, 961-1063 (1984).
3.   E. A. Brener, M. B. Geilikman, and D. E. Temkin, "Kinetics of block pattern enlargement," *Zh. Éksp. Teor. Fiz.*, **90**, No. 5, 1819-1829 (1986).
4.   I. M. Lifshits and V. V. Slezov, "On the kinetics of diffusional decomposition of supersaturated solid solutions," *Zh. Éksp. Teor. Fiz.*, **35**, No. 2, 479-492 (1958).
5.   K. A. Jackson and J. D. Hunt, "Lamellar and rod eutectic growth," *Trans. Am. Inst. Min., Metall. Pet. Eng.*, **236**, No. 8, 1129-1142 (1966).
6.   J. Christian, *Theory of Transformations in Metals and Alloys*, Pergamon Press, New York (1975).
7.   J. S. Langer, "Eutectic solidification and marginal stability," *Phys. Rev. Lett.*, **44**, No. 15, 1023-1026 (1980).
8.   B. E. Sundquist, "The edgewise growth of pearlite," *Acta Metall.*, **16**, No. 12, 1413-1427 (1968).
9.   J. S. Kirkaldy, "Predicting the pattern of lamellar growth," *Phys. Rev. B*, **30**, No. 12, 6889-6895 (1984).
10. V. Datye, R. Mathur, and J. S. Langer, "Mode selection in a caricature of eutectic solidification," *J. Stat. Phys.*, **29**, No. 1, 1-16 (1982).
11. V. Datye and J. S. Langer, "Stability of thin lamellar eutectic growth," *Phys. Rev. B*, **24**, No. 8, 4155-4169 (1981).
12. E. A. Brener, M. B. Geilikman, and D. E. Temkin, "Eutectic pattern selection," *Zh. Éksp. Teor. Fiz.*, **92**, No. 1, 165-178 (1987).
13. H. E. Cline, "Interlamellar spacing in directionally solidified eutectic thin films," *Metall. Trans. A*, **15**, No. 6, 1013-1017 (1984).

EFFECT OF AN IMPURITY
ON THE NONEQUILIBRIUM VACANCY CONCENTRATION
IN A GROWING CRYSTAL

V. V. Voronkov

1. INTRODUCTION

Vacancies play a significant role in the capture of a substitutive impurity by a growing crystal. First, the impurity can transfer from the medium (melt) into the surface layer of the crystal by filling surface vacancies. The reverse reaction occurs simultaneously, i.e., surface impurities enter the melt, leaving behind vacant sites. Second, the impurity migrates into the subsurface region by a mechanism involving vacancies. This affects the surface impurity concentration, i.e., the extent of impurity exchange between the crystal and melt. Thus, the impurity capture coefficient K (ratio of captured concentration N to the melt impurity concentration N_m) should depend strongly on the vacancy concentration C. If quantity C assumes its equilibrium value C_e, then the coefficient K will be constant at sufficiently small N_m, when it is possible to neglect the existence of clusters of a few impurity atoms. In fact, a distinctly nonlinear capture (i.e., K as a function of N_m) is observed in many cases. This appears at small N_m but disappears with increasing N_m [1, 2]. This effect indicates that the vacancy concentration C is actually nonequilibrium and subject to the impurity effect. The dependence of C on N_m causes a corresponding dependence of K on N_m. When the impurity concentration N becomes sufficiently large by comparison with C, then the vacancies do not even exhibit a noticeable effect on the quantity of captured impurity and K becomes constant. Aside from the dependence of K on concentration, additional evidence exists showing the strong influence of small impurity additions on the vacancy concentration or, more precisely, on the total accumulation of intrinsic point defects. Thus, anomalous changes of density and lattice constants [3] were observed in gallium arsenide crystals. Also, a considerable change of the electrophysical properties was noticed upon introduction of isovalent (i.e., electrically neutral) impurities of boron [4], indium, and antimony [5]. For example, an impurity of boron causes a sharp increase in the concentration of antistructural Ga_{As} acceptors. This is limited by the quantity of interstitial gallium Ga_i [6]. Such an effect can be explained by the lowered initial concentration of gallium vacancies Ga_v (filled by boron atoms) and the corresponding increase of the residual concentration of Ga_i after their recombination with Ga_v.

The interface model will be used below for quantitative description of a nonequilibrium nonlinear vacancy—impurity system. All atomic layers of the crystal except the surface layer at the boundary with the melt are considered equivalent. They are characterized by a constant equilibrium vacancy concentration C_e and a constant equilibrium impurity distribution coefficient K_e. Analogous values of C_{se} and K_{se} for the surface layer, according to model estimates [7], can significantly exceed the corresponding volume values. The crystal growth is considered to be layered, advancing due to tangential movement in steps of monatomic height h. The distribution of steps over the face depends on their generation mechanism and can be equidistant (for a spiral source) or random (for two-dimensional formation of new layers). The equidistant distribution will be examined below since nonlinear vacancy–impurity effects for it are displayed more clearly [1]. Layered growth is characterized by a frequency ω with which the steps proceed through a given point of the face. This frequency is related to the growth rate V by the simple relation $V = \omega h$. As a rule, it has a value $\sim 10^5$ sec^{-1}. The vacancy–impurity ratio is near equilibrium if all frequencies with which the vacancies and impurity atoms are mixed are large in comparison with ω. Otherwise, nonequilibrium effects arise. This work is concerned with these.

2. NONEQUILIBRIUM VACANCIES IN A CRYSTAL WITHOUT IMPURITIES

The advancing step covers a new portion of the crystal surface layer with a definite vacancy concentration C^* which in general differs from the equilibrium value C_{se}. We will consider that vacancy diffusion in the crystal occurs sufficiently quickly and that an almost equilibrium (within each layer) vacancy distribution with concentrations C (in volume layers) and C_s (in the surface layer) is established even at a small distance from the step source. The equilibrium relation $C_s/C = C_{se}/C_e$ is important between adjacent layers. In other words, the ratio of actual and equilibrium concentrations (denoted henceforth as S) is identical for all layers:

$$S = C/C_e = C_s/C_{se}. \tag{1}$$

The vacancy solution can be supersaturated ($S > 1$) or unsaturated ($S < 1$). A deviation from equilibrium causes a surface current of substance Q_s from the melt into the crystal. This is caused by a difference between the forward current of atoms which are filling the surface vacancies and the reverse current of atoms which are exiting from the surface sites into the melt. The current Q_s is proportional to the deviation of C_s from C_{se} and can be written in the form

$$Q_s = \nu^+ q_s (C_s - C_{se}), \tag{2}$$

where ν^+ is the frequency of surface vacancy filling (i.e., the probability of filling per unit time) and q_s is the surface density of lattice sites. The fraction of defect sites is implied by the vacancy (or impurity) concentration in this layer. Therefore, the concentration must be multiplied by the corresponding site density (for example, the number of surface vacancies per unit area equal to $q_s C_s$) for a transition to measurable (surface or volume) values.

The actual vacancy concentration in the crystal would be equal to the concentration C^*, introduced by steps, in the absence of a surface volume current (i.e., at $\nu^+ = 0$). The total vacancy current from the front into the crystal VqC (where q is the volume density of lattice sites) would be equal to the current VqC^* introduced by steps. Actually, the first of these currents is smaller than the second by the value Q_s. Substituting the expression (2) into this conservation of matter law and considering the vacancy ratio (1) and the relation $q_s = qh$ to be constant, we find the S value for the growing crystal:

$$S_0 = \frac{C^* + C_{se} \nu^+/\omega}{C_e + C_{se} \nu^+/\omega}. \tag{3}$$

The index $\langle\langle 0 \rangle\rangle$ denotes the crystal without impurities.

If the filling of the surface vacancies is limited only by the mobility of melt atoms, then the frequency of filling ν^+ would be $\sim 5 \cdot 10^{11}$ sec^{-1} [1]. This greatly exceeds the frequency of step propagation ω so that the vacancy concentration practically does not differ from equilibrium [the vacancy ratio S in (3) is near 1]. In this case, even an impurity cannot affect the vacancy concentration since the impurity vacancy filling provides only a small modification to the intrinsic filling (due to the low concentration of impurity in the melt).

Significant nonequilibrium arises only in the case where surface vacancy filling by intrinsic melt atoms is coupled with overcoming a barrier sufficiently high (more than 1 eV) that the frequency v^+ is lowered to a value $\leq \omega$. Then, even if the steps introduce an equilibrium vacancy concentration ($C^* = C_{se}$), the vacancy solution turns out to be supersaturated since the surface equilibrium concentration C_{se} exceeds the volume value C_e. The actual vacancy concentration C is included in the range from C_e to C_{se} and approaches C_{se} at small v^+/ω, in agreement with (3).

Besides surface vacancy filling, there exists another mechanism which prevents a deviation of C from C_e. That is recombination of vacancies with intrinsic interstitial atoms. In order to evaluate this effect, a diffusion current of vacancies from the front into the crystal volume, which is caused by recombination, must be added to the filling current Q_s. This current is considered negligibly small by comparison with Q_s in this work, i.e., the characteristic length over which recombinational decay of the concentration C occurs is considered sufficiently large.

At a low frequency v^+, the impurity filling of surface vacancies can become more extensive than the intrinsic under the condition that the barrier for impurity filling is much lower (or absent altogether). This leads to a change in the nonequilibrium vacancy concentration C at a sufficiently large impurity concentration in the melt N_m.

3. IMPURITY VACANCY FILLING

The concentration of impurity N^* in new parts of the surface layer, which are created in steps, is proportional to the concentration of impurity in the melt: $N^* = K_s N_m$, where the coefficient of capture by the step K_s in general differs from the equilibrium distribution coefficient K_{se} for the surface layer. In other words, N^* differs from the equilibrium impurity surface concentration $N_{se} = K_{se} N_m$.

The impurity concentration in the surface layer N_s at the initial moment (immediately after step passage) is equal to N^*. Then, N_s changes due to exchange with the melt (with part of the surface vacancies) and diffusional exchange with the bulk crystal. We initially neglect the second process, considering it to be slow (the effect of diffusion will be examined below). The impurity current from the melt into the surface vacancies is equal to $w^+ q_s C_s$, where w^+ is the frequency of impurity vacancy filling which is proportional to the impurity concentration in the melt. The net impurity surface current W_s from the melt into the crystal is equal to the differences of the filling currents and the reverse current. The reverse current is caused by exiting of the surface impurity into the melt. Since W_s is zero at equilibrium between surface impurity concentrations and vacancies, i.e., at $N_s/C_s = N_{se}/C_{se}$, then the following expression is valid for W_s:

$$W_s = w^+ q_s \left(C_s - C_{se} N_s/N_{se} \right). \tag{4}$$

N_s as a function of time is described by the kinetic equation

$$d\left(q_s N_s \right)/dt = W_s, \tag{5}$$

the solution $N_s(t)$ of which exponentially approaches SN_{se}. After an interval of time $\tau = 1/\omega$, the surface layer becomes embedded (assumes a volume state) by the next step and the final impurity concentration in the crystal N is equal to $N_s(\tau)$. Thus, we obtain for the impurity capture coefficient $K = N/N_m$

$$K = K_s \exp\left(-w^-/\omega \right) + SK_{se} \left[1 - \exp\left(-w^-/\omega \right) \right], \tag{6}$$

where $w^- = w^+ C_{se}/N_{se}$. This value, according to (4), represents the transition frequency of the surface impurity atoms into the melt and is constant for a given impurity. The quantity K as a function of the vacancy concentration, i.e., S, is important only under the condition that w^- is not too small by comparison with ω. In particular, if $w^- \gg \omega$, then the impurity concentration SN_{se} is quickly established in the surface layer. This corresponds to equilibrium relative to interface exchange at a given S, in this case $K = SK_{se}$.

We use the law of conservation of total vacancy and impurity concentration for the transition from the $K(S)$ function obtained to a concentration function of $K(N_m)$. The total current of these defects into the crystal $Vq(C + N)$ differs from that current contributed by the steps $Vq(C^* + N^*)$ due to intrinsic filling of surface vacancies, i.e., it is smaller by the value Q_s. Impurity filling of vacancies does not change the total concentration. This conservation law accounting for (1) and (2) produces a linear relation between K and S at a given impurity concentration in the melt N_m,

$$(K - K_s) N_m = (S_0 - S)(C_e + C_{se} v^+/\omega), \tag{7}$$

where S_0 is determined by (3) and corresponds to the crystal without impurities. We find the nonequilibrium vacancy concentration C from the two equation system (6) and (7) as a function of N_m:

$$S = \frac{C}{C_e} = \frac{\varphi K_s N_m + C^* + C_{se} v^+/\omega}{\varphi K_{se} N_m + C_e + C_{se} v^+/\omega}. \tag{8}$$

This formula is a generalization of the previous expression (3). The value $1 - \exp(-w^-/\omega)$ is designated by φ. It is near 1 at rapid impurity exchange. The quantity S as a function of N_m, which is found using (6), resolves the question about the concentration dependence of the capture coefficient K.

According to (8), the vacancy ratio S at increasing N_m is smoothly changed from S_0 to the limiting value $S_\infty = K_s/K_{se}$. We now examine in detail the limiting case of a large impurity concentration. It follows from (7) that the limiting capture coefficient K_∞ coincides with the coefficient of capture by the step K_s. This means that the initial impurity concentration introduced by the steps $N^* = K_s N_m$ becomes so large that it changes little as a result of interfacial transfer with a part of the vacancies so that the impurity surface concentration N_s is near N^*. The surface vacancy concentration C_s due to an increasing frequency of impurity filling w^+ no longer depends on the initial added concentration C^* and is completely controlled by the impurity. The quantity C_s assumes a value corresponding to equilibrium relative to interfacial exchange, i.e., $C_s/N_s = C_{se}/N_{se}$. From here it follows that the value S_∞ is equal to K_s/K_{se}.

The strong dependence of the vacancy concentration C and the capture coefficient K on N_m is important when S_∞ differs greatly from S_0. The case $S_\infty < S_0$ corresponds to filling of initial vacancies by impurity. The coefficient $K(N_m)$ exceeds the limiting value $K_\infty = K_s$ since the impurity entering the surface vacancies from the melt is added to that initially included by the steps. The case $S_\infty > S_0$ corresponds to the generation of new vacancies as a result of expulsion of a part of the included impurity from the surface sites. Therefore, $K(N_m) < K_\infty$. Indeed, the inequality $K(N_m) > K_\infty$ is fulfilled for all known experimental curves. This corresponds to filling of initial vacancies by impurity ($S_\infty < S_0$). In particular, if $S_\infty << S_0$, an intermediate impurity concentration range N_m exists in which the vacancy ratio S is already greatly decreased by comparison with the initial value S_0 but is still large relative to the limiting value S_∞ (i.e., the capture coefficient K is still large by comparison with K_∞). Then it follows from (7) that in the range $S_\infty << S_0$ the impurity concentration in the crystal $N = KN_m$ remains almost constant:

$$N \approx C^* + C_{se} v^+/\omega. \tag{9}$$

This means that the impurity almost completely fills both the vacancies introduced by the steps (with concentration C^*) and the vacancies generated due to exiting of intrinsic surface atoms into the melt (with frequency $v^- = v^+ C_{se}$). Thus, vacancy filling is a basic mechanism of impurity entrance into the crystal (i.e., the impurity concentration introduced by steps remains relatively small). An intermediate plateau on the $N(N_m)$ curve is actually observed [1, 2]. This is strong evidence in favor of the mechanism of impurity filling of surface vacancies which leads to a sharp decrease in their concentration. The inequality $S_\infty << S_0$ (a condition for existence of the intermediate plateau) can be fulfilled due either to extensive supersaturation of the starting vacancy solution ($S_0 >> 1$) or to undersaturation of the impurity solution ($K_s << K_{se}$), i.e., to weak impurity capture by steps. Data for impurities in germanium [1] show a predominance of the second factor. However,

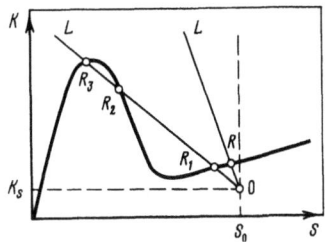

Fig. 1. Nonlinear impurity capture coefficient K as a function of vacancy ratio S caused by diffusional relaxation of embedded impurity.

in other cases extensive filling of vacancies by impurity can be caused by their high nonequilibrium concentration in the initial state ($S_0 \gg 1$).

4. DIFFUSIONAL IMPURITY RELAXATION BY A VACANCY MECHANISM

At the initial moment after passage of the step past a given point of the face, the impurity concentration in the second layer N_2 is equal to $N_s(\tau)$. This was formed in the first (surface) layer at the moment of step arrival and was embedded by this step. When the impurity fills a vacancy, the concentration $N_s(t)$ increases with time. Therefore, the ratio of impurity concentrations in the adjacent layers is $N_s/N_2 < 1$.

Meanwhile, the equilibrium ratio of these concentrations, equal to K_{se}/K_e and designated henceforth by κ, can be large in comparison to 1, i.e., the quantity of impurity in the second layer turns out to be in great excess. Therefore, diffusional relaxation of the embedded impurity begins. It migrates (by a vacancy mechanism) to the surface and then back into the melt. This relaxation was not considered above; however, it can be significant at small growth rates or at high vacancy concentrations. As a result, the impurity concentration finally captured by the crystal N becomes much smaller than the initially embedded concentration $N_s(\tau)$. The vacancy concentration as a function of N_m can also change dramatically. For quantitative illustration of these effects we assume that 1) the surface impurity is in equilibrium with the melt, i.e., $N_s = SN_{se}$; and 2) diffusional exchange between volume layers of the crystal can be neglected. The first assumption, in agreement with the previous section, is valid at $W^- \gg \omega$. The second assumption is based on the surface vacancy concentration being greater than the volume ($C_s \gg C$). Therefore, diffusional jumps of the second layer most often occur into the surface rather than into the adjacent volume layer. If the frequency of impurity jumps from the second layer to the surface under equilibrium conditions is designated f, then in the general case this frequency (proportional to the vacancy concentration) is equal to fS. The total current of the embedded impurity in the surface layer reverts to zero at the equilibrium ratio of concentrations ($N_s/N_2 = \kappa$). Therefore, it can be written in the form

$$W = fS\, q_s\, (N_2 - N_s/\kappa). \tag{10}$$

The quantity N_2 as a function of time is described by a kinetic equation which is analogous to (5), except that it is necessary to replace N_s by N_2 and W_s by W. Solution of $N_2(t)$ exponentially approaches N_s/κ. After the time interval $\tau = 1/\omega$, the next step converts the second layer into the third and relaxation ceases. Therefore, the impurity concentration which is captured by the crystal N is equal to $N_2(\tau)$. For the capture coeffi-

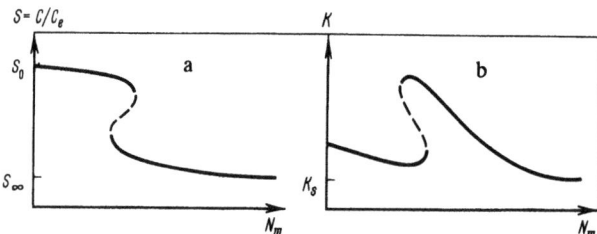

Fig. 2. Example of indeterminate vacancy concentration (a) and impurity capture coefficient (b) as a function of impurity content in the melt.

cient $K = N/N_m$, we obtain

$$K = S K_e \left[1 + (\kappa - 1) \exp\left(-S f / \omega\right)\right]. \tag{11}$$

If the κ parameter (the ratio of equilibrium distribution coefficients for the surface layer and the bulk crystal) exceeds the value $e^2 + 1 \approx 8.4$, then the $K(S)$ function is no longer smooth but passes initially through a maximum and then through a minimum (Fig. 1). The decreasing portion of the $K(S)$ curve is caused by impurity diffusion from the volume to the surface being accelerated with an increase of S (i.e., vacancy concentration). Therefore, although the initially embedded impurity concentration $N_s = SN_{se}$ increases with S, the concentration N maintained in the crystal falls. Diffusional relaxation is insignificant at small S and N remains near N_s (note the initial straight portion in Fig. 1, for which $K = SK_{se}$). At large S, diffusion becomes so rapid that equilibrium between adjacent layers is established, i.e., $N = N_s/\kappa$ (for this linear portion $K = SK_e$).

The behavior of vacancies and impurity as a function of N_m is described by (7) and (11), which is most simply analyzed graphically in the plane of variable S and K. The line L, which passes through the fixed center O with coordinates S_0 and K_s and is inclined to the vertical (counterclockwise) by an angle $\arctan[N_m/(C_e + C_{se}\nu^+/\omega)]$, is obtained geometrically from (7). This angle increases monotonically together with N_m from 0 to $\pi/2$. The desired solution, i.e., the S and K value at a given N_m, corresponds to the intersection of the line L with the curve $K(S)$. From Fig. 1 it is clear that the qualitative path of the $S(N_m)$ and $K(N_m)$ functions is wholly determined by the position of the center O relative to the $K(S)$ curve. This, in agreement with (11), depends on two dimensionless parameters fS_0/ω and S_0K_e/K_s and in the general case can be arbitrary. The particular case shown in Fig. 1 is an example of a rather complicated dependence of S and K on concentration. Initially, while the slope of L is not very large, the intersection R, corresponding to the solution in Fig. 1, moves to the left. This corresponds to the smooth decline of both S and K values. When the concentration N_m exceeds a certain critical value, three intersection points appear, R_1, R_2, and R_3. This means that the vacancy–impurity system has three different stationary states. In this region of concentrations, the $S(N_m)$ and $K(N_m)$ functions become indeterminate and are described by three branches (Fig. 2). The dashed branch, which corresponds to the middle point R_2 in Fig. 1, is unstable [1], i.e., the curves in Fig. 2 are in fact discontinuous. If the center O (Fig. 1) is shifted to the left under the descending portion of the curve, then S and K as functions of N_m become univariant and continuous: the vacancy concentration falls monotonically and the capture coefficient passes through a maximum. On the other hand, if the center O is shifted upwards so that the horizontal line $K = K_s$ intersects all three portions of the $K(S)$ curve, then the dependence on concentration becomes even more complex than in Fig. 2. Actual examples are examined in [1].

Disregarding the fact that diffusional relaxation greatly complicates the behavior of the vacancy–impurity system, a plateau which is described by (9) can occur as before on the function $N(N_m)$ since the existence of a plateau is the result of the general conservation law (7) if the vacancy concentration falls dramatically ($S \ll S_0$) while the capture coefficient remains larger than the limiting value $K_\infty = K_s$.

5. DIFFUSIONAL IMPURITY RELAXATION BY AN INTERSTITIAL MECHANISM

If the impurity is purely substitutive, i.e., has a negligibly small interstitial component, then the concentration of impurity $N(N_m)$ which is captured at the crystallization front is the final, i.e., the experimentally observed, value. This concentration was calculated above and is characterized by the capture coefficient $K(N_m)$. However, if the concentration of interstitial impurity atoms N_i is sufficiently large (but small by comparison with N), interstitial relaxation of the captured impurity can play a substantial role. If N exceeds the equilibrium volume concentration $N_e = K_e N_m$, then the excess impurity is partially displaced into the interstice by the intrinsic interstitial atoms, after which it diffuses through the interstices back into the melt. As a result, the concentration N which is captured on the front is decreased to some (now already final) value N_f. Henceforth, in order not to introduce new symbols, we will use N to represent the variable concentration of substitutive impurity which decreases along the z axis (directed away from the crystal front) from the initial value KN_m to N_f. The impurity concentrations N, N_i and the intrinsic interstitial atoms C_i in this section will be considered to have dimensions equal to the number of atoms in a unit volume of crystal.

The reaction rate for direct displacement of impurity from lattice sites is equal to $\beta N C_i$, where the kinetic coefficient β in the simplest case of a diffusion-limited reaction is determined by the expression $\beta = 4\pi r D_i$ [8], where D_i is the diffusion coefficient of intrinsic interstitial atoms and r is the interaction radius of two defects, equal in magnitude to the interatomic distance. The rate of the reverse reaction (transfer of interstitial impurity into lattice points by displacement of intrinsic atoms from them) is proportional to N_i. The net rate of impurity transfer J from the lattice points into the interstice is equal to the difference of the forward and reverse reaction rates. This reverts to zero with equilibrium between the concentrations of reagents which is described by the law of mass action: $N C_i / N_i = C_{ie}/\gamma$, where C_{ie} is the equilibrium concentration of intrinsic interstitial atoms and γ is the equilibrium ratio of interstitial and lattice point impurity components. Consequently,

$$J = \beta(N C_i - N_i C_{ie}/\gamma). \tag{12}$$

Stationary diffusion of interstitial impurity (with diffusion coefficient D) is described by the equation

$$D\frac{d^2 N_i}{dz^2} - V\frac{dN_i}{dz} + J = 0. \tag{13}$$

The net impurity current, interstitial and lattice point, along z remains constant:

$$V(N + N_i) - D dN_i/dz = \text{const}. \tag{14}$$

In this equation, conservation is neglected by diffusion of the displacing (lattice point) impurity, i.e., its diffusion coefficient is considered small by comparison with Vl, where l is the characteristic scale of the concentration profile $N(z)$.

If the impurity concentration is high enough, the concentration of intrinsic interstitial atoms C_i remains (at a length $\leq l$) near the equilibrium value C_{ie} due to rapid exchange with the melt at the front and rapid diffusion of these atoms in the crystal. In this case, the concentration profiles $N(z)$ and $N_i(z)$ are described by (13) and (14) which have an exponentially decreasing solution $N = N_f + A \exp(-z/l)$. The profile of $N_i(z)$ is described by the analogous expression with coefficients N_{if} and A_i but is increasing ($A_i < 0$). Limiting concentrations of N_f and N_{if} are found at equilibrium, i.e., $N_{if} = \gamma N_f$. At the front ($z = 0$), the concentration $N_i(0)$, due to rapid exchange of the interstitial impurity with the melt, has the equilibrium value $N_{ie} = \gamma N_e$, whereas $N(0)$ is determined by kinetic impurity capture and is equal to KN_m. Substituting these expressions for

$N(z)$ and $N_i(z)$ into (13) and (14) and using the initial conditions at the front, we find the characteristic length l and the unknown coefficients including the limiting concentration N_f, in which we are interested. The final impurity capture coefficient $K_f = N_f/N_m$ is equal to

$$K_f = (K + \xi K_e)/(1 + \xi),\qquad(15)$$

where $\xi = \gamma D/Vl$. Noticeable interstitial relaxation, i.e., a decrease of K_f by comparison with K, occurs only at $\xi \geq 1$. In this case, the following formula is justified for ξ (based on the equilibrium fraction of interstitial impurity γ being small):

$$\xi = \Gamma/2 + \sqrt{(\Gamma/2)^2 + \Gamma},\qquad(16)$$

where $\Gamma = \gamma D\beta C_{ie}/V^2$. If the dimensionless combination $\Gamma \geq 1$, then also $\xi \geq 1$, i.e., noticeable interstitial relaxation occurs. Setting $V \approx 3 \cdot 10^{-3}$ cm/sec and $D \approx 10^{-5}$ cm^2/sec for estimation, using the upper diffusion estimate for β given as $4\pi r D_i$, and setting $D_i C_{ie} \approx 10^{11}$ cm$^{-1} \cdot$ sec^{-1} (which corresponds to a self-diffusion coefficient $\sim 10^{-12}$ cm^2/sec, typical for semiconductor crystals), we find that the condition for noticeable interstitial relaxation can be fulfilled even with very small interstitial impurity component, at $\gamma \geq 10^{-5}$.

Although the examined interstitial impurity relaxation does not affect the vacancy concentration C (which is formed as a result of kinetic processes which occur directly at the crystallization front), it has importance for interpretation of nonlinear impurity capture which is caused by vacancy filling. The process of impurity capture in this case becomes rather unique: initially the impurity fills vacancies and enters the crystal at a rather high concentration (on the order of the concentration of the initial vacancies or higher), but then almost all the impurity is displaced into the interstices and returns to the melt. The effect of the impurity on the vacancies as a result is anomalously strong: the quantity of vacancies filled by impurity (i.e., which disappear) is much larger than the final quantity of impurity N_f which enters the crystal.

A plateau in the function of the initially captured concentration $N(N_m)$, which is determined by (9), may disappear on the $N_f(N_m)$ curve. It is retained under the condition that the capture coefficient K in the region of the plateau exceeds ξK_e, but in this case the plateau is $1 + \xi$ times lower than the value of (9).

6. SURFACE-CATALYTIC EFFECTS

Nonequilibrium in the vacancy solution, which is caused by a small frequency of intrinsic vacancy filling ν^+, leads to a principally important consequence besides the decline in vacancy concentration C as a result of their filling by impurity. As noted above, a small ν^+ value is caused by a high barrier for reaction of intrinsic surface vacancy filling. However, this barrier can be lowered substantially if some defect, either an impurity atom or another vacancy, appears with the vacancies (at the neighboring site). Thus, the frequency ν^+ in the general case is not constant and depends linearly on the concentration of surface defects which catalytically accelerate the intrinsic filling reaction. With respect to the impurity which enters a single-component crystal,

$$\nu^+ = \nu_0^+ + aN_s + bC_s\qquad(17)$$

where ν_0^+ is the vacancy filling frequency in the absence of neighboring defects and the coefficients a and b characterize the effectiveness of the corresponding catalyst (impurity atom or vacancy). If the interfacial impurity exchange occurs rapidly ($w^- \gg \omega$), then the impurity surface concentration N_s is almost constant and equal to $SK_{sc}N_m$. The concentration of surface vacancies C_s according to (1) is equal to SC_{se}. Thus, ν^+ is expressed through S and N_m. Substituting (3) into the conservation law (7) and considering ν^+ as a function of S and N_m, we express N_m through S:

$$N_m = \frac{C^* - SC_e - C_{se}(S - 1)(\nu_0^+ + bSC_{se})/\omega}{K(S) - K_s + S(S - 1)aC_{se}K_{se}/\omega}.\qquad(18)$$

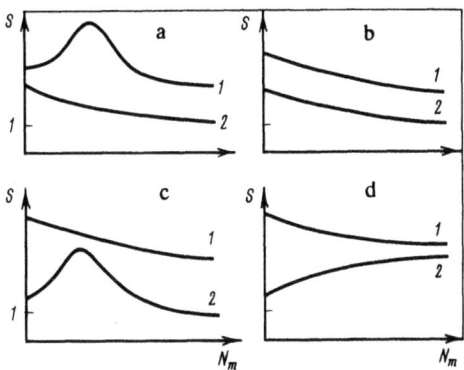

Fig. 3. Change of vacancy concentrations in two sublattices upon doping a crystal of a binary compound. a-d correspond to different sets of catalytic coefficients. 1) sublattice with impurity and 2) the adjacent sublattice.

Analysis of this relation shows that although catalytic effects (i.e., terms with coefficients a and b) can markedly change the path of the $S(N_m)$ curve, they result in no principally new consequences. For example, the vacancy relation S, as before, smoothly changes from S_0 to S_∞ in the absence of diffusional relaxation of an embedded impurity when the $K(S)$ function is described by the linear expression (6). The S_∞ value is near the equilibrium value of 1 at sufficiently large a. If $K(S)$ is not smooth as a result of diffusional relaxation (Fig. 1), then the increase of a leads to simplification of the $S(N_m)$ function, i.e., to disappearance of the indeterminate form (Fig. 2).

The surface-catalytic effects become much more significant for crystals of binary compounds. The mixture which forms the given sublattice and the vacancies of this sublattice in the absence of a catalytic effect of defects on ν^+ are described exactly the same way as for elemental crystals. If the vacancy filling frequency of the sublattice grows catalytically in the presence of a defect neighboring the vacancy (belonging to another sublattice), then the impurity and both types of vacancies form a single interrelated system.

We will label the first sublattice with the impurity index 1 and the second sublattice with 2. The filling frequency ν_2^+ is determined by a formula which is analogous to (17) but now depends on the surface concentration in the adjacent (first) sublattice, i.e., on the impurity concentration N_s and the vacancy concentration C_{s1}:

$$\nu_2^+ = \nu_{20}^+ + a_2 N_s + b_2 C_{s1} . \tag{19}$$

Vacancy filling by the impurity leads to a decrease of C_{s1}. Therefore, introduction of impurity can in principle lead to both an increase of ν_2^+ (due to appearance of the $a_2 N_s$ term) as well as a decrease of ν_2^+ (due to an increase of the $b_2 C_{s1}$ term). The vacancy concentration C_{s2} (the vacancy ratio S_2), which is determined by an expression similar to (3), changes as a result. This, in turn, leads to a decrease of intrinsic filling frequency ν_1^+, which in the case examined depends only on C_{s2}:

$$\nu_1^+ = \nu_{10}^+ + b_1 C_{s2} . \tag{20}$$

Thus, the two crystal sublattices form a system with an inverse connection: the impurity catalytically influences the vacancies of the neighboring sublattice which by the same mechanism influences vacancies of the first sublattice, i.e., the entry of the impurity itself. The impurity—vacancy system is completely described by a system of two conservation equations similar to (7) for each of the sublattices [while the left part of (7) is different from zero only for the first sublattice], by (19) and (20) for frequencies of intrinsic vacancy filling, and by the known dependence of the capture coefficient K on S_1. In particular, the problem for the linear function (6) leads to a quadratic equation for S_1. In view of the large number of unknown constants which determine the path of the $S_1(N_m)$ and $S_2(N_m)$ functions, we will limit ourselves further to qualitative analysis of possible characteristic cases. In the crystal without impurities, both vacancy subsystems are assumed to be supersaturated ($S_1 > 1$ and $S_2 > 1$).

1. The frequency $\nu_2{}^+$ is much more dependent on the impurity than on the vacancy, i.e., $a_2 \gg b_2$. Then, doping leads to an increase of the frequency ν_2^+ and to a corresponding decline in S_2 (up to the limiting equilibrium value of 1). Therefore, the frequency ν_1^+ decreases, which by itself would lead to an increase of S_1 if there were no competitive vacancy filling by impurity. Thus, two qualitatively different cases are possible here: a) If the catalytic coefficient b_1 is sufficiently large, S_1 initially increases as a result of a decrease of ν_1^+. However, the catalytic effect can be saturated (the frequency ν_1^+ tends toward a constant limit). Therefore, the S_1 value at a sufficiently large impurity concentration begins to decrease as a result of impurity vacancy filling. The quantities S_1 and S_2 as functions of N_m are shown schematically in Fig. 3a. b) If the coefficient b_1 is small, then the predominant effect from the very start is impurity vacancy filling, and S_1 falls smoothly (Fig. 3b).

2. The frequency ν_2^+ depends more strongly on the vacancies than on the impurity, i.e., $b_2 \gg a_2$. Therefore, impurity vacancy filling decreases the frequency ν_2^+, at least while the impurity concentration N_s is not too large by comparison with C_{s1}. This leads to an increase of S_2 and, consequently, of the frequency ν_1^+, which also decreases the vacancy concentration of the first sublattice, initially caused by impurity filling. The net path of the concentration curves depends on the value of the coefficient a_2. Two characteristic cases are possible here. a) The impurity catalytic effect becomes noticeable (a_2 is not too small) at large concentrations and the frequency ν_2^+ begins to increase and S_2 falls (to a limit of 1) so that the $S_2(N_m)$ curve has a maximum (Fig. 3c). b) The coefficient a_2 is negligibly small so that ν_2^+ falls monotonically to a value which corresponds to the limiting vacancy concentration of the first sublattice. Therefore, S_2 also monotonically approaches the limiting value (Fig. 3d).

A variable vacancy concentration as a function of impurity (Fig. 3a and b), which is caused by the joint influence of impurity filling and the surface-catalytic effect, means that the effect of the impurity on the system of native point defects is strongest at some intermediate concentration.

7. CONCLUSIONS

A low value of the intrinsic filling frequency of surface vacancies ν^+ leads to the initial (at the crystallization temperature) vacancy concentration C becoming nonequilibrium and strongly dependent on the impurity concentration. Two mechanisms of impurity effect on C exist, the impurity vacancy filling and catalytic increase of the ν^+ frequency. A catalytic dependence of ν^+ on vacancy concentration can show a strong indirect effect on C. Diffusional relaxation of embedded impurity steps can also have this effect through the vacancy mechanism. Simultaneous influence of both of these effects leads in a number of cases to variable and even indeterminate (discontinuous) dependence of C on impurity concentration. The concentration of impurity captured at the front N is, as a result, an even more complicated function of its content in the melt N_m. The last effect, i.e., nonlinear impurity capture, is a simple experimental indicator which shows that the vacancy nonequilibrium is factual for a given growing system. Therefore, doping can help control the vacancy concentration. A similar nonlinear capture is exhibited well for impurities in GaAs and Ge, whereas it is relatively weak for impurities in Si [9]. Linear impurity capture still does not mean that C maintains its equilibrium value. The other possibility consists in that impurity vacancy filling occurs sufficiently slowly and weakly affects the captured impurity concentration but can greatly affect the vacancy concentration.

REFERENCES

1.   V. V. Voronkov, "Impurity atoms and vacancies on the surface of a growing crystal: nonlinear kinetic effects," *Kristallografiya*, **31**, No. 3, 426-433 (1986).

2. V. V. Voronkov, I. V. Stepantsova, R. I. Gloriozova, and Yu. N. Bol'sheva, "Impurity capture coefficients in gallium arsenide as functions of concentration," Abstracts of Papers of the VIIth Conf. on Processes of Growth and Synthesis of Semiconducting Crystals and Films [in Russian], Novosibirsk (1986), Vol. 2, pp. 208-209.

3. V. T. Bublik, M. G. Mil'vidskiĭ, and V. B. Osvenskiĭ, "Nature and behavior of point defects in doped single crystals of A^3B^5 compounds," *Izv. Vyssh. Uchebn. Zaved. Fiz.*, No. 1, 7-22 (1980).

4. L. B. Ta, H. M. Hobgood, and R. N. Thomas, "Evidence of the role of boron in undoped GaAs grown by liquid encapsulated Czochralski," *Appl. Phys. Lett.*, 41, No. 11, 1091-1093 (1982).

5. E. V. Solov'eva, N. S. Rytova, M. G. Mil'vidskii, and N. V. Ganina, "Electrical properties of gallium arsenide doped with isovalent impurities (GaAs:Sb, GaAs:In)," *Fiz. Tekh. Poluprovodn.*, 15, No. 11, 2141-2146 (1981).

6. V. V. Voronkov, Yu. N. Bol'sheva, R. I. Gloriozova, L. I. Kolesnik, and O. G. Stolyarov, "Intrinsic point defects in GaAs: distinct dependence of residual concentrations on melt composition," *Kristallografiya*, 32, No. 1, 208-214 (1987).

7. V. V. Voronkov, "Impurity capture coefficient from a melt as a function of surface step angle and growth rate," in: *Growth of Crystals*, Vol. 11, Consultants Bureau, New York (1979).

8. T. R. Waite, "Theoretical treatment of the kinetics of diffusion-limited reactions," *Phys. Rev.*, 107, No. 2, 463-470 (1957).

9. Yu. M. Shashkov and V. P. Grishin, "Effective distribution coefficients of aluminum and phosphorus during growth of silicon needles from a supercooled melt," *Izv. Akad. Nauk SSSR, Neorg. Mater.*, 3, No. 5, 755-759 (1967).

EFFECT OF SOLUBLE AND INSOLUBLE ADDITIVES
ON THE CRYSTALLIZATION OF METALS FROM THE MELT

D. E. Ovsienko

1. INTRODUCTION

Impurities are known to play an important role in hardening processes and are widely used in metallurgy for improving the quality of castings. However, the nature and mechanism of their effect have been insufficiently studied. This impedes development of a theory of modification. Therefore, the choice of modifying additives is made empirically in the majority of cases. Elucidation of the effect of soluble and insoluble additives on the kinetics of formation and growth of crystals and on their morphology is crucial from the viewpoint of primary structure modification since formation of a casting is determined namely by these factors. Investigation of these questions has a rather lengthy history. A definite contribution to progress on this problem has been made by the studies carried out at the Metal Physics Institute of the Academy of Sciences of the Ukrainian SSR. Fundamental concepts about the role and mechanism of the effect of impurities on crystallization processes and formation of cast structures have been based on these studies. The present work is also concerned with these.

2. EFFECT OF SOLIDS ON CRYSTAL AND
STRUCTURE FORMATION PROCESSES

2.1. Kinetics of Nucleation on a Solid

2.1.1. Theory. According to classical theory [1-3], arbitrary or homogeneous nucleus formation of a new phase in the bulk of an old one is examined as a fluxional process. The formation rate of crystallization centers (cc) is given by the following expression:

$$J = K_0\, e^{-U/kT}\, e^{-W_0/kT},\qquad (1)$$

where W_0 is the work for critical nucleus formation, equal in its spherical form to

$$W_0 = 16\,\pi \left(\frac{M}{\rho}\right)^2 \left(\frac{T}{q}\right) \frac{\sigma^3}{\Delta T},\qquad (2)$$

$$K_0 = Z i_k\, (a\,\sigma/9\,\pi kT)^{1/2}\, n\, (kT/h) \approx 10^{33}.\qquad (3)$$

Here U is the activation energy, σ is the surface tension at the crystal–melt interface, ΔT is the undercooling, k is Boltzmann's constant, M is the molecular weight, ρ is the nucleus density, q is the heat of fusion per mole-

31

cule, T_0 is the equilibrium temperature of the solid and liquid phases, T is the absolute temperature, i_k is the number of atoms per critical nucleus surface, n is the number of atoms per cm^3, h is Planck's constant, and Z is the Zel'dovich nonsteady-state factor.

For heterogeneous nucleation (nucleation on a solid surface), the work for nucleus formation is equal to

$$W_s = W_0 \left(\frac{1}{2} - \frac{3}{4} \cos\theta + \frac{1}{4} \cos\theta^3 \right) \equiv W_0 \, f(\theta), \tag{4}$$

where θ is the contact angle which is expressed through surface energies σ_{s-1} (solid–liquid), σ_{s-n} (solid–nucleus), and σ_{n-1} (nucleus–liquid) as follows:

$$\cos\theta = \frac{\sigma_{s-1} - \sigma_{s-n}}{\sigma_{n-1}}. \tag{5}$$

The rate of cc nucleation on the solid is

$$J_s = K_s \, e^{-U/kT} \, e^{-W_0 f(\theta)/kT}, \tag{6}$$

$$K_s = K_0 \, \frac{n_s}{n} \, [f(\theta)]^{1/6}. \tag{7}$$

where K_s is the quantity of nucleus atoms which contact 1 cm^2 of surface.

Comparison of (5) and (2) shows that nuclei should appear on the substrate at significantly smaller undercoolings than in the melt volume at $\theta < 180°$ and $W_s < W_0$, i.e., a distinct lowering of the metastable limit should occur. Physical requirements which the values of σ_{s-1} and σ_{n-s} should satisfy so that the work for critical nucleus formation on the substrate is decreased by comparison with homogeneous nucleation result in crystal chemical requirements relative to the lattice. In particular, σ_{s-n} can be stated in the form

$$\sigma_{s-n} = \gamma + \alpha (\delta - \epsilon), \tag{8}$$

where γ is the chemical term which characterizes the magnitude and type of chemical bond and $\alpha(\delta - \epsilon)$ is the structural term which depends on the degree of discrepancy in parameters which join the planes of the nucleus and substrate $\delta = \Delta a/a$, and the shift modulus ϵ. Accordingly, undercooling ΔT of the melt at the surface depends on

$$\Delta T = \frac{C}{\Delta S_v} \delta^2, \tag{9}$$

where C is the elasticity constant and ΔS_v is the specific entropy of fusion.

2.1.2. Experiment. Results of an experimental investigation of cc nucleation on solid isomorphous surfaces are reviewed in [4, 5]. Thus, during study of undercooled hydroquinone crystallization it was found that addition of calcite (CaCO$_3$) or siderite (FeCO$_3$) crystals causes cc nucleation of hydroquinone at undercoolings of 4 and 18°K, respectively, by comparison with pure hydroquinone ($\Delta T = 37$°K). Oriented cc nucleation is observed in some cases. The differences of parameters which join the planes of the nucleus and substrate $\Delta a/a$ and $\Delta b/b$ for CaCO$_3$ are 13 and 9, and for FeCO$_3$, 19 and 7%. This determines the catalytic activity of these minerals. The width of the metastable limit [ΔT] of heterogeneous nucleation, i.e., the undercooling range in which the expectation time for appearance of crystals τ changes from several hours to 1-3 sec, i.e., practically from ∞ to 0, is relatively narrow and is not dependent on superheating of the melt but is different for different surfaces. Thus, it is ~ 1°K for the hydroquinone–CaCO$_3$ system, as seen from Table 1.

Besides this, at small ΔT of 3.8-4°K, the τ_i values differ greatly from each other and from τ_{av}. Knowing τ_{av}, the cc nucleation rate on the solid surface $I_s = N/(S\tau_{av})$ (S is the area of the solid, N is the quantity of centers on the surface) can be plotted as a function of $1/T\Delta T^2$. (It is possible to set $N = 1$ for a small value of

Table 1. τ_i as a Function of ΔT for Liquid Hydroquinone Containing a CaCO$_3$ Single Crystal

ΔT, K	3.6	3.8	4.0	4.2	ΔT, K	3.6	3.8	4.0	4.2
τ_i, sec	1,5 h	540	41	2			170	140	2
	No cc	306	110	3			1310	192	
		478	130	1			710	55	
		332	20	3			70	20	
		10	230	2	τ_{av}, sec				
		570	129	1			477	106	1,8

Table 2. Data on the Catalytic Effect of Various Solids

System	Maximal undercooling ΔT, °K	Width of metastable limit [ΔT], °K	K_s	σ, ergs/cm^2	Sample volume, cm^3
Pure hydroquinone	37	1.6	10^{29}	23.5	0.06
Hydroquinone – FeCO$_3$	18	2.5	–	10	0.06
Hydroquinone – ZnCO$_3$	15	1.5	10^{11}	9	0.06
Hydroquinone – CaCO$_3$	4.4	1.0	10^5	8.6	0.05
Pure bismuth	46	6	10^{11}	25	0.05
Oxidized bismuth	13	1.5	10^{11}	10	0.05
Pure lead	8	1	–	6	0.04
Oxidized lead	3.5	0,5	–	4	0.04
Oxidized potassium	1.5	0.25	10^4	0.4	0.03
Pure sodium	3.5	0.15	10^{22}	2	0.03
Oxidized sodium	2.5	0.15	10^6	1	0.03
Pure iron	280	–	–	–	10
Pure iron	340	30	10^{16}	160	10^{-6}
Oxidized iron	100–200	35	10^{15}	140	10^{-6}
Fe + MgO	30		–	–	10
Fe + BeO	130	–	–	–	10
Fe + ZrO$_2$	140	–	–	–	10
Fe + Al$_2$O$_3$	280	–	–	–	10
Steel X18H9 [7]	205	–	–	–	10
Steel X18H9–ZrC, ZrN, TiC, TiN [7]	10–20	–	–	–	10

$S < 1$ mm^2.) Thus, the experimental function ln $I_s(\Delta T)$ follows a linear law (Fig. 1). This agrees with (6) and indicates a fluxional character of cc nucleation. The K_s (from the intercept of the straight line with the ordinate axis) and σ_1 (from the slope) values for the hydroquinone–CaCO$_3$ system found from this function are 10^5 and 8.6 ergs/cm^2, respectively. For ZnCO$_3$, they are 10^{11} and 9 ergs/cm^2, and for FeCO$_3$, $\sigma_1 = 10$ ergs/cm^2. The values found for K_s, as seen from Table 2, are many orders of magnitude smaller than expected from theory, $K_s = 10^{26}$. This discrepancy is related to disregard in the theory of the localized character of cc nucleation, which is caused by surface defects (dislocations, fissures, etc.) that initiate cc nucleation. The fact that the K_s coefficient for the hydroquinone–CaCO$_3$ system is much smaller than for the hydroquinone–ZnCO$_3$ system is explained by the density of defects which initiate cc nucleation being much smaller for the CaCO$_3$ single crystal than for the ZnCO$_3$ polycrystal.

Very good reproducibility characterizes crystallization on these surfaces. It does not depend either on superheating of the melt or on the number of repeated recrystallizations. If these minerals are added initially to the melt, then undercooling as a function of superheating dissapears completely. However, the surface acquires additional activity after prolonged dwelling of these samples in a hypercrystalline state and cc begin to be formed at lower undercoolings of ~1°K (Fig. 2). The rate of the activation process is larger the higher the

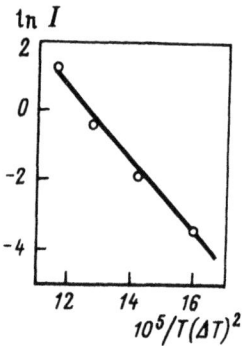

Fig. 1. Plot of ln I as a function of $1/T(\Delta T)^2$ for the hydroquinone–$CaCO_3$ system.

temperature and the nearer the crystal chemical relation of the partners. The acquired activity is easily eliminated with superheating of the melt by 30-40°K above the melting point of hydroquinone, while the natural activity is preserved even at very large superheatings by 200°K. It should be mentioned that the effects of activation and deactivation have been described earlier by V. I. Danilov [6], who related the activation effect to formation of a molecular-contact layer, which has an elevated melting point. The deactivation effect was examined as the reverse process, which consists of the destruction of this layer by superheating of the liquid.

Oxides, nitrides, carbides, etc. containing the same or different metals comprise the naturally active impurities. These have various effects on the kinetics of cc nucleation, as seen in Table 2. For example, MgO is the most active and Al_2O_3 the least active among the foreign oxides in the case of iron. These cause cc nucleation in large samples (100 g) at undercoolings of 30 and 280°K, respectively (Table 2). Qualitatively the same effect is observed upon crystallization of 100-200 μm drops on these same substrates. However, the undercoolings are significantly larger in this case, while the activity differences are smaller. The maximal undercoolings are 300 and 340°K on MgO and Al_2O_3, respectively.

The cc nucleation in all cases has a heterogeneous character. This is evident from the undercooling as a function of the superheating and the small K_s value of $\sim 10^{15}$. The reasons for the indicated differences in catalytic surface activity upon crystallization of large and small (drops) volumes is not clear. It may in part be related to the decreased probability of cc nucleation due to diminished contact area (by 10^6-10^7 times) of liquid with substrate by comparison with the melt in a crucible. However, this can increase ΔT only by tens of degrees and not by hundreds. It is also possible that a part of the drops impinge on defectless portions of the substrate with relatively small defect densities and a very small contact area (hundredths of a mm^2). Nucleation of cc occurs here at larger undercoolings than on the portions which contain dislocations and other defects. Besides this, lowered activity for the drops can be caused by poor molecular contact due to the contamination of the surface, the appearance of surface effects, etc. This is only conjecture and requires experimental verification. However, the fact itself that the catalytic activity of the solid is lowered upon metal drop crystallization is important since it demonstrates the possibility of attainment of large undercoolings in small volumes even in the absence of solid particles.

2.2. Effect of Solid Particles on Structure Formation

2.2.1. Nucleation Effects. The fact that solid particles in a melt can change the basic kinetics of crystallization and affect the formation of a cast metal structure lies at the foundation of the theory of metal and alloy modification. The effectiveness of such a modifier depends not only on their nature but also on the crystallization conditions, in particular, on the cooling rate. At slow cooling rates, even a large quantity of particles in the melt can be ineffective as a result of the probability nature of cc nucleation. This can easily be discerned

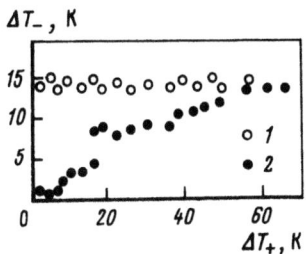

Fig. 2. Supercooling ΔT as a function of superheating ΔT_+ for the system hydro-quinone–$FeCO_3^+$. 1) Direct addition of $FeCO_3$ to the melt; 2) after 25 days stora ge as crystals at room temperature.

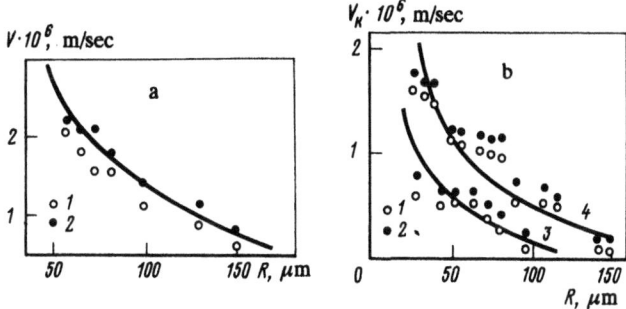

Fig. 3. Rates of repulsion (1) and incorporation (2) as a function of nickel sphere radius: for camphene crystals (a), for salol (b), the (110) face (3), the (111) face (4), calculation from (10) (a), and calculation from (11) (b).

from Table 1. It follows from the data for τ_i that only the upper metastable limit is reached with a sufficiently slow cooling rate. For example, of 10 particles (which is equivalent to 10 successive observations for a single particle) in the melt, a nucleus can form only on one particle for which the expectancy time is minimal (10 sec). It must be considered in this case that from the moment of appearance of the first crystal the probability of cc formation on other particles is suppressed due to a rapid growth rate and release of latent heat of crystallization which raises the melt temperature. At large cooling rates, which ensure attainment of the lower metastable limit, when the τ_i are small and similar, nucleation on a multitude of particles should be important. This forms a multitude of casting grains. However, large melt cooling rates from high temperatures cannot be developed for large volumes and such a situation cannot be realized. Pouring of the metal into a cold mold makes this partially possible when various casting zones are formed. Experiments showed [7] that grain separation in the region next to the wall (frozen) occurs using such a method for steel containing high-melting nitride and carbide particles. High cooling rates are attained in this zone. As a result, a tapering of the columnar crystals occurs.

The most effective influence of solid particles in formation of a cast structure appears with suspended metal castings [8], where they are introduced into a melt stream and can act as a microcondenser, creating local undercoolings and having a nucleating effect.

2.2.2. Interaction of Solid Particles with a Growing Crystal. Morphological Effect. The effect of solid particles on the morphology of a growing crystal under conditions where their nucleation effect is not realized

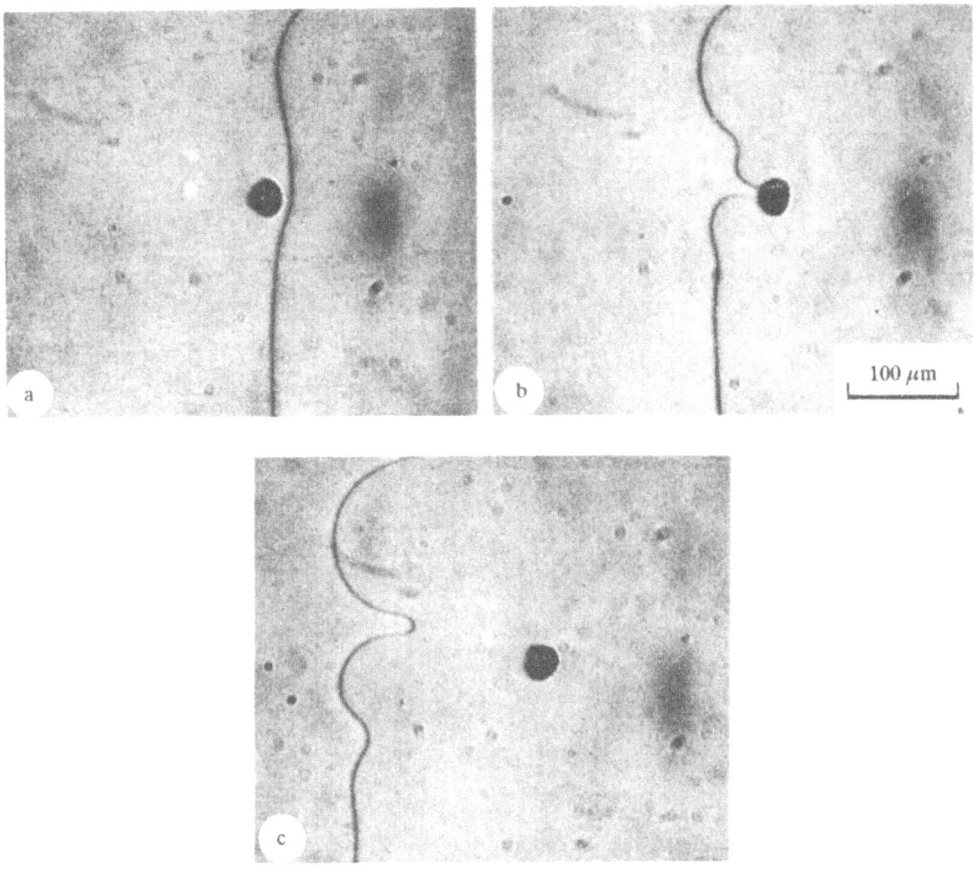

Fig. 4. Lycopodium particle incorporation by a planar succinonitrile crystallization front ($V = 6 \cdot 10^{-7}$ m/sec).

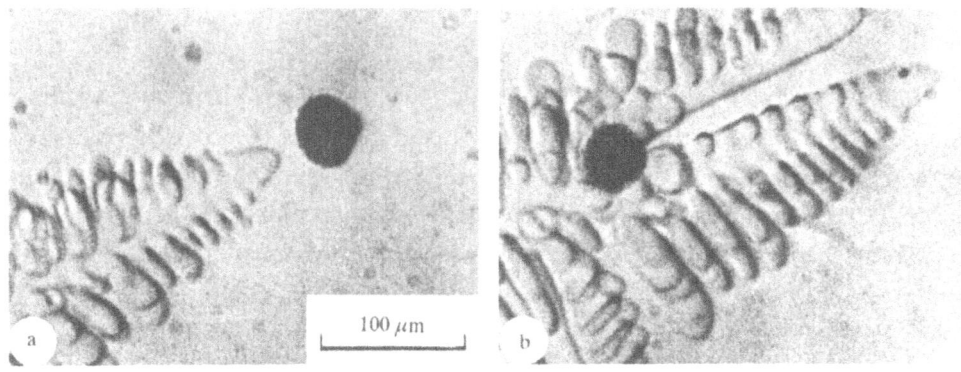

Fig. 5. Succinonitrile dendrite splitting upon encountering a lycopodium particle ($V = 9 \cdot 10^{-5}$ m/sec).

(positive temperature gradient, low melt undercooling [9-12]) was observed besides the above effects. Morphological changes are produced during incorporation of solid particles by the growing crystal and are exhibited in a loss of planar front stability and a splitting of cells and dendrite branches. In [9-12], solid particle incorporation conditions by crystals, growing normally and in layers, were studied in detail for model transparent substances. The critical growth rates V_{cr} above which particle incorporation occurs and below which their repulsion occurs, and V_{cr} as a function of particle size were determined. It should be pointed out that these studies were largely stimulated by the formulation of a theory [13] developed for a normal growth mechanism and giving the following expression for V_{cr}:

$$V_{cr} = 0.1 \, (\alpha \, B_3^2)^{1/3}/\eta \, R^{4/3}, \tag{10}$$

where α is the surface energy, B is the cleaving pressure constant, η is the melt viscosity, and R is the particle radius.

The experimental function $V_{cr}(R)$ for crystals of camphene growing by a normal mechanism and containing spherical nickel particles, shown in Fig. 3a, has the form $V_{cr} = A/R^{1.4}$ and agrees satisfactorily with (10). The rates are 20-25 times smaller for cyclohexanol, which has a melt viscosity 30 times larger, than those for camphene with the same nickel particle size. This also agrees with (10).

The V_{cr} values for salol and benzophenone crystals growing by a layered mechanism with the same nickel particles are many times larger (Fig. 3b) than for camphene crystals regardless of the fourfold increase of their viscosities by comparison with camphene. However, a substantial anisotropy of V_{cr} is observed. It is larger for the (111) face than for (110) (Fig. 3b). Such a situation was explained as being related to unique solid particle incorporation features which were not considered in deriving (10). While evaluating the experimental results, Temkin [13] developed a theory for application to a layered growth mechanism and obtained the relation [12]

$$V_{cr} = \left(\frac{\Delta S \, \Delta T B_3^2}{\Omega}\right)^{1/3} /6 \, \eta \, R, \tag{11}$$

where Ω is the molecular volume, ΔS is the entropy of fusion, and ΔT is the melt undercooling, which satisfactorily describes the experimental functions $V_{cr}(R)$ (Fig. 3b).

Distortions created during solid particle incorporation at the crystallization front were also found to differ both in form and in their further behavior as a function of growth mechanism. Interaction of the particle with the face growing in layers leads to a facial distortion which is completely "healed" after incorporation of the particle. A curved depression which is retained due to isotropic growth even after complete inclusion of the particle into the crystal (Fig. 4) is formed with particle incorporation by a smooth face (normal growth mechanism). Cells, and then dendrites, which undergo morphological changes upon encountering particles, begin to form with increased melt undercooling and a correspondingly increased growth rate. If the dendrite radius is similar to the particle radius, splitting occurs (Fig. 5). Such breakdown of dendrites becomes substantial and is accompanied by a change of their orientation with high particle density.

The presence in succinonitrile of small amounts of impurity (for example, water) which lower the liquidus temperature by 0.2-0.3°K leads to such a drastic lowering of V_{cr} that particle incorporation is observed only at the lowest measurable growth rates. This agrees qualitatively with theory [14].

Incorporation of a solid particle and subsequent crystal growth with the included particle is accompanied by formation of an impurity channel which, in contrast to the pure substance (Fig. 4b), trails from the particle to the front and persists until hardening is completed. A system of channels or fibrous divisions is formed with a multitude of solid particles. The quantity and size of these is determined by the solid particle distribution. These formations have a character analogous to that of the cellular-fibrous type of division which is formed at a definite (critical) value of structural undercooling. A cellular-fibrous structure is not formed under our experimental conditions in the presence of solid particles. Therefore, solid particles facilitate a premature loss of planar front stability and creation of cells. The reason lies in local elevation of concentration near the particles, which occurs during their approach to the front due to "opaqueness" of the solid particle relative to diffusion of impurity atoms into the melt volume. The excessive impurity concentration around the particle is confirmed by the formation of local regions of melting around the particle upon fully isothermal sample crystallization.

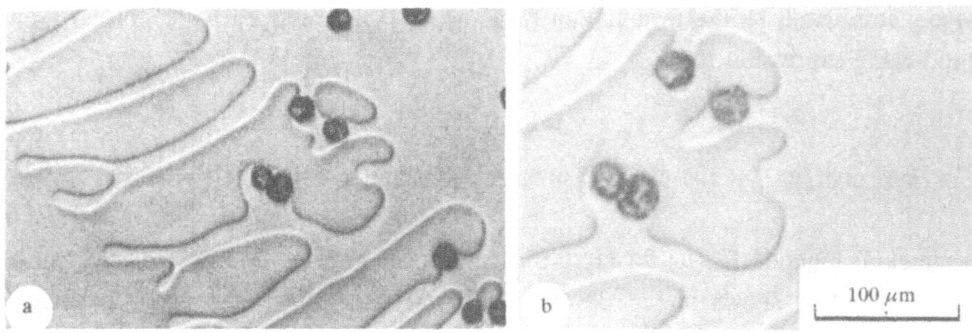

Fig. 6. Dissolution and division of a dendrite branch between two lycopodium particles in cyclohexanol with added water.

Thus, solid particles can be a source of local concentration inhomogeneities of impurity atoms which not only facilitate a premature loss of planar front stability, particle incorporation, and dendrite breakdown, but also seem to cause division of dendrite branches. Exactly such a case is shown in Fig. 5. Here particles in the melt are located at sufficiently close distances and the elevated concentration fields overlap. The dendrite branch which extends into this region at first breaks down and then is dissolved. Division of part of the dendrite stem occurs as a result. This then continues with independent growth (Fig. 6).

The processes described occur at the crystallization front and can lead to both a dispersion of dendrite structure and creation of new grains in the casting.

Besides these effects, which are related to melt crystallization, solid particles can also affect formation of secondary structure by generation of polygons, recrystallization, and phase transitions which occur in the solid [15].

3. CRYSTALLIZATION OF PURE METALS

3.1. Undercooling and the Problem of Homogeneous Nucleation

The magnitude of metal melt undercooling, as was shown, depends on the catalytic activity of the solids (crucible walls, coatings, dispersed particles). This activity appears differently in small and large volumes. Complete removal of solid surfaces from the liquid must occur for homogeneous nucleation. This is an exceedingly complex problem which is aggravated even more by the absence of reliable methods for control of trace quantities of these impurities. In this sense, the established principles of cc nucleation on solids and the criteria formulated on their basis [4-6] of homogeneous crystallization takes on special importance. They result in requirements of complete absence of activation and deactivation effects, and undercooling as a function of superheating and increased undercooling as a function of the degree of purification, as well as other enumerated effects which are related to crystallization on solids.

Various methods [5, 16, 22, 23, 25] for microscopic (10-1000 μm drops) and bulk (0.1-500 g) samples were developed for removal of insoluble impurities and attainment of large undercoolings of liquid metals. In the first case, a powder melting method on a neutral substrate in an inert atmosphere (argon, helium) or under a protective layer (glass, rosin, etc.) as well as drop–emulsion and insular methods were applied. Their total merit is the possibility in principle of producing drops which are free from solid particles since the quantity of drops can be larger than the number of solid particles in the starting sample. For bulk samples, melting in a crucible in an inert atmosphere under a glass layer or in a glass sleeve was applied. These assist removal of foreign particles and prevent interaction with the crucible. Methods that do not use crucibles also exist. These include suspended melting and dropping through a long vertical tube in a neutral atmosphere.

Table 3. Combined Data on Maximal Undercoolings of Pure Metals

| Metal | mp, °K | ΔT of 10-100 μm drops [3, 16] | ΔT, °K, from new data | | | | | ΔT/T_f | ΔT_cr = ΔH/C_p, °K |
| | | | 10-100 μm drops | Reference | Bulk sample | | | | |
					1-10 g	100-500 g	Reference		
Ag	1234	227	252	[22]	–	250	[18]	0.18; 0.20	363
Al	933	130	160	[23]	–	–	–	0.14; 0.17	329
Au	1336	230	–	–	190	–	[19]	0.17	436
Bi	544	90	227	[23]	–	–	–	0.71; 0.42	357
Co	1768	330	470	[4, 5]	310	–	[19]	0.19; 0.27; 0.17	383
Cu	1356	236	277	[22]	210	218	[25, 17]	0.17; 0.15; 0.16	413
Fe	1809	286	500	[21]	420	280	[20, 5]	0.16; 0.28; 0.23; 0.16	329
Ga	303	76	174	[23]	–	–	–	0.25; 0.58	202
Ge	1210	227	296	[22]	200	–	[19]	0.19; 0.25; 0.16	1333
Cd	591	–	110	[23]	–	–	–	0.19	–
Hg	234	80	88	[23]	–	–	–	0.34; 0.38	83
In	430	–	110	[23]	–	–	–	0.26	110
Mn	1517	308	–	–	–	–	–	0.26	318
Nb	2741	–	525	[24]	–	–	–	0.19	875
Ni	1726	319	480	[4, 5]	365	305	[24, 5]	0.18; 0.21	445
Pb	600	80	153	[23]	–	–	–	0.13; 0.26	157
Pd	1825	332	–	–	310	–	[19]	0.18; 0.17	482
Pt	2042	370	–	–	–	–	–	0.18	639
Sb	904	135	210	[23]	–	–	–	0.15; 0.23	633
Sn	505	118	187	[23]	–	–	–	0.23; 0.37	235
Te	723	–	236	[23]	–	–	–	0.33	929
Zr	2130	–	425	[24]	–	–	–	0.20	575

Using microvolume methods, Turnbull and Cech [3, 16] first carried out systematic studies of undercoolings ΔT of a number of metals (for drops of 50 μm diameter). The relative maximal undercoolings T/T_f which were reached by them for various metals, with the exception of mercury (Table 3), change little (from 0.14 to 0.25) and average $0.18T_f$. These undercoolings remained for a long time as the limiting ones and were accepted as those corresponding to homogeneous nucleation. However, almost the same undercoolings for bulk samples of 10-500 g for Ag, Au, Cu, Co, Fe, Ni, and Pd (Table 3) have been obtained at comparable cooling rates in more recent studies (when melting under slag glass which removes oxides and other impurities). Suspended melting gave even larger values, 420°K for Fe and 480°K for Ni. Significantly larger undercoolings $(0.3-0.6)T_f$ than in [16] were attained in microvolumes (at cooling rates of 5-20°K/min) using more modern methods [5, 21-24]. Therefore, values of undercoolings of ~$0.2T_f$ can be confirmed as corresponding not to homogeneous nucleation, as suggested earlier, but to heterogeneous, which was indicated earlier in [4, 5] and later in [2, 3].

More detailed studies of cc nucleation kinetics, development of necessary process characteristics, and proof of homogeneous nucleation are required for revealing the character of nucleation at elevated undercoolings of $(0.3-0.6)T_f$. Regardless of disagreement with respect to the problem of experimental observation of homogeneous crystallization, the very attainment of larger melt undercoolings played a large role in the formulation of new studies for elucidating the mechanism of soluble additive effects, structure-formation processes at large undercoolings, and other important aspects.

3.2. Effect of Undercooling on Crystal Morphology and Casting Structure

In a number of previous works [26-28 among others], principles of crystal morphology and casting grain structure changes were established depending on the magnitude of melt undercooling which preceded its hardening. Experiments were done on model pseudometallic organic substances (with $L/kT < 2$) in plane-

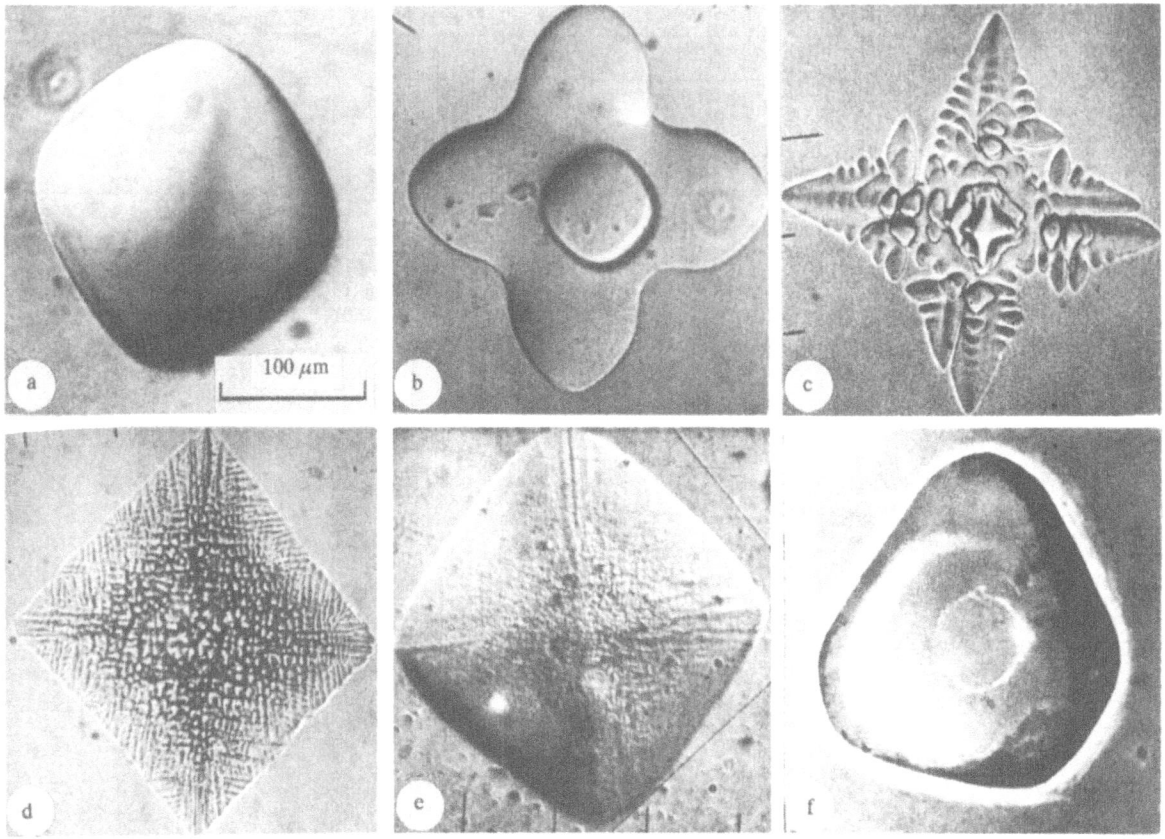

Fig. 7. Morphological changes of a cyclohexanol crystal with increased melt undercooling ΔT: a-f, $\Delta T \leq 0.1$, 0.3, 0.45, 2.35, 3.5, and 10.2°K, respectively.

parallel cuvettes and on metals upon melting in crucibles under a glass layer or in an inert atmosphere. The following principles were found for the example of cyclohexanol (Fig. 7). At small undercoolings <0.1°K, stable growth of a curved crystal form with a smooth surface occurs. At $\Delta T \sim 0.3$°K, the smooth boundary loses stability and forms protuberances after attainment of critical sizes which are smaller than at $\Delta T = 0.1$°K. These transform into dendrites upon increasing ΔT to 0.45°K. With further increase of ΔT, the dendrite branches become finely divided and at $\Delta T = 2.35$°K at a definite stage of growth the dendritic crystal assumes the macroscopically quasifacial (square) form through geometrical selection. At $T = 3.5$°K, the dendritic form degenerates into a needlelike one which is characterized by the presence of a system of radially directed thin needles without side branches. Rounding of the crystal occurs at this stage. At large undercoolings of 10°K, the crystals adopt a curved and macroscopically smooth form. For cyclohexanol, this undercooling by 10°K is higher than the critical $T_{cr} = \Delta H/C_p = 8.5$, i.e., $\Delta T/\Delta T_{cr} = 1.2$.

Analogous features, with the exception of the last form, are also observed for metals. A well-developed dendritic structure (Fig. 8a) is always observed in castings which are hardened at small undercoolings. With increased undercooling, it is finely divided and is retained until certain undercoolings above which it transforms into a needlelike structure similar to Fig. 8b. Thus, a tendency toward such a transition is observed in pure nickel in castings which are crystallized at melt undercoolings of ~140°K, i.e., at $\Delta T/\Delta T_{cr} = 0.3$ it was similar to the corresponding value for cyclohexanol, $\Delta T/\Delta T_{cr} = 0.4$. A macroscopically curved form, similar to cyclohexanol at $\Delta T/\Delta T_{cr} = 1.2$, could not be detected in metals, possibly due to insufficient undercoolings ($\Delta T < \Delta T_{cr}$) but mainly due to disruption of the primary structure, which occurs as a result of secondary processes, phase transitions (Fe), or recrystallization (Ni). Thus, the primary structure is completely disrupted for pure Ni at $\Delta T \sim 140$°K and a fine-grained structure appears (Fig. 8c). This, as was shown in [25], is caused by recrys-

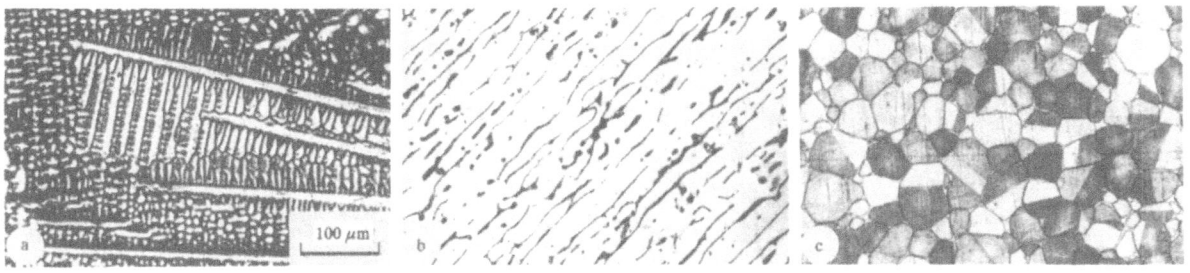

Fig. 8. Structures of castings at various melt undercoolings: Fe + 5 mass % Si, $\Delta T = 30°K$ (a), Fe + 5 mass % Si, $\Delta T = 270°K$ (b), and Ni, $\Delta T = 200°K$ (c).

tallization and not by an increase of the number of nuclei in the melt, as was proposed earlier. It should be noted that a slight increase of the number of grains in the casting was observed for undercoolings which correspond to dendritic growth due to truncation (by convectional currents) or fusion (upon recalescence) of dendritic branches.

4. EFFECT OF SOLUBLE ADDITIVES

Earlier [29, 30], the effect of soluble impurities was explained by their influence on the value of crystal–melt interfacial surface energy, σ_{n-l}, and the activation energy, i.e., on parameters contained in the basic equation of cc nucleation (1). In particular, it was assumed that the surface-active substances, which lower σ_{n-l}, should lead to decreased work of nucleus formation according to (2), and consequently, to an increase of cc nucleation rate and correspondingly to a fractionation of the casting grains. A requirement for realization of homogeneous nucleation lies at the basis of this theory. However, as shown above, such a process is never achieved under industrial conditions. Thus, the necessity of further studies of the nature of the influence of soluble additives arose. This was largely stimulated by the possibility of producing large undercoolings of metals. Special attention in these investigations was focused on elucidation of the effect of soluble additives on the cc nucleation kinetics, their morphology, and the cast structure.

The elements Sn [31], Fe [32], and Ni [33] with various additives in quantities which form solid solutions were studied. The cc nucleation kinetics were studied in detail only for Sn. The range of undercoolings for 50-80 g samples at cooling rates of 5-7°K/sec were determined for the remaining metals. Experimental results are given in Table 4. It follows from these data that all soluble impurities, independent of their surface activity at the melt–vapor interface, practically do not affect undercooling of metals, with the exception of Te in Sn. However, a definite effect of these additives (Fig. 9a) is revealed on the kinetic curves of $J(\Delta T)$ for Sn. This appears at a decreased cc nucleation rate and over a broadened $[\Delta T]$ range in which the rate changes from negligibly small to a perceptible value. Additives of Bi show a larger effect than those of In. The values of the kinetic coefficients K_s with In and Bi additives remain just as small as for Sn, implying a heterogeneous cc nucleation mechanism. The decrease of the kinetic coefficient K alone can be treated as a lowered atomic mobility which is related to formation at the interface of an impurity layer. This hinders entrance of atoms of the primary substance from the melt. This, more likely than not, is manifested not at the critical nucleus formation stage, but during its growth. It can be assumed that nucleus growth at a given ΔT will be retarded with time as the impurity accumulates at the crystallization front while it remains smooth. A loss of stability with formation of cells will occur at some critical concentration. This leads to a jump in the growth rate. Retarded growth up to the stability loss of the planar front leads, in turn, to an increased expectancy time τ_{min} and, consequently, to decreased J.

Table 4. Data for Undercooling of Sn, Fe, and Ni Containing Soluble Additives

System	Surface activity of additives at the liquid–vapor interface	Undercooling range [ΔT], °K	K_s	σ_l, ergs/cm^2
Sn	–	40–45	10^{13}	24
Sn + 1 at. % In	Neutral	43–46	10^{11}	23
Sn + 1 at. % Bi	Increased	39–42	10^8	21
Sn + 1 at. % Te	The same	5–8	–	–
Fe		260–280	–	–
Fe + (0.1–5) mass % Si	Decreased	260–290	–	–
Fe + (0.02–0,2) mass % Mn	The same	270–290	–	–
Fe + (1–5) mass % Sn		280–300	–	–
Fe + (1–8) mass % Mo	Increased	260–280	–	–
Ni	The same	280–300	–	–
Ni + 3,7 mass % Si	Decreased	270–300	–	–
Ni + 5 mass % W	Increased	270–290	–	–

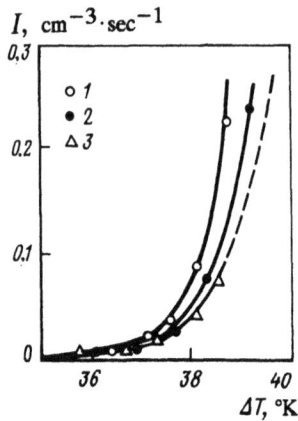

Fig. 9. Plot of $I(\Delta T)$ for pure tin (1) and with additives of 1 at. % In (2) and 1 at. % Bi (3).

The accuracy of the above was confirmed by direct experimental observations of the shifting of the phase boundary and the change of its morphology with time in a planar sample cooled at a defined rate after partial melting. Results of these observations are given in Fig. 10. The growth rate for Sn and Sn with 1 at. % Te with time, i.e., as the temperature is lowered and the crystallization front advances from its initial state, is seen to increase continuously. A positive gradient existed in our experiments in a melt. The crystallization front in pure Sn remained smooth for almost all hardening times. Formation of a cellular form as a result of traces of some impurities occurred only for hardening of the last portions. This function has a variable character for Sn with additives of 1 at. % In and Bi. Initially the growth rate increases similarly to that of Sn. Then, it slows gradually and falls precipitously after passage of a certain time. The last is accompanied by a loss of stability of the planar front and formation of many cellular-type outcroppings which improve heat transfer in comparison with the planar front. The decrease of growth rate in the initial stage is related to the accumulation of In and Bi impurities at the crystallization front. These interfere with approach of the primary atoms to the crystal. The loss of stability of the planar crystallization front and the distinct decrease of growth rate related to it are connected with the presence of concentration undercooling. The value of this depends on the impurity distribution coefficient. The Bi impurities lead to more effective growth rate deceleration at the initial stages and to an

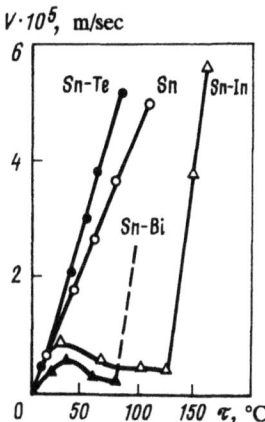

Fig. 10. Effect of soluble additives on tin crystal growth rate.

earlier loss of stability which is accompanied by formation of needlelike protrusions and a sharp decrease of growth rate than for the In additives. This fact is explained by the Bi distribution coefficient in tin ($k = 0.22$) being smaller than that for In ($k = 0.56$). The absence of an effect of Te additives at the growth stages where the retarding action of In and Bi appear is caused by the distribution coefficient of Te in Sn being larger than unity and, consequently, the melt layer which adjoins the phase boundary is depleted in Te.

The fact that an additive of 1 at. % Te leads to a sharp lowering of the undercooling from 5-8°K, as seen from Table 4, is explained by a catalytic effect of tin telluride (SnTe) particles and not by an intrinsic Te effect as for the soluble additive [31].

It should be noted that in the experiments of [34], which used drops of 350 μm, the influence of soluble additives (1 at. %) on the cc nucleation kinetics was also not observed. Thus, for pure Sn, the undercooling range [ΔT] was 51-77°K, for Sn with additives of Cu, Zn, Cd, In, Te, Pb, and Bi [ΔT] = 54-78°K, and for Pb and Pb with additives of Sn and Te these [ΔT] ranges are 13-28 and 12-24°K, respectively.

Thus, soluble additives which do not form chemical compounds do not affect melt undercooling of the basic metal independent of whether they increase or decrease the surface tension at the melt—vapor (base metal) interface [30].

In our experiments, and more so under industrial conditions, cc nucleation has a heterogeneous character. Here we discuss only the effect of soluble impurities on the behavior of solid surfaces or particles, although we did not detect such an effect. Nevertheless, it is in principle possible since the impurity atoms adsorbed on the particle—substrate surface can change the catalytic properties of the latter, changing the critical undercooling which is necessary for cc nucleation. However, the effect of insoluble additives and realization of homogeneous nucleation can hardly be expected to be removed completely since the catalytic effect of solid particles has a large sphere of influence [35]. Consequently, the question of the effect of surface-active impurities on the process of homogeneous nucleation, which lie at the foundation of adsorption theories of modification [29], is very controversial. To a large degree, this relates to the role of surface-active impurities in the formation of a cast structure since under industrial conditions homogeneous nucleation is never achieved. Soluble additives, not affecting or even weakly affecting the cc nucleation kinetics, markedly strengthen further growth of cc, slowing this process and changing the crystal morphology which appears upon loss of stability of the planar front, development of cellular or dendritic growth, etc. In turn, reinforcement of the dendritic structure, which is accompanied by emergence of concentration inhomogeneities, facilitates fusion of dendritic branches and a corresponding increase in the number of grains in the casting. The presence of soluble additives, as shown in the Fe—Si, Ni—Si, and other systems [25-28], leads to a broadening of the temperature region for existence of the needlelike structure (Fig. 9b) and to a shift of the critical undercooling of the finely divided grain (Fig. 9c),

ΔT_{cr}, to lower temperatures. For example, an additive of 3.7 at. % Si in the case of Ni causes an increase of ΔT_{cr} from 140 to 225°K. Occurrence of these processes is not related to the surface activity of the soluble additives and is determined mainly by the degree of their incorporation by the growing crystal and the presence of concentration undercooling. Exactly this condition should be one of the important factors for building a theory of modifying additives of this type.

Finally, we should dwell shortly on the interaction of the growing crystal with the liquid inclusions that are artificially added to the melt. In [10, 12], the morphological effect of such inclusions, which is manifested differently depending on their solubility and temperature, was discovered in the transparent substance succinonitrile. Thus, insoluble liquid mercury and gallium drops in liquid succinonitrile cause the same effects as insoluble solid particles upon encountering the planar crystallization front or dendrite branches. Soluble inclusions, for example, camphene drops, have a different influence. A layer of variable concentration from pure succinonitrile (mp = 58°C) to pure camphene (mp = 48°C) is formed around the camphene drop upon partial dissolution, in agreement with the phase diagram (eutectic type). The width of this layer depends on the rate of dissolution and residence time. The dendrite, falling in such a variable concentration region, undergoes various changes depending on temperature. At $T_e < T < T_L$ (T_e is the eutectic temperature), the dendrite at first slows its growth as it nears this region. Then, an extensive dissolution of the dendrite branches and formation of curved crystals around the undissolved camphene drop occurs.

If crystallization occurs at $T < T_e$, i.e., in the monophasic region, then the dendrite structure around the inclusion degenerates into a dispersed structure of an apparently eutectic type. A collection of various structural and phase states in agreement with their phase diagrams can be expected in more complicated systems in the layer of variable concentration around the liquid inclusion. Besides this, the picture can change significantly upon increasing the undercooling below T_e as a result of morphological changes (for example, degeneration of the dendrite form into a needlelike one) and due to appearance of new metastable phases. In principle, similar structural changes should also be important around partially soluble solid particles. However, in the case of high catalytic particle activity which causes cc nucleation at small undercoolings, crystallization will occur only in the biphasic region, i.e., at $T_e < T < T_L$.

Further studies are necessary for resolution of these and other questions. Nevertheless, the extensive structural dispersion around liquid inclusions demonstrates the possibility in principle of effective influence on the cast structure by introduction into the melt of a doping additive as a large quantity of fine drops. This possibility was verified by metals [10] in the Al–Cu system by addition to the crystallizing Al melt of drops (ϕ 2 mm) with composition Al + 10 mass % Cu in a quantity such that the total Cu content in the casting was 4.5 mass %. The ingot prepared by this method (250 cm³) consisted of a multitude of randomly located fine dendritic grains together with a columnar hyperdendritic structure in an ingot of the same composition (Al + 4.5 mass % Cu) which was prepared under the same conditions but with uniform Cu distribution.

Naturally, special studies of the structure, phase composition, and properties, as well as resolution of a number of technical difficulties, are necessary for practical application of this method to actual metallic alloys.

5. CONCLUSIONS

The presence of solid particles in a melt can exert nucleation and morphologic effects upon interaction with a growing crystal.

Crystallization center nucleation on active surfaces has a fluxional character. Mutually oriented nucleus and substrate, a distinct metastable boundary, absence of a dependence on melt superheating, and reproducibility of results are characteristic for this. The catalytic activity of surfaces is determined by the crystal chemical relation of the planes which join the nucleus and substrate. As a result of molecular contact with crystals of the principal substance, such surfaces acquire additional activity which causes cc nucleation at small undercoolings and can be completely removed at relatively small melt superheatings.

The presence of solid particles in the melt can fundamentally change the crystallization kinetics and by this change the casting grain structure. However, such a possibility can be realized only with definite crystallization conditions. In particular, at high cooling rates, which can be reached in the layer next to the wall upon pouring into a cold mold or upon suspension of a metal casting when added to a metal stream, solid particles can play the role of a microcondenser and have a nucleating effect. Besides the catalytic influence, solid particles incorporated into the growing crystal can have an effect on the morphology of the surface at the interface. This causes a loss of stability of the planar front and splitting of cells and dendritic branches. Their presence substantially affects the occurrence of secondary processes in the solid phase, i.e., phase transitions, recrystallization, and polygon formation.

Establishment of principles and a mechanism of cc nucleation on solid surfaces allows criteria for separation of homogeneous nucleation from heterogeneous to be formulated. They lead to the absence of effects related to nucleation on solid particles. A conclusion was reached based on an analysis of these and literature data that with undercooling of liquid metals of $\sim 0.2 T_p$ crystallization occurs not homogeneously as assumed earlier but on foreign particles. This demonstrates the necessity for further studies of the homogeneous crystallization problem. Homogeneous crystallization was confirmed to be unrealized and to give no contribution to the formation of a cast structure under industrial conditions.

The possibility of production of large undercoolings of liquid metals allowed the mechanism of action of soluble additives to be better understood. In a number of systems it was found that soluble additives which do not form chemical compounds with the main metal and the impurities contained in it have practically no influence either on the kinetics or on the mechanism of cc nucleation independent of their surface activity relative to the melt. However, they can have an effect on the growth and morphology of crystals, thereby affecting the formation of a cast structure. The effectiveness of impurity influence is determined by the degree of its incorporation by the growing crystal and the creation of structural undercooling.

Local concentration nonuniformities in the melt can play a large role in the formation of a casting.

REFERENCES

1. M. Volmer, *Kinetik der Phasenbildung*, Leipzig (1939).
2. Ya. I. Frenkel', Kinetic Theory of Liquids [in Russian], Nauka, Leningrad (1975), Vol. 3.
3. J. Holloman and D. Turnbull, "Formation of nuclei during phase transitions," in: *Advances in Metal Physics* [Russian translation], Metallurgizdat, Moscow (1959), Vol. 1, pp. 304-367.
4. D. E. Ovsienko, "On crystallization center nucleation in undercooled liquid metals," in: *Current Problems in Crystallography* [in Russian], Nauka, Moscow (1975), pp. 127-149.
5. D. E. Ovsienko, "On homogeneous crystallization center nucleation in liquid metals," *Growth of Crystals* [in Russian], Izd. Erevan Univ., Erevan (1975), Vol. 2, pp. 11-25.
6. V. I. Danilov, Structure and Crystallization of Liquids [in Russian], Naukova Dumka, Kiev (1956).
7. Yu. Z. Babaskin, D. E. Ovsienko, L. A. Rostovskaya, and G. A. Alfintsev, "Modification of steel by high-melting particles and its effect on the structure and properties of castings," *Liteinoe Proizdvod.*, No. 2, 17-19 (1975).
8. S. S. Zatulovskii, *Suspension Casting* [in Russian], Naukova Dumka, Kiev (1981).
9. D. E. Ovsienko, G. A. Alfintsev, B. A. Kirievskii, G. I. Gershtein, and G. P. Chemerinskii, "Dendrite splitting upon incorporation of solid particles," *Metallofizika*, 5, No. 6, 110-111 (1983).
10. D. E. Ovsienko, O. P. Fedorov, B. A. Kirievskii, G. I. Gershtein, and G. P. Chemerinskii, "Effect of solid and liquid inclusions in a melt on crystallization front morphology," *Metallofizika*, 8, No. 4, 64-71 (1986).
11. D. E. Ovsienko, O. P. Fedorov, B. A. Kiriyevsky, G. I. Gershtein, and G. P. Chemerinsky, "The effect of solid and liquid inclusions on morphology of crystallization front," *Cryst. Res. Technol.*, 21, No. 6, 721-727 (1986).
12. D. E. Ovsienko, D. E. Temkin, O. P. Fedorov, and G. P. Chemerinskii, "Interaction of solid particles with crystals growing from a melt," *Kristallografiya*, 32, No. 5, 1246-1252 (1987).
13. D. E. Temkin, A. A. Chernov, and A. M. Mel'nikova, "Theory of solid inclusion incorporation during crystal growth from a melt," *Kristallografiya*, 21, No. 4, 652-660 (1976).
14. D. E. Temkin, A. A. Chernov, and A. M. Mel'nikova, "Incorporation of foreign particles by a crystal growing from a melt with impurities," *Kristallografiya*, 22, No. 1, 27-35 (1977).
15. M. B. Movchan, "Solid-phase modification of the primary structure of cast alloys by dispersed nonmetallic particles," Author's Abstract of Candidate's Dissertation, Technical Sciences, Kiev (1983).

16. D. Turnbull and R. E. Cech, "Microscopic observation of the solidification of small metal droplets," *J. Appl. Phys.*, **21**, No. 8, 804-810 (1950).

17. G. L. F. Powell and L. M. Hagen, "Undercooling of copper and copper–oxygen alloys," *Trans. Metall. Soc. AIME*, **242**, No. 10, 2133-2138 (1968).

18. G. L. F. Powell and L. M. Hagen, "The influence of oxygen content on the grain size of undercooled silver," *Trans. Metall. Soc. AIME*, **245**, No. 2, 407-412 (1969).

19. J. Fehling and E. Scheil, "Untersuchung der Unterkuhlbarkeit von Metallschmelzen," *Z. Metallk.*, **53**, No. 9, 593-600 (1962).

20. D. W. Gomersal and S. X. Shirashi, "Undercooling phenomena in liquid metal droplets," *J. Aust. Inst. Met.*, **10**, No. 5, 220-222 (1965).

21. A. I. Dukhin, "Crystallization of metals and alloys in small volumes," in: *Problems in Metal Science and Metal Physics* [in Russian], Metallurgizdat, Moscow (1959), Vol. 6, pp. 9-33.

22. V. P. Skripov and V. P. Koverda, *Spontaneous Crystallization of Undercooled Liquids* [in Russian], Nauka, Moscow (1984).

23. J. H. Perepezko, "Nucleation in undercooled liquids," *Mater. Sci. Eng.*, **65**, 125-135 (1984).

24. C. Merton, M. Flemings, and Y. Schiohara, "Solidification of undercooled metals," *Mater. Sci. Eng.*, **65**, 157-170 (1984).

25. D. E. Ovsienko, M. V. Laputskii, O. Nechiporenko, and A. Medvedovskii, "Effect of aluminum on undercooling and spheroidization of metal drops," in: *Metal Physics* [in Russian], Naukova Dumka, Kiev (1971), No. 3, pp. 62-64.

26. D. E. Ovsienko, V. P. Kostyuchenko, V. V. Maslov, and G. A. Alfintsev, "Effect of undercooling on the structure of a nickel ingot," in: *Kinetics and Mechanism of Crystallization* [in Russian], Nauka i Tekhnika, Minsk (1973), pp. 75-81.

27. D. E. Ovsienko, G. A. Alfintsev, and V. V. Maslov, "Kinetics and shape of crystal growth from melt for substances with low L/kT values," *J. Cryst. Growth*, **26**, 233-238 (1974).

28. D. E. Ovsienko and G. A. Alfintsev, "Crystal growth from the melt. Experimental investigation of kinetics and morphology," in: *Crystal Growth. Properties and Applications*, Springer, Berlin–Heidelberg (1980), pp. 119-170.

29. V. K. Semenchenko, *Surface Effects in Metals and Alloys* [in Russian], Gostekhizdat, Moscow (1957).

30. N. L. Pokrovksiĭ and T. G. Smirnova, "Growth of tin grains in the presence of surface-active and -inactive impurities," in: *Growth and Imperfection of Metallic Crystals* [in Russian], Naukova Dumka, Kiev (1966), pp. 82-88.

31. D. E. Ovsienko, V. V. Maslov, and G. A. Alfintsev, "Mechanism of effect of soluble impurities on tin crystallization," *Kristallografiya*, **22**, No. 5, 1042-1050 (1977).

32. D. E. Ovsienko, V. V. Maslov, and G. A. Alfintsev, "Effect of some soluble additives on undercooling and structure of an iron ingot," in: *Problems of Steel Casting: Works of the VIth Conf. on Casting* [in Russian], Metallurgiya, Moscow (1976), pp. 60-62.

33. D. E. Ovsienko, V. V. Maslov, and G. A. Alfintsev, "Effect of silicon additives on undercooling and structure of a nickel ingot," *Izv. Akad. Nauk SSSR, Metally*, No. 4, 92 (1974).

34. T. Takahashi and W. A. Tiller, "Nucleation experiments of some alloys with low eutectic temperatures," *Acta Metall.*, **17**, No. 5, 651-656 (1969).

35. G. I. Distler, A. N. Lobanov, V. P. Vlasov, O. K. Mel'nikov, and N. S. Grindina, "Novel method of single crystal growth," *Dokl. Akad. Nauk SSSR*, **215**, No. 1, 91-94 (1974).

Part II

GROWTH OF CRYSTALS FROM MELTS
AND HIGH-TEMPERATURE SOLUTIONS

SYNTHESIS AND ELECTROPHYSICAL PROPERTIES
OF SUPERIONIC CONDUCTORS $Li_3M_2(PO_4)_3$ (M = Fe, Sc, Cr)

A. B. Bykov, L. N. Dem'yanets, S. N. Doronin, A. K. Ivanov-Shits,
O. K. Mel'nikov, B. K. Sevast'yanov, V. A. Timofeeva,
and A. P. Chirkin

Compounds with high ionic conduction (solid electrolytes) attract the attention of investigators with respect to both scientific interest (the nature and mechanism of ionic transport) and the promise of practical use (storage batteries, fuel elements, sensitive gas detectors and analyzers, etc.).

In spite of extensive searches, the number of superionic conductors (with conductivity at $300°C \geq 10^{-2}$ $\Omega^{-1} \cdot cm^{-1}$) is rather small. Individual compounds and solid solutions based on the structural types AgI, $NaAl_{11}O_{17}$ (β-alumina), β-$LiAlSiO_4$ (eucryptite), Li_4GeO_4 (Li_4SiO_4), and $Na_5TRSi_4O_{12}$ represent the most well-known examples.

Formation of the three-dimensional open framework $Me_2T_3O_{12}$ is characteristic for a large number of oxygen-containing compounds. Octahedra (MO_6) and tetrahedra (TO_4) can be distinguished in these structures. In such a mixed framework [1, 2], M—O—M or T—O—T bonds are not found. Each M atom is surrounded by six O atoms which belong to the vertices of six TO_4 tetrahedra. Each T atom is surrounded by four O atoms which belong to the vertices of MO_6 octahedra. The degree of skeletal deformation and the size of the void channels in the structure, and as a result the properties of the compound, change depending on the type of M, T cation, and the type and quantity of compensator cations A^{+n} which fill the framework.

The presently well-known Na-superionic conductors [3-5] are compounds with a mixed framework $M_2T_3O_{12}$.

At the start of this study, information in the literature [6] consisted of the preparation of the rhombohedral modification of $Li_3Fe_2(PO_4)_3$. However, data on the complete family of materials based on $Li_3M_2(PO_4)_3$ (M = Fe, Sc, Cr) were lacking.

Existing structural motifs for oxygen-containing compounds, whose framework contained cations of one (or several) types, were analyzed by us in the search for new superionic conductors with lithium ion conductivity. Experiments on the transport properties of these compounds and compounds which were prepared by isomorphic replacement with preservation of the structural type were carried out in parallel.

A detailed study of a new class of superionic conductors $Li_3M_2(PO_4)_3$ (M = Fe, Sc, Cr) resulted from this work. The study took the following directions: choice of synthetic conditions and growth of single crystals

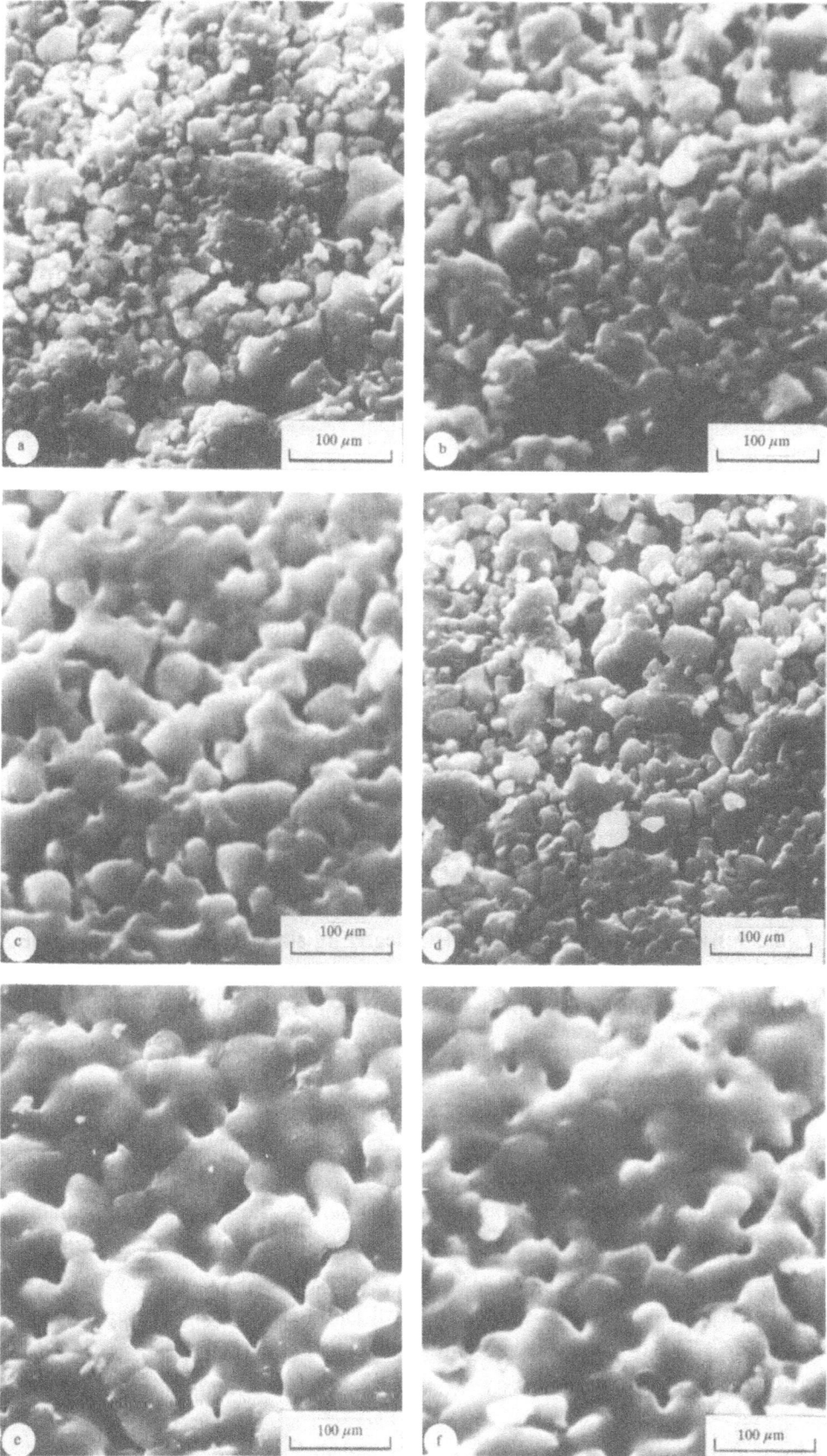

Fig. 1. $Li_3Sc_2(PO_4)_3$ microstructure as a function of thermal treatment procedure: (a) annealing time 2 h, annealing temperature T = 900°C; (b) 5 h, 900°C; (c) 10 h, 900°C; (d) 2 h, 1000°C; (e) 5 h, 1000°C; (f) 10 h, 1000°C; (g) 2 h, 1100°C; (h) 5 h, 1100°C; (i) 10 h, 1100°C; (j) 2 h, 1200°C; (k) 5 h, 1200°C; and (l) 10 h, 1200°C.

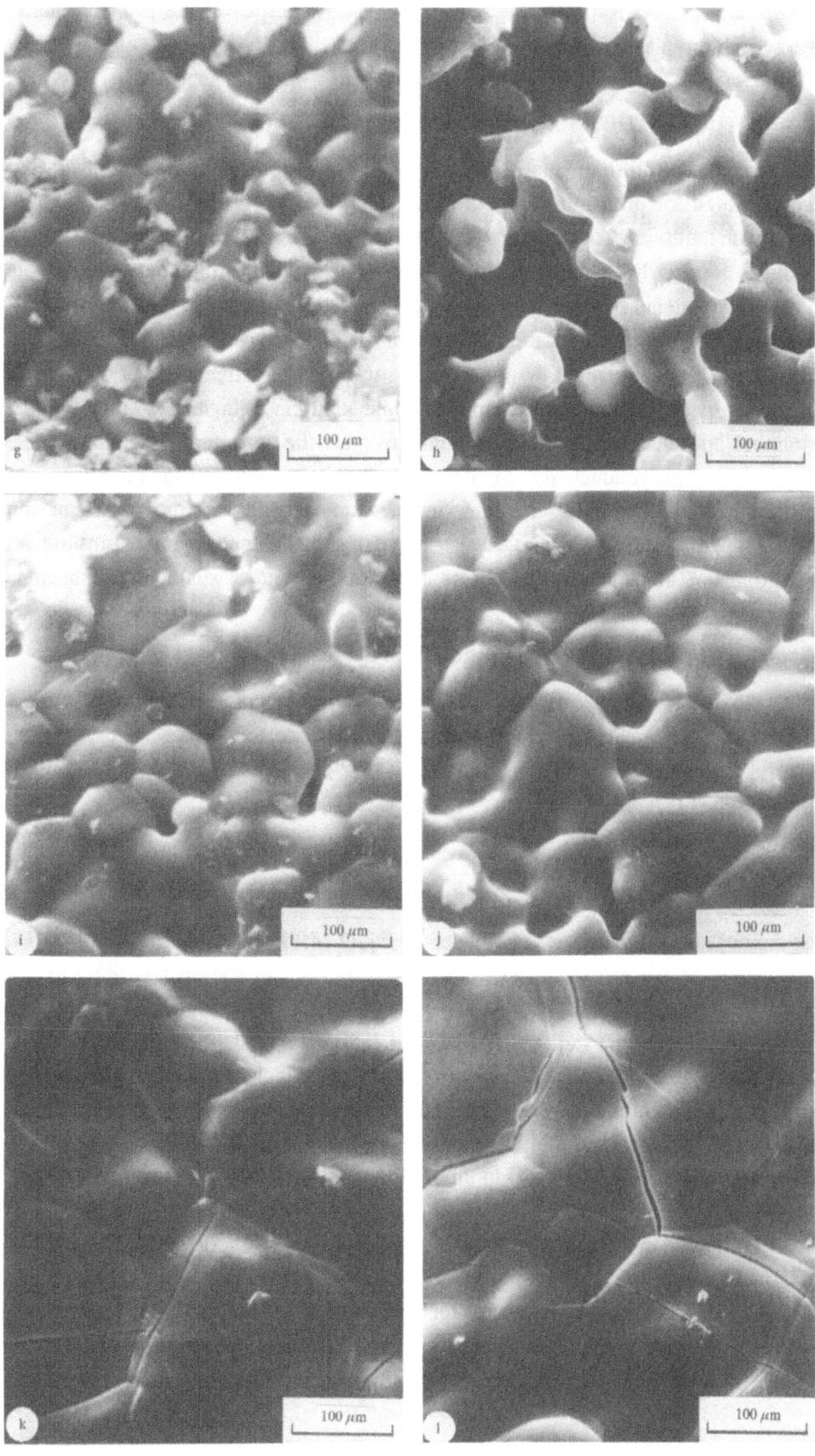

of individual solid solutions, investigation of the electrophysical properties of the prepared materials, and study of the effect of isomorphic replacement in the $Li_3M_2(PO_4)_3$ structural type on the electrophysical properties.

1. SYNTHESIS AND GROWTH OF $Li_3M_2(PO_4)_3$ CRYSTALS

Solid-phase synthesis, hydrothermal synthesis, crystallization from solution in a melt, and crystallization from a melt were employed for preparation of poly- and single crystals of the Li, M-phosphates.

The variety of methods used is explained by the formulation of the problem and the specificity of the examined compounds. The polycrystalline samples were necessary for preliminary studies of the structural, electrophysical, and electrochemical properties as well as for determination of the isomorphic relations in this group of compounds. The electrophysical properties for the majority of solid electrolytes were studied on polycrystalline samples. This is explained by the difficulty of preparing large single crystals of these compounds which are mainly incongruently melting multicomponent solid solutions. Single-crystalline samples were required for precision structural and electrochemical measurements since the structural features of the majority of the compounds indicate anisotropic conductivity. Single crystals were also required for determination of the conductivity value in different crystallographic directions and a detailed study of other physicochemical properties. Peculiarities of the Li, M-phosphates (ideal cleavage, highly aggressive melts, phase transitions, etc.) entailed a search for various routes of single crystal growth for mixed lithium and trivalent metal phosphates.

1.1. Solid-Phase Synthesis

The starting mixture for solid-phase synthesis was prepared by homogenization of a mechanical mixture of carbonates, phosphates, and oxides of the corresponding elements. Variants of the starting compositions, taken in ratios corresponding to the stoichiometry $Li_3M_2(PO_4)_3$, were

$Li_2CO_3 + NH_4H_2PO_4 + M_2O_3$;
$Li_2CO_3 + MPO_4 + NH_4H_2PO_4$;
$Li_3PO_4 + MPO_4$;
$Li_2CO_3 + MPO_4$.

The mixture was homogenized in a spherical grinder in liquid (acetone, isopropyl or ethyl alcohol) for 3-5 h. The temperature regime for the course of the solid-phase reaction was chosen according to DTA data for the starting mixture.

The ceramic was prepared by stepwise sintering of the mixture. The temperature of the first annealing (400-600°C) corresponded to the temperature for removal of gaseous decomposition products from the starting mixture (CO_2, NH_3). The mixture was held at this temperature for 2-4 h. After repeated mixing, the samples were placed into muffle furnaces and the temperature was raised to 900-1200°C depending on the sample composition. The completion of reaction was monitored by x-ray (DRON-2 diffractometer) and thermal (DTA-TG analyzer) methods.

Monophasic samples, corresponding to the composition $Li_2M_2P_3O_{12}$ (M = Sc, Fe, and Cr), were obtained after thermal treatment.

A ceramic with high density was necessary for studies of electrophysical and electrochemical properties. The microstructure of samples prepared by various thermal treatment procedures was examined in order to reveal the influence of such treatment and the choice of optimal conditions for preparation of high quality ceramics.

Fresh chips of tablets which were pressed and annealed by various procedures were covered with a 100 Å layer of gold and studied with a JSM-2 scanning microscope. The typical microstructure of polycrystalline

Table 1. Calculated X-Ray Powder Patterns of Synthesized Phosphates

Li$_3$Fe$_2$(PO$_4$)$_3$		Li$_3$Sc$_2$(PO$_4$)$_3$		Li$_3$Cr$_2$(PO$_4$)$_3$		Li$_3$Fe$_2$(PO$_4$)$_3$		Li$_3$Sc$_2$(PO$_4$)$_3$		Li$_3$Cr$_2$(PO$_4$)$_3$	
d/n, Å	I/I_0	d/n, Å	I/I_0	d/n, Å	I/I_0	d/n, Å	I/I_0	d/n, Å	I/I_0	d/n, Å	I/I_0
6.84	10	–	–	–	–	3,01	20	3,07	18	2,97	26
6.02	6	–	–	–	–	2,95	8	3,03	12	2,91	15
5.44	21	5,59	13	5.38	17	2.78	13	2,87	9	2,75	25
4.29	100	4,39	65	4.24	100	2.70	13	2,76	29	2.67	19
4.06	7	4,16	6	4.04	10	2.67	21	2,73	16	2,63	31
3.84	30	3,96	30	3,80	37	2.48	25	2,55	22	2,44	40
3.65	74	3.76	82	–	–	2.47	27	2,52	14	2,03	7
3.63	60	3,73	100	3.60	76	1.920	33	1,972	13	1,897	21
3.35	13	3,43	24	3.32	19	1.897	10	1,951	21	1,879	23
3.25	30	3.32	44	3.22	52	1.639	6	–	–	–	–
3.04	40	3,13	40	3,02	65	1,617	10	1,631	7	1.603	28
						–	–	1.587	6	1.596	35

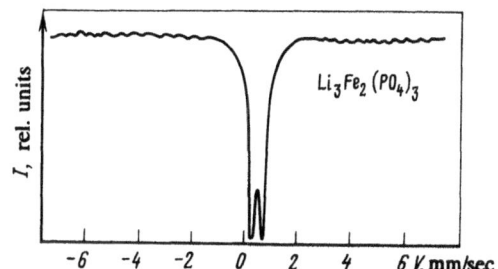

Fig. 2. Mössbauer spectrum of Li$_3$Fe$_2$(PO$_4$)$_3$ at T = 300 K.

Li$_3$Sc$_2$(PO$_4$)$_3$ samples is shown in Fig. 1. An increased duration of annealing at constant temperature evidently leads to partial coalescence of grains. An elevated thermal treatment temperature leads to the same effect. Samples with optimal microstructure (large single-crystalline regions, small number of intergrain interfaces) were prepared at an annealing temperature of 1200°C for 5-10 h. The average grain size in the lithium–scandium phosphate at T = 900°C was 3 μm. At T = 1200°C, it was 30 μm. Samples containing scandium are easily recrystallized. This allows single crystals of 0.3 mm size, which are suitable for single crystal x-ray structural studies, to be prepared with extended annealing.

1.2. Properties of Li$_3$M$_2$(PO$_4$)$_3$ Compounds

Calculated x-ray powder patterns of Li$_3$M$_2$(PO$_4$)$_3$, where M = Sc, Fe, and Cr, are given in Table 1.

Partial reduction of Fe^{+3} is possible during synthesis of Li$_3$Fe$_2$(PO$_4$)$_3$. The valence of iron ions in the synthesized compound was verified by Mössbauer spectroscopy. The Mössbauer spectrum of Li$_3$Fe$_2$(PO$_4$)$_3$ (Fig. 2) consists of a symmetric doublet. The isomer shift δ and the quadrupole splitting Δ (relative to α-Fe) are 0.4095 and 0.315 mm/sec, respectively, at room temperature. These values of δ and Δ are typical for trivalent high-spin iron ions. Based on these results, Fe^{+2} ions can be said to be absent up to 0.1 mole %.

With increased temperature, Li$_3$M$_2$(PO$_4$)$_3$ undergoes reversible phase transitions. The nature of these is illustrated in Fig. 3.

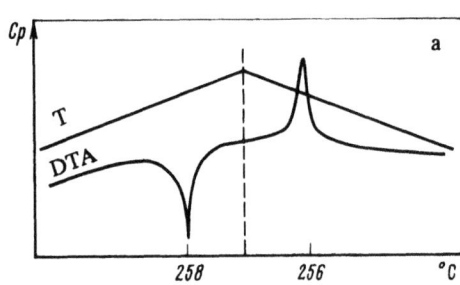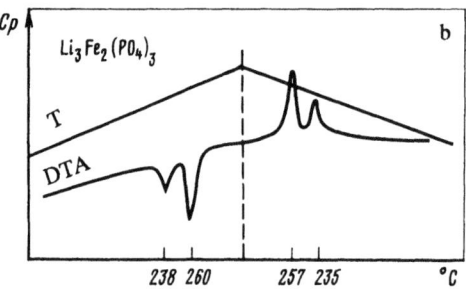

Fig. 3. Differential thermal analysis of $Li_3Sc_2(PO_4)_3$ (a) and $Li_3Fe_2(PO_4)_3$ (b). Thermal effects upon heating (left curve) and cooling (right curve) are shown.

Table 2. X-Ray Diffraction Data for the $Li_3M_2(PO_4)_3$ Compounds [6-13]

Composition	Unit cell constants				Space group
	a	b	c	angle	
α-$Li_3Sc_2(PO_4)_3$	8.853(2)	12.273(2)	8.802(2)	$\gamma = 90.00(2)$	$P2_1/n$
α-$Li_3Fe_2(PO_4)_3$	8.562(2)	12.005(3)	8.612(2)	$\gamma = 90.5(2)$	$P2_1/n$
α-$Li_3Cr_2(PO_4)_3$	8.453	8.547	12.520	$\beta = 125.15$	$P2_1/c$
β-$Li_3Sc_2(PO_4)_3$	8.858(1)	12.327(2)	8.816(1)	$\gamma = 90.00(2)$	$P2_1/n$
β-$Li_3Fe_2(PO_4)_3$	8.588(1)	12.112(2)	8.638(1)	$\gamma = 90.19(2)$	$P2_1/n$
γ-$Li_3Sc_2(PO_4)_3$	8.828(1)	12.399(2)	8.823(1)	–	$Pcan$
γ-$Li_3Fe_2(PO_4)_3$	8.592(2)	12.129(2)	8.637(1)	–	$Pcan$
R-$Li_3Fe_2(PO_4)_3$	8.310	–	22.06	–	$R3c$ ($R\bar{3}c$)

DTA data for Li, Fe-phosphate have been published [7, 8]. Our results agree with those in [8] on phase transitions in $Li_3Fe_2(PO_4)_3$ upon heating. One transition (Fig. 3) is detected for Li, Sc-phosphate. It should be noted that the temperatures of the phase transitions can change by several degrees depending on the thermal treatment procedure and the presence of microimpurities (various starting reagents). The $Li_3M_2(PO_4)_3$ compounds have several polymorphic modifications [6-13]. The three-dimensional MT-framework ($Fe_2(SO_4)_3$ type), which consists of M-octahedra and T-tetrahedra, provides the basis for these. This framework practically does not change during the phase transitions. Various structural variants (space group $P22_12_1$, $R\bar{3}c$, $P2_1/n$, and $Pcan$) are preferred for individual modifications. The complexity of unambiguous determination of the true space group is related to the effect of microtwinning. This aspect was examined in detail in [13]. The monoclinic space group $P2_1/n$ was identified for the low-temperature α-phases of $Li_3M_2(PO_4)_3$. Unit cell constants are given in Table 2. A transition to the β-phase, which is characterized by the same space group but with a different distribution of mobile ions, is observed with increased temperature. The furthest increase of temperature leads to appearance of a rhombic superionic γ-phase with space group $Pcan$ (Table 2).

1.3. Hydrothermal Synthesis

Experiments on hydrothermal crystallization in multicomponent phosphate systems for preparation of single $Li_3M_2(PO_4)_3$ crystals were carried out on the following systems: LiH_2PO_4–Fe_2O_3–H_3PO_4–H_2O, LiH_2PO_4–Sc_2O_3–H_3PO_4–H_2O, $Li_3Fe_2(PO_4)_3$–$NH_4H_2PO_4$–H_2O, $Li_3Sc_2(PO_4)_3$–$NH_4H_2PO_4$–H_2O, $Li_3Fe_2(PO_4)_3$–LiH_2PO_4–H_2O, $Li_3Sc_2(PO_4)_3$–LiH_2PO_4–H_2O, LiH_2PO_4–$FePO_4$–H_3PO_4–H_2O, and $Li_3Fe_2(PO_4)_3$–H_3PO_4–H_2O.

Table 3. Phase Formation in the $Fe_2O_3-NH_4H_2PO_4-LiF-V_2O_5$ System

Solution component			Crystallizing phase	Solution component			Crystallizing phase
LiF	$NH_4H_2PO_4$	V_2O_5		LiF	$NH_4H_2PO_4$	V_2O_5	
4.0	3,2	–	Fe_2O_3 ; $Li_3Fe_2(PO_4)_3$	5,6	4,9	3,9	$Li_3Fe_2(PO_4)_3$
6,0	3,2	–	Fe_2O_3	3,7	5,6	1,8	$LiFeP_2O_7$
6,0	3,2	3,0	Fe_2O_3 ; $Li_3Fe_2(PO_4)_3$				

Note. Fe_2O_3 = 1 mole.

Experiments were performed under temperature gradient conditions. The experimental temperature was 300–450°C. The temperature gradient was 1.5–2 deg/cm. The pressure was $(5-15) \cdot 10^7$ Pa. The concentration of solvent and ratio of starting solid components were varied in the experiments. The solid products of hydrothermal reaction were washed in running water and analyzed by x-ray powder diffraction and thermal analysis.

In the first two of the above systems, formation of $Li_3M_2(PO_4)_3$ was not observed in the studied range of variable parameters. Stable products of the hydrothermal crystallization were iron and scandium orthophosphates, $FePO_4$ and $ScPO_4$; lithium phosphate, Li_3PO_4; $LiFePO_4OH$; and oxides and hydroxides of iron and scandium. Due to this, the experimental scheme was changed and the starting oxides and salts were replaced with a mixed Li, M-phosphate which was prepared by preliminary solid-phase synthesis. Thus, if formation of solid products in the first case proceeded due to exchange reactions and synthesis, then in the second case (the four following systems) an attempt was made to prepare single $Li_3M_2(PO_4)_3$ crystals by direct crystallization. In this case, separate single $Li_3M_2(PO_4)_3$ crystals were actually prepared but as a metastable phase which with extended duration of the experiment transformed into stable lithium and trivalent element orthophosphates. The hydrothermal $Li_3M_2(PO_4)_3$ samples had structural properties which were analogous to those of samples prepared by solid-phase synthesis. Thus, the studies carried out showed that in this case the hydrothermal method was not optimal for preparation of large $Li_3M_2(PO_4)_3$ crystals.

1.4. Crystallization from Solution in a Melt

Experiments on solution–melt crystallization were performed in fluoride–oxide fluids. A mixture of $LiF-V_2O_5$ was used as solvent. The lithium fluoride acted as both solvent and a source of Li ions for the crystallizing compound. Data on phase formation in the $Fe_2O_3-NH_4H_2PO_4-LiF-V_2O_5$ system are given in Table 3. Crystallization of Fe_2O_3 and $LiFeP_2O_7$ besides that of the desired compound is seen to be possible.

Growth of $Li_3M_2(PO_4)_3$ crystals was done in standard platinum crucibles [number (100-13)] with a solution–melt volume of 100–150 cm^3 by localization of nucleus formation on a rotating platinum rod of 6 mm diameter at an average system cooling rate of 0.5 deg/h. The temperature of initial crystallization in all cases did not exceed 850°C. The crystallization temperature range was usually 800–650°C.

Study of phase formation in the $M_2O_3-NH_4H_2PO_4-LiF-V_2O_5$ systems allowed the crystallization temperature stability range of phosphates with these compositions to be determined and the conditions for crystallization of monophasic $Li_3Sc_2(PO_4)_3$, $Li_3Fe_2(PO_4)_3$, $Li_3(Fe, Sc)_2(PO_4)_3$, and $Li_3(Sc, Cr)_2(PO_4)_3$ to be found. The habit of the crystals depends substantially on the crystallization conditions (rate of system cooling) and the concentration of crystal-forming components. Prismatic crystals are formed at lower concentrations and smaller cooling rates. Platelets occur at higher concentrations and larger cooling rates (Fig. 4). Crystals of $Li_3Fe_2(PO_4)_3$ as elongated yellowish-brown prisms of $2 \times 2 \times 10$ mm size were prepared by successive dilution of the solution–melt by vanadium and phosphorus oxides. Crystals of $Li_3Sc_2(PO_4)_3$ and $Li_3Sc_{2-x}Cr_x(PO_4)_3$ of

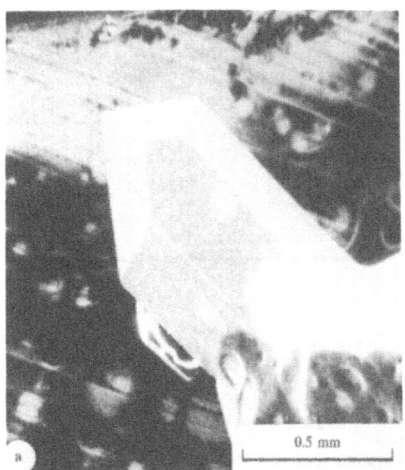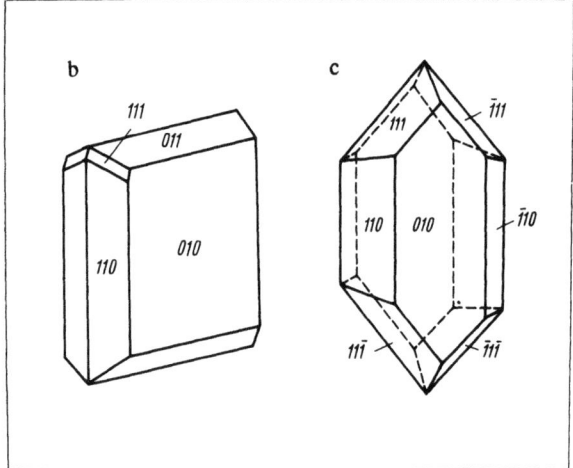

Fig. 4. $Li_3Fe_2(PO_4)_3$ single crystals prepared from a solution–melt (a) and simplified habits of platelet (b) and prismatic (c) crystals.

$1 \times 4 \times 4$ mm size change color depending on chromium content from colorless to green. These crystals contain inclusions of mother solution and were not used for precise electrophysical measurements.

1.5. Crystallization from the Melt

DTA results showed that $Li_3Sc_2(PO_4)_3$ crystals melt congruently at 1490°C. This gave hope for possible preparation of crystals by direct crystallization from the melt. This was all the more necessary since the other methods did not produce $Li_3Sc_2(PO_4)_3$ crystals of the required quality. However, even the first experiments showed that mixtures of composition $Li_3Sc_2(PO_4)_3$ gradually decompose to starting material with extended storage as a melt according to the equation

$$Li_3Sc_2(PO_4)_3 \rightarrow Li_3PO_4 + Sc_2O_3 + P_2O_5.$$

Disturbance of the melt stoichiometry results from this reaction, and the quality of the growing crystals is correspondingly degraded.

Crystals of $Li_3Sc_2(PO_4)_3$ can be prepared by the Czochralski and Kyropoulos methods on an apparatus with high-frequency heating at sufficiently high growth rates. Either Pt or Ir crucibles of 100-150 cm³ capacity were used as containers. Growth occurred in air (Pt containers) or helium (Ir containers). A Pt rod was used as a seed in the first experiments. In subsequent ones, single-crystalline blocks prepared by crystallization from a melt were used. The rate of rotation of the seed rod was 15-20 rpm. The rate of withdrawal was 1-5 mm/h.

The prepared boules consisted of several single-crystalline blocks which had ideal cleavage in two directions. The block size of $5 \times 10 \times 10$ mm allowed samples which were suitable for precision electrophysical measurements to be prepared.

The compound $Li_3Fe_2(PO_4)_3$, as well as $Li_3Sc_2(PO_4)_3$, melts incongruently at 1060°C. However, the melt in air undergoes slow decomposition with extended storage by the following reaction:

$$2Li_3Fe_2(PO_4)_3 \rightarrow 4LiFePO_4 + 2LiPO_3 + O_2\uparrow.$$

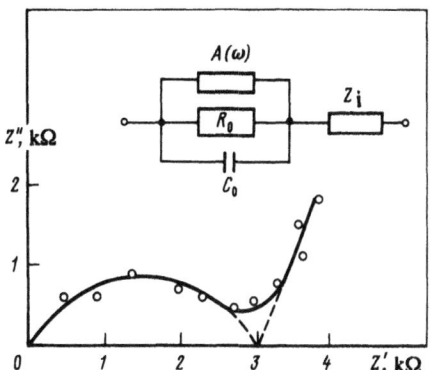

Fig. 5. Hodograph of impedance of the electrochemical cell $C|Li_3M_2(PO_4)_3|C$. R_0 is the active resistivity, C_0 is the geometric capacitance, $A(\omega)$ is the frequency-dependent element, and Z_i is the element which describes the impedance of the electrode/sample interface.

Excess pressure of pure oxygen > 120 atm is necessary for prevention of this decomposition. This presents serious demands for the growth apparatus which at present cannot be satisfied.

The experiments carried out by us on growth of $Li_3Fe_2(PO_4)_3$ crystals from a melt in an oxidizing atmosphere showed that the oxygen pressure ($3 \cdot 10^7$ Pa) attained in the apparatuses was insufficient for suppression of the decomposition reaction cited above.

Another substantial difficulty, which arises during growth of single Li, Sc- and Li, Fe-phosphate crystals, should be pointed out. This is the presence of phase transitions which occur upon lowering the temperature from the crystallization temperature to ambient. These transitions occur with a significant change of unit cell volume. This leads to large stress and cracking of the prepared crystals.

2. ELECTROPHYSICAL PROPERTIES OF SUPERIONIC CONDUCTORS OF THE $Li_3M_2(PO_4)_3$ FAMILY

2.1. Experimental Method

The $Li_3M_2(PO_4)_3$ materials have primarily ionic conduction, as noted in [6, 7, 10]. Therefore, measurements were made in alternating current for determination of total resistivity of samples. The experimental setup, described in [14], permitted experiments to be carried out over a wide temperature range from 100 to 900 K with a temperature control of ≈ 0.5 K. Measurements can be taken in vacuum (≈ 6 Pa) or in an atmosphere of inert argon which has been purified from traces of oxygen and water.

Two types of samples were used in the course of the experiments, ceramic polycrystalline samples and single-crystalline. Polycrystalline samples as tablets of 5-10 mm diameter and 1-3 mm thickness were prepared from uniform homogenized starting material by cold pressing with uniaxial compression of $\approx (7-10) \cdot 10^3$ kg/cm^2. Since the studied materials are not plastic, the pressed samples were annealed at temperatures 50-100° below the melting point. As a result, the tablets experienced shrinkage and the sample acquired ceramic properties. Various additive—fillers (polyvinyl alcohol type) were applied in a number of cases for preparation of ceramics. This allowed samples with a density up to ~90% of theoretical to be prepared.

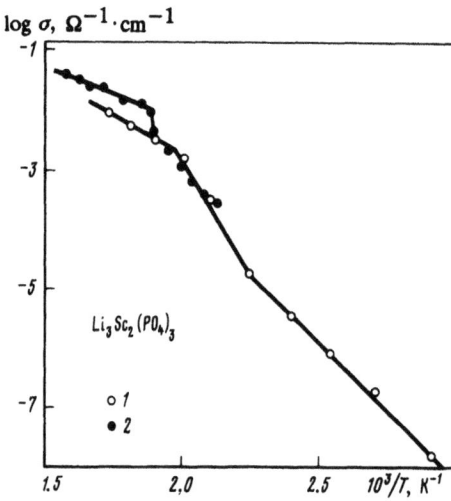

Fig. 6. Conductivity of $Li_3Sc_2(PO_4)_3$ as a function of temperature. 1) Total conductivity of the ceramic and 2) of a single crystal (in the direction $\parallel C$).

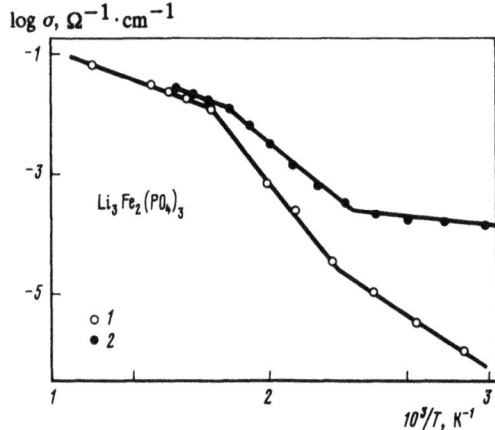

Fig. 7. Conductivity of $Li_3Fe_2(PO_4)_3$ as a function of temperature. 1) Total conductivity; 2) conductivity at a frequency of 500 MHz.

An electrochemical cell with inert electrodes was used for recording frequency characteristics. The electrodes were irreversible relative to the type of ions which participated in the charge transfer. Carbon, gold, nickel, and silver, which were deposited as pastes or a sprayed layer on ends of the ceramic or single crystals, were used as inert electrodes. The single-crystalline samples were oriented before measurements along the three principal crystallographic axes.

Impedance of samples was measured using E8-2 (frequency range from 0.1 to 20 kHz) and P-5021 ($\Delta f = 0.02$-200 kHz) alternating current bridges and a BM-507 ($\Delta f = 0.005$-500 kHz) impedance meter. Corrections for spurious impedance of the supply wires were introduced during measurements at high frequencies ($f > 100$ kHz).

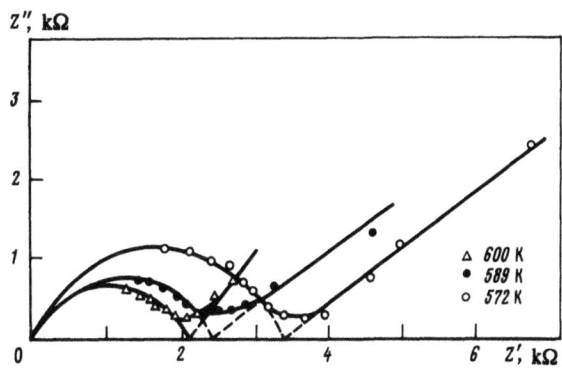

Fig. 8. Hodographs of impedance of the electrochemical cell
$C \mid Li_3Sc_2(PO_4)_3 \mid C$ for the single-crystalline sample.

2.2. Electric Conductivity of Ceramic Samples

The measured resistivity during study of polycrystalline samples is frequently the sum of resistivities of the single-crystalline grains themselves (bulk resistivity) and the intergrain interfaces. Hodographs of the impedance of polycrystalline ceramics at frequencies from 5 to $5 \cdot 10^5$ Hz are shown in Fig. 5 as an example. Analysis of the hodographs [i.e., the functions $Z''(Z')$] shows that the bulk resistivity of the sample cannot be separated in the case examined. The equivalent electrical diagram which correctly describes the observed path of the hodograph has the form shown in the inset of Fig. 5. Here R_0 is the total (summed) resistivity of the sample, $A(\omega)$ is the frequency-dependent element whose impedance can be described as $Z_A = A(j\omega)^\alpha$ [15, 16], and Z_i is the element which characterizes the processes which occur at the electrode/sample interface. Final analysis of the measurement results of cell frequency characteristics was carried out with the assumption that the equivalent electrical diagram (Fig. 5) adequately describes the cell over the whole working temperature range. Thus, the active resistivity, which characterizes the resistivity of the studied sample, is found from the intersection of the hodographs with the abscissa. The conductivity (σ) of the material is then determined from the sample geometry.

Figures 6 and 7 show the conductivities of lithium–scandium and lithium–iron phosphates as functions of temperature. A linear path of electric conductivity on inverse temperature is observed at high temperatures, i.e., σT as a function of $1/T$ is described by the relation

$$\sigma T = \sigma_0 \exp(-E_\sigma/kT),$$

where E_σ is the activation energy of conductivity and σ_0 is the preexponential factor.

A significant lowering of σ with a change of activation energy is observed upon lowering the temperature to the vicinity of the $\gamma \leftrightarrow \beta$-phase transition from the superionic phase. As noted earlier [13], the β-modification of $Li_3M_2(PO_4)_3$ is stable in a limited temperature range. Therefore, one more break, related to the transition to the α-modification, is visible on the $\sigma(T)$ curves with further lowering of temperature. It should be pointed out that the $\beta \leftrightarrow \alpha$-transition is not exhibited very markedly and apparently depends mainly (from the DTA results) on the prior history of the sample (its thermal treatment) and its quality (presence of impurities, cracks, etc.).

Figure 7 also shows measurements of ceramic conductivity at a frequency of 500 MHz. The resistivity of intergrain interfaces in this case can logically be assumed to be insignificant (in any case, substantially less than

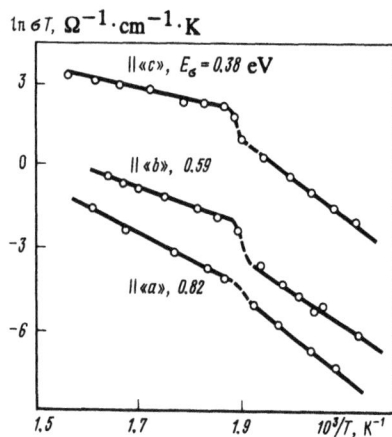

Fig. 9. Conductivity of single $Li_3Sc_2(PO_4)_3$ crystals along the three crystallographic axes as functions of temperature. Numbers by the curves are activation energies of conductivity.

for measurements at frequencies up to 500 kHz). Thus, the measured resistivity characterizes the bulk properties of the sample. As seen from Fig. 7, the largest deviation in the results is encountered in the α- and β-modifications whereas for the superionic γ-phase the conductivity value determined by various methods differs insignificantly.

2.3. Electric Conductivity of Single Crystals

A sample in the shape of a right-angled block of dimensions 5 × 3.5 × 3 mm was cut from a single-crystalline block of $Li_3Sc_2(PO_4)_3$ for measurement of electric conductivity. The sides of the block were oriented along the three principal crystal axes. Electrodes of silver or graphite pastes were attached to the polished surfaces of the sample.

Figure 8 shows the impedance hodographs of the cell $C|Li_3Sc_2(PO_4)_3|C$ for a sample oriented along the b axis. In this case, as for the ceramic samples, the path of the hodograph is described by the equivalent electric diagram (Fig. 5). However, R_0 now corresponds to the pure bulk resistivity of the crystal since intercrystallite interfaces are absent.

The conductivities, measured between 480-700 K, as functions of temperature are shown in Fig. 9 [17]. It should be noted that the two main features of the behavior of σ are inherent to the superionic conductor $Li_3Sc_2(PO_4)_3$. First, the transition to the superionic state is accompanied by a sharp jump in σ. Second, clearly exhibited anisotropy of conductivity is observed both before and after the superionic $\gamma \leftrightarrow \beta$-phase transition. The transition to the superionic state occurs at $\approx$525 K. At this same temperature, an endothermic effect is observed on the forward path (increasing temperature) of the DTA curve (Fig. 3). The observed transition is also accompanied by rearrangement of the crystal lattice with a symmetry change from $P2_1/n$ to $Pcan$ and a redistribution in the population of the lithium atom positions [11-13]. Only a change of the activation energy of conductivity is noticed in the $\sigma(T)$ curves for polycrystalline ceramic samples (Figs. 6 and 7, as well as [6]) in the vicinity of the phase transition. However, the presence of an abrupt change of conductivity is also characteristic for the single crystals. The absence of a jump in electric conductivity is apparently related to a washout of the transition during measurements on polycrystalline samples. This was seen for other solid electrolytes. Besides this, the jump is weakly exhibited during measurements along the a axis. This also muddles the picture of the transition in ceramic samples.

Table 4. Possible Isomorphic Replacements in $Li_3(Fe, Sc)_2(PO_4)_3$ (Ionic Radii of the Variously Charged Ions Are Given)

Li^{1+} 0.68 (3)		(Fe^{3+}, Sc^{3+}) 0.67 0.83 (2)				P^{5+} 0.35 (3)			O^{2-} 1.36 (12)
Na^{1+} 0.98	Mg^{2+} 0.74	Al^{3+} 0.57	Mg^{2+} 0.74	Ti^{4+} 0.64 / Nb^{5+} 0.66	As^{5+} 0.47	Si^{4+} 0.39	Mo^{6+} 0.65	Fe^{1-} 1.33	
Cu^{1+} 0.98	Zn^{2+} 0.83	Ga^{3+} 0.62	Ca^{2+} 1.04	Zr^{4+} 0.82 / Ta^{5+} 0.66	V^{5+} 0.40	Ge^{4+} 0.44	W^{6+} 0.65		
Ag^{1+} 1.13	Cu^{2+} 0.80	In^{3+} 0.92	Zn^{2+} 0.83	Hf^{4+} 0.82 / Sb^{5+} 0.62	Nb^{5+} 0.66				
		Y^{3+} 0.97	Cd^{2+} 0.99	V^{4+} 0.61 / Bi^{5+} 0.74	Ta^{5+} 0.66				
		TR^{3+} 0.8–1.02	Cu^{2+} 0.80	Na^{4+} 0.67	Sb^{5+} 0.62				
		Cr^{3+} 0.64	Ni^{2+} 0.74	Mo^{4+} 0.68	Bi^{5+} 0.74				
		Mn^{3+} 0.70	Mn^{2+} 0.91	Sn^{4+} 0.67					
		Co^{3+} 0.64		Pb^{4+} 0.76					

Table 5. Solid Solutions with Heterovalent Isomorphism (Starting Compound $A_3^{+1}M_2^{+2}(PO_4)_3$; B, K, and T are ionic replacements)

Replacement	Solid solution	Replacement	Solid solution
$2A^{1+} \rightarrow B^{2+} + Vac$	$A_{3-2x}^{1+}B_x^{2+}M_2^{3+}(PO_4)_3$	$M^{3+} + T^{5+} \rightarrow B^{2+} + K^{6+}$	$A_3^{1+}B_x^{2+}M_{x-2}^{3+}K_x^{6+}T_{3-x}^{5+}O_{12}$
$A^{1+} + M^{3+} \rightarrow 2B^{2+}$	$A_{3-x}^{1+}B_{2x}^{2+}M_{2-x}^{3+}(PO_4)_3$	$3A^{1+} + M^{3+} \rightarrow B^{6+} + Vac$	$A_{3-3x}^{1+}M_{2-x}^{3+}B_x^{6+}(PO_4)_3$
$T^{5+} \rightarrow B^{4+} + A^{1+}$	$A_{3+x}^{1+}M_2^{3+}B^{4+}T_{3-x}^{5+}O_{12}$	$2T^{5+} \rightarrow K^{4+} + B^{6+}$	$A_3^{1+}M_2^{3+}K_x^{4+}B_x^{6+}T_{3-2x}^{5+}O_{12}$
$M^{3+} \rightarrow A^{1+}$	$A_{3+x}^{1+}M_{2-x}^{3+}P_3O_{12-x}$	$2M^{3+} \rightarrow K^{4+} + B^{2+}$	$A_3^{1+}B_x^{2+}M_{2-2x}^{3+}K_x^{4+}(PO_4)_3$
$M^{3+} + A^{1+} \rightarrow B^{4+}$	$A_{3-x}^{1+}M_{2-x}^{3+}B_x^{4+}(PO_4)_3$	$2M^{3+} \rightarrow B^{5+} + A^{1+}$	$A_{3+x}^{1+}M_{2-2x}^{3+}B_x^{5+}(PO_4)_3$

The highest value of conductivity is noted during measurements along the c axis, as seen in Fig. 9. This agrees with conclusions based on examination of structural features of lithium–scandium phosphate [11-13]. Transfer of ions along this same direction in the superionic phase is achieved with the least difficulties. This follows from comparison of the values of activation energy E_σ, shown in Fig. 9.

2.4. Isomorphic Replacement in the $Li_3M_2(PO_4)_3$ Structural Type

A change of various physicochemical properties of compounds is often observed upon isomorphic replacements. The effect of composition of materials based on $Li_3M_2(PO_4)_3$ on the electric conductivity can be related either to a distortion of the initial lattice, or to a change of the degree of occupancy of crystallographic

Table 6. Conductivity of Compounds Based on Scandium and Iron Phosphates with Several Isomorphic Substitutions

Composition	T, °C	R, $\Omega^{-1} \cdot cm^{-1}$	Composition	T, °C	R, $\Omega^{-1} \cdot cm^{-1}$
$Li_3Sc_{1.8}In_{0.2}P_3O_{12}$	470	$1.78 \cdot 10^{-2}$	$Li_{2.9}Sc_{1.7}Zr_{0.2}P_3O_{12}$	470	$6.58 \cdot 10^{-2}$
$Li_3Sc_{1.8}Cr_{0.2}P_3O_{12}$	470	$8.58 \cdot 10^{-2}$	$Li_3Sc_2P_{2.8}Mo_{0.2}O_{12}$	470	$5.3 \cdot 10^{-4}$
$Li_3Sc_{2.9}Y_{0.1}P_3O_{12}$	470	$7.94 \cdot 10^{-2}$	$Li_{3.3}Fe_2P_{2.7}O_{0.3}O_{12}$	320	$1.5 \cdot 10^{-2}$
$Li_3Sc_{1.8}Y_{0.2}P_3O_{12}$	470	$5.62 \cdot 10^{-2}$	$Li_{3.1}Fe_2P_{2.9}Si_{0.1}O_{12}$	320	$3.8 \cdot 10^{-2}$
$Li_3Sc_2P_3O_{12} \cdot 0.1Y_2O_3$	470	$9.43 \cdot 10^{-2}$	$Li_{2.8}Fe_2P_{2.9}Si_{0.1}O_{12}$	320	$1.0 \cdot 10^{-2}$
$Li_{2.8}Fe_{1.8}Ti_{0.2}P_3O_{12}$	470	$5.01 \cdot 10^{-2}$	$Li_2MgScP_2MoO_{12}$	450	$1.99 \cdot 10^{-2}$
			$Li_3Sc_{1.8}Zn_{0.2}P_{2.8}Mo_{0.2}O_{12}$	470	$3.2 \cdot 10^{-2}$

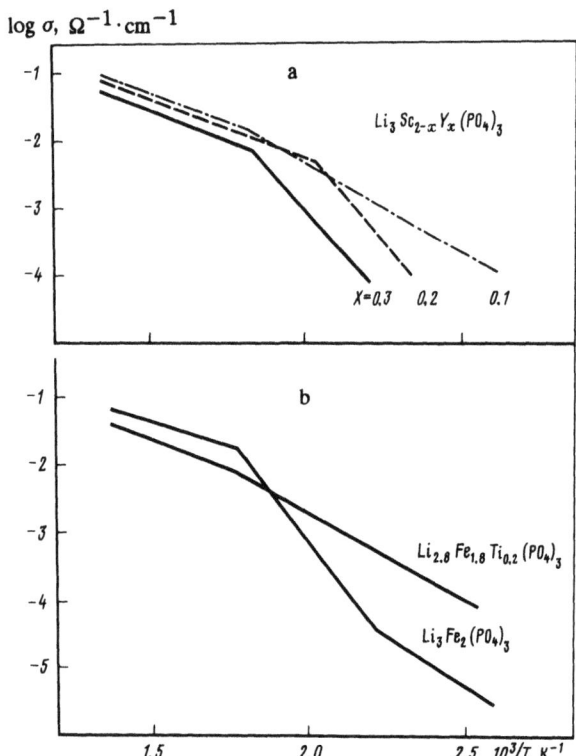

Fig. 10. Conductivity in solid solutions based on $Li_3M_2(PO_4)_3$ as functions of temperature. a) Replacement $Sc^{+3} \rightarrow Y^{+3}$; b) replacement $Li^+ + Fe^{+3} \rightarrow Ti^{+4}$.

positions by mobile ions, or to a redistribution of atoms among the vacant crystallographic positions. Eventually, this leads to increased possibility of ionic transport within the frameworks examined.

Cases of complete isomorphism based on $Li_3Sc_2(PO_4)_3$ are rather limited. The replacements Sc—Fe, Sc—Cr, and Sc—V are known. Partial isomorphism presumes an enormous number of heterovalent replacement variants (Table 4). We studied isomorphic replacements in systems with complete isomorphism. These systems are characterized by the formation of solid substitution solutions over the whole range of compositions. The change of crystal lattice constants of the studied materials obeys Vegard's law.

The electric conductivity of all solid solutions with a high-temperature superionic γ-phase depends to a small degree on the type of substituting atom and varies between $1 \cdot 10^{-2}$-$5 \cdot 10^{-2}$ $\Omega^{-1} \cdot cm^{-1}$ at 600 K.

A larger number of variants is realized upon heterovalent substitution since formation of both vacancies and interstitial ions is possible in the structures. A whole series of substitution variants in the initial composition $A_3^{+1}M_2^{+3}(PO_4)_3$ is given in Table 5. In this situation, replacement of all elements of the starting matrix by atoms of various valences is examined.

Measurements of electric conductivity of several solid solutions are shown in Table 6 and Fig. 10. We note that marked improvement of conductivity (in the superionic phase) is not observed in most cases. The weak influence of isomorphism on the electrophysical characteristics of the high-temperature phase, the values of σ and E_a, is apparently related to the fact that the concentration of mobile ions in the superionic state depends weakly on the doping additive. Also, the small possible structural distortions practically do not affect the potential relief. This result is characteristic not only for materials based on $Li_3M_2(PO_4)_3$ but also for a large number of other superionic materials whose structure is characterized by the presence of isolated tetrahedral anions [18].

The conductivity can vary within small limits of lattice saturation [18] for heterovalent replacements which lead to a change of concentration of the monovalent mobile ions.

We also note that introduction of a whole series of additives leads to a change of the phase transition temperature $T_{\beta \to \gamma}$ (Fig. 10). This allows the region of existence of the high-temperature superionic phase to be widened.

3. CONCLUSIONS

Studies on composition–synthesis–property diagrams with incorporation of structural data allowed a new family of promising superionic conductors based on $Li_3M_2(PO_4)_3$ compounds, where M = Sc, Fe, and Cr(V), to be discovered and investigated in detail. Their chemical composition and features of the thermal behavior of the $Li_3M_2(PO_4)_3$ compounds give rise to significant difficulties for preparation of single crystals of these compounds. The difficulties are caused by the high reactivity of the melt, a possible change of valence at elevated temperature, the presence of microtwinning, phase transitions, etc. Special requirements in this plan are also placed on preparation methods of single crystals. Narrow ranges of concentrations in the hydrothermal and solution–melt methods, pressure of oxygen and precise temperature procedures in the melt methods, etc., are noted.

Methods of solid-phase synthesis for preparation of dense ceramics were developed as a result of the completed studies. Crystallization of Li-, Sc-, (Fe)-phosphates in hydrothermal media, from solution in a melt, and from a melt, were investigated for choice of optimal conditions for growth of single crystals. $Li_3Sc_2(PO_4)_3$ and $Li_3Fe_2(PO_4)_3$ single crystals with sizes up to 0.5 cm^3 and of good quality were grown.

The electrophysical properties of mixed phosphates were studied. Conductivities as functions of temperature were obtained for various compositions. Two main features of the behavior of σ, inherent to the superionic conductor $Li_3Sc_2(PO_4)_3$, were detected. The transition to a highly conducting phase is accompanied by an abrupt change of σ, the value of which depends on the choice of crystallographic direction. This effect was not observed in the ceramic, possibly due to wash-out caused by the average orientation of the microcrystals or due to metastability of the β-phase (microtwinning effect). Anisotropy of conductivity is exhibited both before and after the phase transition. This feature agrees with data of precise structural studies which attests to further promise of quantitative comparison of electrophysical and structural parameters with directed change of phosphate phase compositions.

REFERENCES

1. V. V. Ilyukhin, A. A. Voronkov, and V. K. Trunov, "Crystal chemistry of mixed frameworks. Features of the $M_2T_3O_{12}$ framework," *Koord. Khim.*, **7**, No. 11, 1603-1612 (1981).

2. R. G. Sizova, V. A. Blinov, A. A. Voronkov, V. V. Ilyukhin, and N. V. Belov, "Refined $Na_4Zr_2(SiO_4)_3$ structure and its place in the series of mixed frameworks with general formula $M_2(TO_4)_3$," *Kristallografiya*, **26**, No. 2, 293-300 (1981).

3. H. Y.-P. Hong, "Crystal structures and crystal chemistry in the system $Na_{1+x}Zr_2Si_xP_{3-x}O_{12}$," *Mater. Res. Bull.*, **11**, No. 1, 173-182 (1976).

4. V. A. Efremov and V. B. Kalinin, "Determination of the crystal structure of $Na_3Sc_2(PO_4)_3$," *Kristallografiya*, **23**, No. 4, 703-708 (1978).

5. B. I. Lazoryak, V. B. Kalinin, S. Yu. Stefanovich, and V. A. Efremov, "Crystal structure of $Na_3Sc_2(PO_4)_3$ at 60°C," *Dokl. Akad. Nauk SSSR*, **250**, No. 4, 861-864 (1980).

6. M. Pintard-Screpel, F. d'Yvoire, and F. C. Remy, "Polymorphism and ionic constants of the phosphate $Na_3R_2(PO_4)_3$," *C. R. Hebd. Seances Acad. Sci., Ser. C*, **286**, No. 13, 381-383 (1978).

7. M. Pintard-Screpel, F. d'Yvoire, and F. Bretey, "Ionic conductivity, phase transitions and structure of 3D framework phosphates $M_3^IM_2^{III}(PO_4)_3$: M^I = Li, Na, Ag; M^{III} = Cr, Fe," Solid State Chem. Proc. Second Eur. Conf., R. Metselaar, H. J. M. Heijligers, and J. Schoonman (eds.), Studies in Inorganic Chemistry, Veidhoven, 7-9 June, 1982 (1983), Vol. 3, pp. 830-839.

8. F. d'Yvoire, M. Pintard-Screpel, F. Bretey, and M. de la Rochere, "Phase transitions and ionic conduction in 3D framework phosphates," *Solid State Ionics*, **9**, No. 10, 851-858 (1983).

9. C. Delmas, R. Olazcuaga, G. le Flem, et al., "Crystal chemistry of the $Na_{1+x}Zr_{2-x}L_x(PO_4)_3$ (L = Cr, In, Yb) solid solution," *Mater. Res. Bull.*, **16**, No. 2, 285-290 (1981).

10. E. A. Genkina, L. N. Dem'yanets, A. K. Ivanov-Shits, et al., "High ionic conduction in the compounds $Li_3Fe_2(PO_4)_3$ and $Li_3Sc_2(PO_4)_3$," *Pis'ma Zh. Tekh. Fiz.*, **38**, No. 3, 257-259 (1983).

11. E. A. Genkina, L. A. Muradyan, and B. A. Maksimov, "Crystal structure of the monoclinic modification of $Li_3Sc_2(PO_4)_3$ at T = 293 K," *Kristallografiya*, **31**, No. 3, 595-596 (1986).

12. I. A. Verin, E. A. Genkina, B. A. Maksimov, and L. A. Muradyan, "Crystal structure of the ionic conductor $Li_3Fe_2(PO_4)_3$ at T = 593 K," *Kristallografiya*, **30**, No. 4, 677-681 (1985).

13. I. P. Kondratyuk, M. I. Sirota, B. A. Maksimov, et al., "X-ray structural study of microtwinning of $Li_3Sc_2(PO_4)_3$ and $Li_3Fe_2(PO_4)_3$ crystals," *Kristallografiya*, **31**, No. 3, 488-494 (1986).

14. A. K. Ivanov-Shits, V. S. Borovkov, V. G. Fadeev, and V. A. Baskakov, "Apparatus for measurement of conductivity of solid electrolytes over a wide temperature range," All-Union Institute of Scientific and Technical Information (VINITI), deposited paper No. 3002-77 (1977).

15. A. K. Jonsher, "Analysis of the alternating current properties of ionic conductors," *J. Mater. Sci.*, No. 13, 553-562 (1978).

16. D. P. Almond and A. R. West, "Comments on the analysis of a.c. conductivity data for single crystal Na-β-alumina at low temperatures," *J. Electroanal. Chem.*, **193**, No. 1/2, 49-55 (1985).

17. L. N. Dem'yanets, A. K. Ivanov-Shits, O. K. Mel'nikov, and A. P. Chirkin, "Anisotropy of electric conductivity and phase transition in single crystals of the superionic conductor $Li_3Sc_2(PO_4)_3$," *Fiz. Tverd. Tela*, **27**, No. 6, 1913-1914 (1985).

18. A. K. Ivanov-Shits and S. E. Sigarev, "Ionic conduction in crystal structures with isolated tetrahedral ions," *Kristallografiya*, **32**, No. 1, 254-259 (1987).

PREPARATION OF SINGLE CRYSTALS
OF THE NONSTOICHIOMETRIC FLUORITE PHASES $M_{1-x}R_xF_{2+x}$
(M = Ca, Sr, Ba; R = RARE EARTH ELEMENTS)
BY THE BRIDGMAN–STOCKBARGER METHOD

B. P. Sobolev, Z. I. Zhmurova, V. V. Karelin,

E. A. Krivandina, P. P. Fedorov, and T. M. Turkina

INTRODUCTION

Research on the preparation and study of single-crystalline $M_{1-x}R_xF_{2+x}$ (R = rare earth element, REE) materials with large deviations from stoichiometry began at the Institute of Crystallography of the Academy of Sciences of the USSR in 1963. The efforts were continued in cooperation with the chemistry staff of Moscow State University. Progress on the quest and preparation of crystals of nonstoichiometric fluorides for quantum electronics up to 1978 was reviewed in dissertations [2]. However, a large part of the data on growth of $M_{1-x}R_xF_{2+x}$ single crystals is scattered over a large number of obscure publications.

The purpose of the present article is to review and correlate all work published by us on the growth of $M_{1-x}R_xF_{2+x}$ fluorite phase crystals.

The uniqueness of this family of crystals and hence the unwavering interest in their preparation and study consists in the gross deviations from stoichiometry which permit fluorite structural types with heterovalent substitution of M^{2+} by R^{3+} (up to 50 mole % RF_3). Changes in the physical and physicochemical characteristics of the crystals are effected and controlled as a result of the gross nonstoichiometry. Mechanical, radiative, thermal, chemical, optical, transport, acoustic, and other structurally sensitive characteristics are affected.

The $M_{1-x}R_xF_{2+x}$ phases are the most striking with respect to nonstoichiometry and most feasible from the viewpoint of technological production among the more than 150 types of known fluorites with gross stoichiometric imbalance [1]. Numerically, they comprise a third of all known types (according to qualitative chemical composition) of fluorite–fluoride nonstoichiometric (two-component) phases.

The scope of our program characteristically includes preparation of single crystals of the $M_{1-x}R_xF_{2+x}$ nonstoichiometric fluorite phases for the whole concentration region covering hundredths of a fraction of a percent to saturation, i.e., ~50 mole % RF_3. Numerous publications by other authors have appeared for REE trifluoride concentrations of 1-3 mole %. Review of these is a separate task. Few articles can be singled out on the preparation of single crystals with high contents of RF_3. The most systematic of these should be mentioned [3-6]. In particular, the high content of heterovalent impurities in the CaF_2 structure is of prime

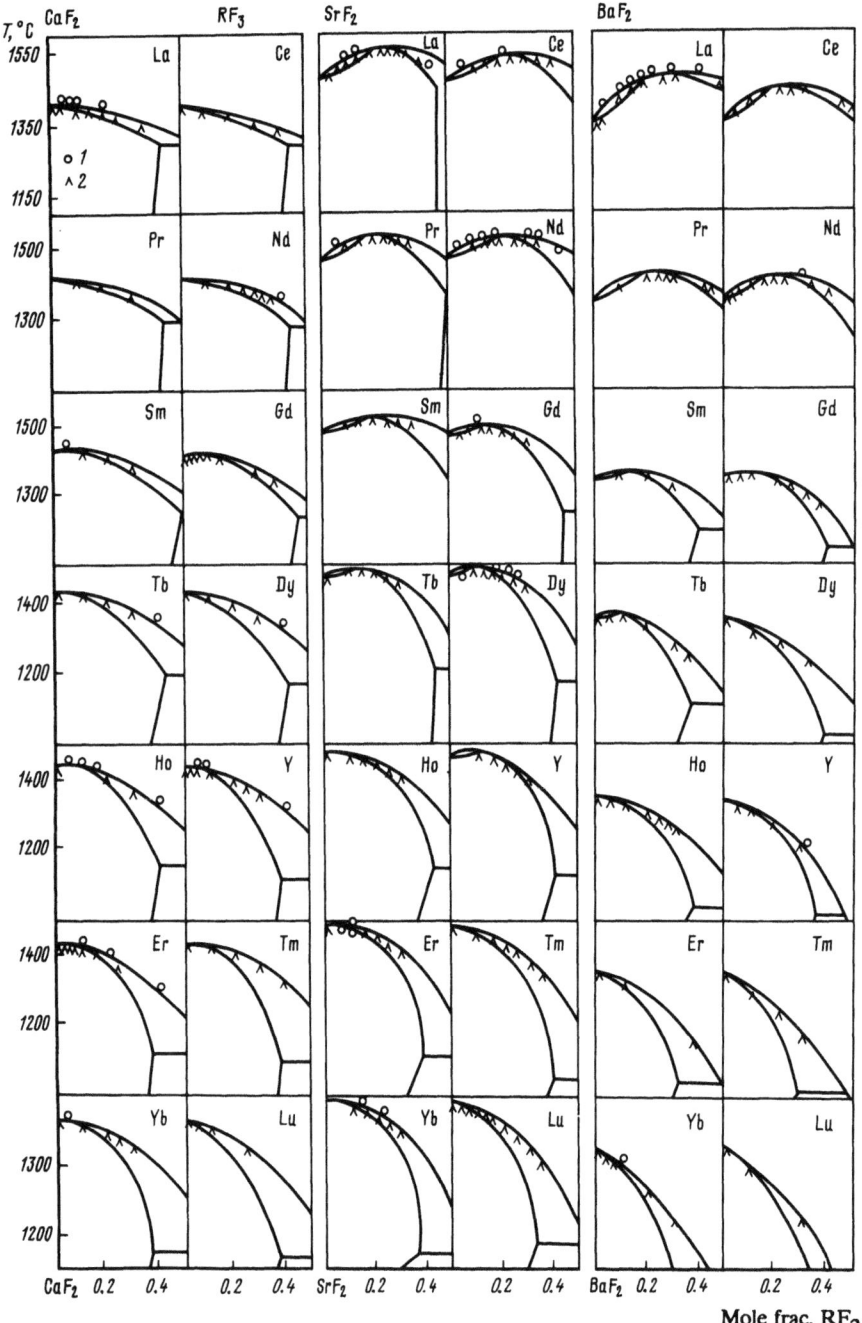

Fig. 1. Parts of the phase diagrams of the MF_2–RF_3 systems from [7-9]. Compositions of single crystals of concentrated solid solutions are shown by conventional symbols. 1) Literature data; 2) our results.

interest since substantial disturbance of stoichiometry is required for significant changes of the properties of such ionic crystals.

The present review considers, primarily, the current state of preparation of single crystals of grossly nonstoichiometric fluorite phases by the Bridgman–Stockbarger method. Previous reviews [7-9] were concerned with new data obtained during study of phase diagrams of MF_2–RF_3 systems. One type (the main) of grossly nonstoichiometric phase, heterovalent solid solutions with the fluorite structure, was chosen from a number of structural types for this review.

Practical problems on the growth of $M_{1-x}R_xF_{2+x}$ single crystals necessitate completion of phase diagrams through work on the thermodynamics and kinetics of crystallization of multicomponent fluoride melts.

Table 1. Equilibrium Distribution Coefficients of Rare Earth Ions R^{3+} upon Crystallization from a Melt of Solid Solutions with the Fluorite Structure in MF_2–RF_3 Systems ($c_0 = 0.5$ mole % RF_3)

Dopant	CaF$_2$		SrF$_2$		BaF$_2$	
	Our data	From [19, 20]	Our data	From [19]	Our data	From [19]
LaF$_3$		0.705		1.98		2,26
CeF$_3$	0.82 ± 0.03	0.82; 0.88		1.945		1.96
PrF$_3$		0.935		1.875		1.795
NdF$_3$		1.0; 1.06		1.82		1.69
PmF$_3$	0.94[+], 1.05[++]		1.77[++]		1.54[++]	
Sn.F$_3$		1.10; 0.98		1.715		1.38
EuF$_3$	1.13[+], 1.12[++]		1.68[+]		1.29[+]	
GdF$_3$	1.14 ± 0.03	1,14; 1.07		1.66		1.195
TbF$_3$	1,09 ± 0.03	1,10	1.36 ± 0.16		1.05 ± 0,03	
DyF$_3$	1.00 ± 0.06		1,2[+]		1.1[+]	
HoF$_3$	0.98 ± 0.03		1,10 ± 0,10		1,14 ± 0.06	
ErF$_3$	1.01 ± 0.04		1.11 ± 0.16		1,20 ± 0,14	
TmF$_3$	1,02 ± 0,03		1.11 ± 0,06		0.80 ± 0,08	
YbF$_3$	1.02 ± 0.03	0.97		1		0,58
LuF$_3$	0.96 ± 0.03		0,98 ± 0,04		0.58 ± 0,16	
YF$_3$	1.07 ± 0.03	1.11		1,375		0,84
ScF$_3$	0,26 ± 0,05					

Note. Experimental value ("+") and extrapolated ("++").

Studies of equilibrium distribution coefficients of all available REE ions in MF_2 during directed crystallization of melts were undertaken since data on the thermodynamics and kinetics of $M_{1-x}R_xF_{2+x}$ fluorite phase crystallizations at small and large x were practically absent in the literature. Estimation of the distribution coefficients for high contents of RF_3 were made from the fusibility disgrams of MF_2–RF_3 systems [7-9]. Problems of the pyro-hydrolysis pressure during crystallization and questions of homogenization of multicomponent fluoride melts as related to crystal growth mechanisms are examined. Formation of axial and radial nonuniformities in single crystals is studied. An analysis of preparation conditions for uniform distribution of dopants is given. Several questions on technical formulation of the growth experiment as applied to nonstoichiometric fluorite phases are discussed.

1. FUSIBILITY DIAGRAMS OF $M_{1-x}R_xf_{2+x}$ SOLID SOLUTIONS AND PROBLEMS OF CRYSTAL PREPARATION

Figure 1 depicts parts of the fusibility diagrams for the MF_2–RF_3 systems at 0-50 mole % RF_3. Fluorite solid solutions $M_{1-x}R_xF_{2+x}$ are formed in this region. EuF_3 is excluded from the available RF_3 compounds since its partial reduction upon reaction with graphite has been demonstrated [10]. Thus, fluorite phases are prepared in 42 systems (individual crystals were synthesized in the CaF_2–TmF_3 system and in the CaF_2, SrF_2, and BaF_2 with ScF_3 systems [11]).

The most characteristic feature of the fusibility of the nonstoichiometric phases is the presence of maxima on the curves for 32 of them (of a total of 42 systems). This feature of thermal behavior has great significance for preparation of single crystals. It is inherent among ionic crystals only for heterovalent solid solutions which are formed by a change of the number of ions in the unit cell [12]. At the maxima, the equilibrium distribution coefficient k_0 is equal to 1 and the two-component melt behaves as a pseudo-one-component melt.

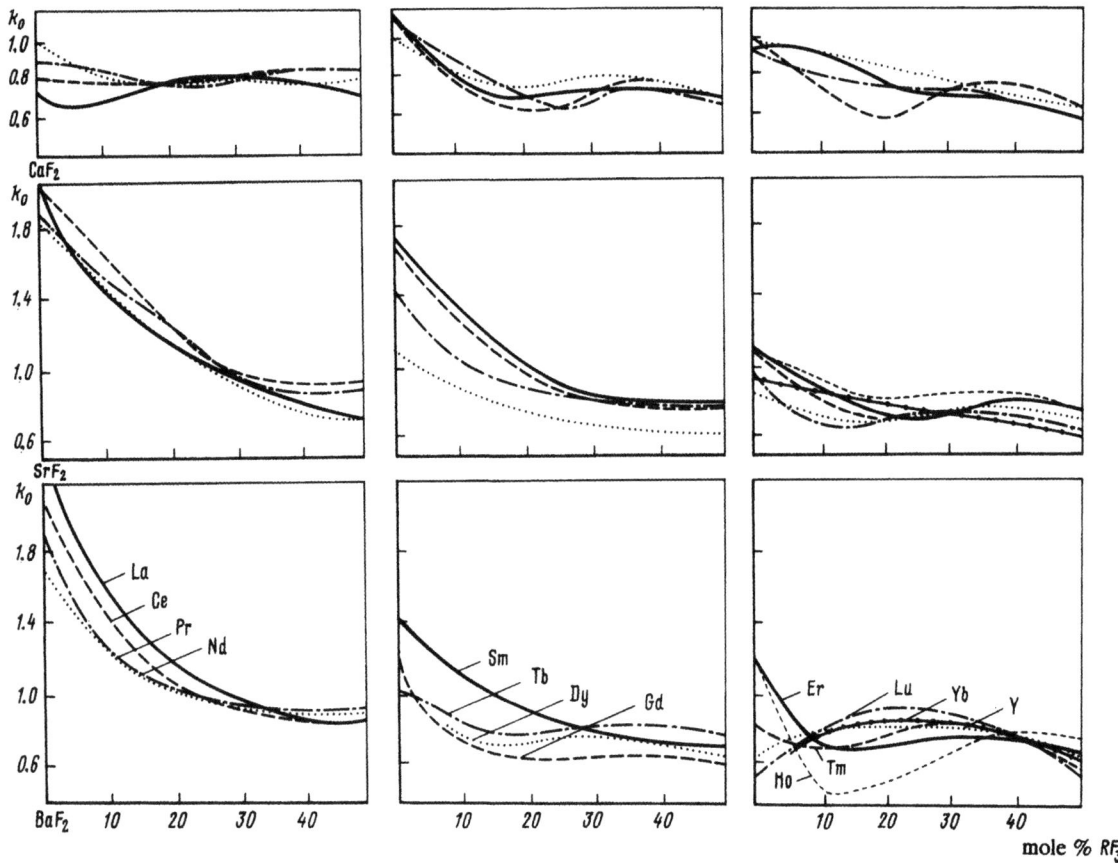

Fig. 2. Equilibrium distribution coefficients of R^{3+} in MF_2 as functions of concentration calculated from phase diagrams [7-9] and determined experimentally for $x = 0.05$ [14-18].

The methodology for study of physical properties of nonstoichiometric crystals, beginning with the pioneering work [13] on spectroscopy of R^{3+} in $M_{1-x}R_xF_{2+x}$, uses properties as a function of concentration for elucidation of the optimal compositions.

The necessity to crystallize melts with large deviations of distribution coefficients k_0 from unity arose during such studies on single-crystalline specimens.

1.1. Distribution Coefficients of R^{3+} during Directed Crystallization of Melts

Determination of these quantities which are fundamental for directed crystallization was carried out by two different methods for various composition ranges.

The k_0 value was determined by extrapolation to zero growth rate for experimentally obtained distribution coefficients [11, 14-18, 25, 29] at RF_3 content up to 0.5 mole %. The obtained k_0 values, given in Table 1, together with the literature data [19, 20] completely describe the behavior of the two-component $M_{1-x}R_xF_{2+x}$ melts at small RF_3 contents. The presumption of independence of distribution coefficients from composition holds rather well in this region. Stability criteria of a planar crystallization front in a known form [21] were applied for estimation of the crystallization parameters which guarantee a very uniform R^{3+} distribution throughout the crystal.

Ultrasmall RF_3 contents (<0.1 mole %) deserve special attention. A sharp change of distribution coefficient is observed in this composition region, as first shown for fluorides [22]. This effect is treated [23] as a capture of impurity by native thermally stimulated defects of the crystal matrix when their concentrations become comparable. The RF_3 concentrations for matrices with fluorite structure at which this effect is important should be rather high due to diffuse phase transitions near the melting point which are characteristic for

MF_2 fluorites. These transitions are associated with partial disordering of the anionic sublattice of all compounds with the fluorite (and antifluorite) structure.

Study of the R^{3+} distribution coefficients at small concentrations enabled refinement of a number of phase diagrams which were obtained by thermal analysis [7-9]. The existence of maxima in four systems, SrF_2-TmF_3, BaF_2-DyF_3, BaF_2-HoF_3, and BaF_2-ErF_3 [11], along with the maxima which were found earlier on the fusibility curves of $M_{1-x}R_xF_{2+x}$ phases was shown convincingly by using radionuclides. Maxima were probable (within limits of experimental error of the equilibrium distribution coefficients) in another four systems.

The k_0 values can be calculated from the fusibility diagrams for the $M_{1-x}R_xF_{2+x}$ phases at RF_3 contents from several mole % up to saturated solid solutions. The liquidus and solidus curves of these are determined by differential thermal analysis.

Figure 2 shows k_0 as functions of concentration which are calculated from the experimental points of the $M_{1-x}R_xF_{2+x}$ phase fusibility diagrams. The curves for high RF_3 contents are satisfactorily extrapolated to k_0 values for $M_{0.995}R_{0.005}F_{2.005}$ which were obtained in [14-17]. The strongest dependences of coefficients on composition are observed for all three MF_2 at 0-20 mole % RF_3. Further increase of the trifluoride concentrations is not accompanied by substantial changes of k_0.

Figures 1 and 2 show that compositions of the maxima in the fusibility curves of the $M_{1-x}R_xF_{2+x}$ phases in systems based on SrF_2 and BaF_2 approach the ordinate of the pure component whereas the melting points at the maxima are lowered as the difference in size of the substituting M^{2+} and R^{3+} cations increases. This function acquires another form in solid solutions based on CaF_2, in which the sizes of the heterovalent cations are closest. In this case, maxima appear in the systems with RF_3 where R is located in the center of the lanthanide series. It disappears for R at the beginning and end of the series. The smallest variation of k_0, either by composition or by REE series, is characteristic for these solid solutions. A number of consequences which must be considered during preparation of $M_{1-x}R_xF_{2+x}$ single crystals ensue from the R^{3+} distribution coefficients as functions of concentration.

1.2. Nonuniform Distribution of Dopant Component in $M_{1-x}R_xF_{2+x}$ Single Crystals

Nonuniformities in the dopant distribution should arise in the general case upon solid solution crystallization from a two-component melt (except for compositions at the maxima of fusibility curves). However, preparation of single crystals of solid solutions which for practical purposes can be considered uniform is possible with judicious choice of crystallization parameters. In this section, forms of nonuniformity which are found in crystals of nonstoichiometric fluorite phases are examined. First, we will discuss processes for homogenization of multicomponent fluoride melts which precede crystallization. Mechanical mixing of fluoride melts in the Bridgman–Stockbarger and zone melting methods is technically complicated. Therefore, the question of effectiveness of natural homogenization in the liquid phase and the kinetics of equilibrium state establishment through the melt volume is of utmost importance.

We now examine results of our study for the example of the two-component CaF_2 and TmF_3 mixture [11]. A crucible with eight cells of various diameter, on the bottom of each of which was loaded 0.5 mole % TmF_3 containing an indicator amount of the radioactive isotope ^{170}Tm, was used. CaF_2 was added on top. The crucible was heated to the melting point of fluorite. The melt was maintained for about 20 min. It was cooled by turning off the furnace. The height of the melt column was 60 mm. The temperature gradient for this length was estimated at ~100°C. The crystal was cut into 15 sections (perpendicular to the boule axis). The specific activity of each section was measured. Results are shown in Fig. 3. Equilibrium distribution of the Tm^{3+} mixture is seen to be attained very quickly in the melt if the cell diameter is more than 6-8 mm. Maintenance for 20 min is insufficient with smaller diameters. However, the melts experienced one hour's maintenance for degassing and fluorination under actual conditions of the growth experiment. Therefore, from

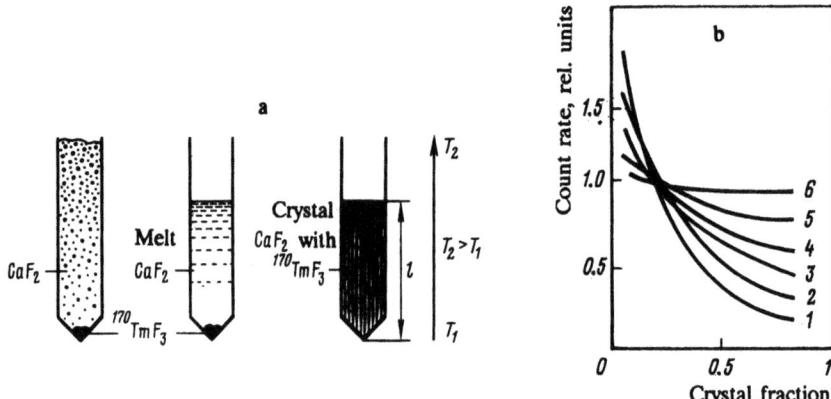

Fig. 3. Experimental setup for study of melt mixing (a) and degree of melt mixing (b) in growth cells.
Diameters: 2, 2.5, 3, 3.5, 4, and 10 mm for curves 1-6, respectively.

a practical viewpoint, the homogenization step of multicomponent fluoride melts does not present any additional demands on the growth process. Such a rapid equilibration of concentrations in the melt column, which is placed in anticonvectional conditions, remained unexplained until the publication of [24]. The reason for mixing under anticonvectional conditions was shown to be the loss of mechanical stability; this was done by mathematical modeling based on general solutions of heat and mass transfer equations. There may be two causes for this effect. One cause could be the lack of a common direction (in the general case) between the directions of the force of gravity and the temperature gradient along the cell growth axis. For this, an axial inclination of 30-40' is sufficient. The second cause may be axially asymmetric heating of the melt, for which an inclination of 5-7° is sufficient. Obviously, neither of these causes can be eliminated under actual conditions.

Thus, standard experimental growth conditions guarantee effective mixing of components for the multicomponent fluoride melts. Such behavior of a dopant in a melt in classical works on crystallization theory is known as a condition of partial melt mixing. Distribution of the dopant along the length of the crystal for this case (with the exception of the initial portion) is described, for example, by the Burton—Prim—Slichter equations [25]. Distribution of the dopant in the initial (nonstationary) portion can be described, to a first approximation, by the Tiller equation [21].

With respect to the preparation of single crystals with an equilibrated distribution of dopant along the length, this means that the crystal will be nonuniform at the initial and final portions for all compositions with $k_{eff} \neq 1$. The length of the central part of the crystal that is characterized by a practically uniform dopant distribution depends on how closely k_{eff} approaches 1.

The distribution of RF_3 dopants along the length of $M_{1-x}R_xF_{2+x}$ single crystals was studied by various methods: x-ray fluorescence with averaging for the large capacity (5-10%) of the boule [15, 26, 27], local x-ray spectral analysis with scanning by a wide probe along the growth axis [28], and radionuclidic methods (both by determination of specific activity and by autoradiography without single crystal destruction) [11, 14, 15]. Numerous experiments showed that not less than 70-80% of the crystal length was found in the portion with equilibrated axial distribution of dopant for the chosen growth conditions for $M_{1-x}R_xF_{2+x}$. This effect correlates with the data of Fig. 2, from which it follows that the distribution coefficients of R^{3+} in $M_{1-x}R_xF_{2+x}$ with a deviation from 1 by ±0.2 encompasses the concentration region 10-50 mole % RF_3. Compositions for 80% of the synthesized single crystals (Fig. 1) fall into this region.

The change of dopant concentration along the crystal axis was studied by integral methods of chemical analysis, averaging over the nonuniform volume in the crystalline boule cross section if applicable. Variation of dopant content in the perpendicular cross sections, radial nonuniformity, was observed besides the axial nonuniformities. A cellular substructure [29] dominates the several varieties of nonstoichiometric fluorite phases.

It should also be indicated that radial nonuniformity is an arbitrary term. In actuality, the cells permeate the whole crystal. Thus, the cellular substructure in the radial cross section becomes fibrous at the axis and affects the axial dopant distribution. Therefore, integral methods of concentration determination are necessary for studying this. Both forms of nonuniformity arise for the same reason, i.e., the incongruent behavior of the two-component melts. However, they are not interrelated. Crystals with axial nonuniformity without noticeable radial effects can be made.

Theoretical analysis of radial nonuniformities has been carried out more than once. Stability criteria of the planar crystallization front have been generally acknowledged and experimentally confirmed. Among these, the criterion of Tiller [21] has the simplest form:

$$\frac{G}{R} > \frac{m(1 - k_0)}{D k_0} c_0,$$ (1)

where G is the gradient temperature, D is the diffusion coefficient, R is the growth rate, m is the slope of the liquidus curve, c_0 is the dopant concentration in the starting melt, and k_0 is the equilibrium distribution coefficient. Loss of stability by such a front also leads to radial nonuniformities with a cellular substructure.

The Tiller criterion for RF_3 contents up to 0.5 mole % was checked for a number of $M_{1-x}R_xF_{2+x}$ crystals (M = Ca, Ba, Sr; R = Sc, Y, La–Lu). Good agreement was noted between the shape of the stability function and the shape of the experimental curve which determines the boundary between regions of presence and absence of cells in coordinates of G/R vs. RF_3 concentration.

Equations that describe stability conditions of the planar crystallization front suggest that k_0 is independent of system composition. This is true, to a first approximation, for a region of small dopant content. However, such an assumption in the general case (Fig. 2) for fluorite solid solutions is unacceptable for the majority of compositions for the two-component MF_2–RF_3 systems (and, naturally, for any others with wide regions of solid solution formation).

The possibility of expanding the Tiller stability criterion for a planar crystallization front to the region of binary system compositions was examined in [29]. In this case, k_0 is a function of the dopant content. According to [21], the reason for loss of shape of a planar crystallization front is concentration overcooling. This effect does not arise if the temperature gradient in the melt at the crystallization front G is larger than the equilibrium crystallization temperature gradient T_L, which is determined by the path of the liquidus curve in the phase diagram:

$$G = \frac{dT}{dx}\bigg|^t_{x=0} > \frac{dT_L}{dx},$$

where x is the bar length. Writing the material balance equation at the crystallization front, we obtain

$$R \Delta c = -D \frac{dc}{dx},$$

where Δc is the jump of concentration at the growth front. Considering that

$$\frac{dT_L}{dx} = \frac{dT_L}{dc} \frac{dc}{dx} = m \frac{dc}{dx},$$

we obtain for the stability criterion

$$\frac{GD}{R} > m\Delta c = F(c).$$ (2)

The first part of this inequality is called the stability function of the planar front [29].

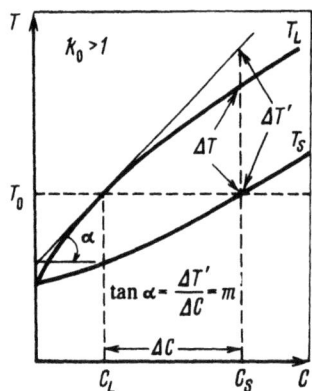

Fig. 4. Portion of a solid solution fusibility diagram ($k_0 > 1$) deduced from the stability function of the planar crystallization front.

If the crystallization process occurs under conditions which are near to equilibrium, the stability function can be calculated from the phase diagram (Fig. 4):

$$F(c) = \Delta T' \approx \Delta T. \tag{3}$$

The stability of directed crystallization of a binary system to concentration supercooling is determined, to a first approximation, by the difference between the temperatures of the beginning and end of melt crystallization, the composition of which corresponds to the crystal composition at the crystallization front. The conclusion of (3) is assumed without accounting for the heat of crystallization for nonstoichiometric fluorite crystals where low enthalpies of fusion are characteristic.

If the combination of principal parameters under actual growth conditions is such that the quantity GD/R is greater than the values of $F(c)$ calculated from experimental phase diagrams, then concentration supercooling is not observed and the single crystals that are prepared are free from cellular substructure. Thus, fusibility diagrams of the $M_{1-x}R_xF_{2+x}$ phases allow quantitative estimation of the growth conditions necessary for synthesis of single crystals of a multicomponent system with enhanced uniformity. We calculated changes of the stability function $F(c)$ for all grossly nonstoichiometric fluorite phases.

At small dopant contents, m and k_0 can be taken as constant. Then, we have a particular case of the Tiller criterion (1) from expression (2).

At the maxima on the fusibility curves of the $M_{1-x}R_xF_{2+x}$ phases, $k_0 = 1$ and $\Delta C = 0$, as is characteristic for grossly disturbed stoichiometry in general. This precludes concentration supercooling. The liquidus and solidus curves have a common tangent and $m = 0$. Thus, we not only have $F(c) = 0$ but its first derivative is also equal to zero, whereas the second derivative is positive. The stability function has a minimum. Small deviations from its composition do not cause a loss of planar crystallization front stability. Finally, uniform $M_{1-x}R_xF_{2+x}$ single crystals with compositions near the maxima are prepared over a larger range of crystallization values and parameters than for the region of small x values [29].

The cellular substructure that arises at large RF_3 contents is morphologically similar to that for small trifluoride concentrations.

In conclusion, we note that analysis of the crystallization trends for nonstoichiometric $M_{1-x}R_xF_{2+x}$ phases from multicomponent melts demonstrates the critical role of technical parameters of the growth experiment (temperature gradient and growth rate) in the preparation of single crystals with a high uniformity of dopant distribution.

2. GROWTH OF $M_{1-x}R_xF_{2+x}$ SINGLE CRYSTALS

2.1. Growth Techniques

Single crystals of alkaline earth fluorides are prepared by Bridgman–Stockbarger, Czochralski, Kyropoulos, and zone melting methods. However, the most extensive and developed methods for industrial production are variants of the Bridgman, Stockbarger, Shamovskii, and Stepanov methods.

The crystallization apparatus used by us basically repeats the one- and two-zone setup which was described earlier. Beginning in 1985, a KRF-1 setup was used. This was developed and built at the Special Design Bureau of the Institute of Crystallography, Academy of Sciences of the USSR. It has two regulated thermal zones which are divided by a changeable diaphram. This allows control of the temperature gradients. Thus, the concentration ranges for preparation of uniform $M_{1-x}R_xF_{2+x}$ single crystals are broadened. Typical temperature gradients in the crystallization zone in these setups were 25-70 deg/cm [27, 29]. Rates of crucible lowering varied in the range 2-22 mm/h [14, 29].

2.2. Chemical Processes during Mixture Preparation and Growth

Fluoride melts at high temperatures are known to be not only aggressive relative to crucible materials, heaters, and shields, but also exceedingly susceptible to pyrohydrolysis [30, 31]. Not one of the known preliminary solid-phase fluorinations of reagents guarantees sufficient purity of melts relative to oxygen for MF_2, much less for RF_3. The main reason for this is the good ability of the fluoride surfaces to absorb moisture.* Even the most oxygen-free powdered fluoride mixtures contain sufficient water to cause pyrohydrolysis with consequences that are unacceptable for optical materials.

The pyrohydrolysis products from MF_2 (M = Ca, Sr, and Ba) are oxides that are practically insoluble in melts. At oxygen contents of ~0.2 mass %, the oxide as fine particles leads to strong light scattering which makes the crystal unsuitable for use (at least in optics). The pyrohydrolysis of MF_2 was studied long ago and in detail. Methods for purification of melts from oxygen were recommended. These included treatment with anhydrous HF (up to bubbling through the melt), with a series of fluorine-containing gases (fluorocarbons, fluorophosgene, boron trifluoride, etc.), and exchange reactions of MO with CdF_2 and PbF_2 (reductants) [32, 35].

Pyrohydrolysis occurs more vigorously for melts of REE fluorides or mixtures of alkaline earth and REE fluorides since RF_3 are much more reactive relative to water vapor. The first step of pyrohydrolysis of RF_3 was shown in [36] to involve $RF_{3-2x}O_x$ oxyfluorides with small x values and not phases of the ROF or $R_4O_3F_6$ type, as was thought earlier. In contrast to MF_2, oxygen partially dissolves in RF_3 melts (see the MF_2–ROF systems [37]). However, such solid solutions are unstable and partially decompose upon cooling. A certain fraction of the oxygen remains dissolved in the single crystals [37]. Thus, the RF_3 content can, up to definite concentrations that depend on the REE type, mask the oxygen impurity in the melt as $RF_{3-2x}O_x$ solid solutions. These are partially preserved for some REE not only by quenching but also by slow cooling.

---

*Other reasons, which occur under conditions where preparation of mixture components does not include a melting step, are analyzed in [31] and others. We are concerned namely with the latter variant of mixture preparation.

Table 2. Ranges of Compositions of Single Crystals of Grossly Nonstoichiometric $M_{1-x}R_xF_{2+x}$ Fluorite Phases Prepared by the Bridgman–Stockbarger Method

R	$Ca_{1-x}R_xF_{2+x}$		$Sr_{1-x}R_xF_{2+x}$		$Ba_{1-x}R_xF_{2+x}$	
	x	Reference	x	Reference	x	Reference
La	0.01–0.35	46–49, 57, 58 66, 75–78, 92	0.01–0.42	29, 47, 51, 66 67, 74, 76, 85, 86	0.005–0.5	27, 47, 57, 73, 74 76, 81, 90, 91
Ce	0.005–0.39	14, 41, 47, 57, 58, 76	0.03–0.35	26, 47, 57, 68, 76, 85, 86	0.005–0.5	47, 57, 76, 89, 93
Pr	0.10–0.3	46, 47, 57, 58, 76	0.1–0.35	57, 68, 76, 85, 86	0.1–0.3	47, 57, 76, 89, 93
Nd	0.1–0.4	46, 47, 57, 58, 76, 79	0.1–0.41	47, 57, 69, 70, 76, 85, 86	0.01–0.32	4, 47, 57, 76, 89
Sm	0.1–0.3	57, 58, 80	0.1–0.35	57, 85	0.1–0.3	57
Gd	0.005–0.35	14. 46–48, 57, 58, 75	0.1–0.3	26, 29, 47, 57, 85–87	0.005–0.35	15, 47, 57, 89, 93
Tb	0.005–0.4	14, 47, 57, 58, 82	0.005–0.3	15, 16, 47, 57, 85	0.005–0.35	16, 47, 57
Dy	0.005–0.4	14, 47, 57, 58 82	0.005–0.3	15, 17, 57, 71, 85	0.005–0.3	15, 17, 57
Ho	0.005–0.4	14, 47, 57, 58, 82, 83	0.005–0.3	15, 17, 57, 85	0.005–0.3	15, 17, 57, 94
Er	0.005–0.4	14, 47, 48, 58, 82, 92	0.005–0.3	15, 17, 47, 71, 85, 88	0.005–0.38	15, 17, 47, 95
Tm	0.005–0.4	14, 47, 57, 58	0.005–0.30	15, 17, 57, 85	0.005–0.3	15, 16, 57
Yb	0.005–0.3	14, 47, 57, 58	0.1–0.35	47, 57, 72, 85	0.10–0.3	4, 47, 57
Lu	0.005–0.25	14, 46, 47, 58, 75, 84	0.005–0.3	15, 16, 26, 29, 51, 85	0.005–0.1	15, 16
Y	0.005–0.4	14, 46–48, 50, 57, 58, 79, 82, 96	0.1–0.3	57, 89	0.05–0.3	4, 57, 93

Anhydrous HF was used as fluorinating agent in the initial years. The fluoride growth method in such an atmosphere was developed jointly at the Institute of Crystallography and the Institute of General and Inorganic Chemistry of the Academy of Sciences of the USSR (Kh. S. Bagdasarov and E. G. Ippolitov) simultaneously and independently of [35]. It has been used by us since 1965 for preparation of crystals of multicomponent compositions [38]. Later, a technically simpler and equally effective method for purification of a melt from traces of oxygen by fluorination with tetrafluoroethylene pyrolysis products was used. This was proposed in [32] and studied in detail in [33]. However, this method has temperature limitations since corrosion of the graphite elements becomes noticeable with its rise [33].

Traditional container materials for fluoride melts are graphite and glassy carbon. Graphite is practically universal for all melts except those which contain Sm^{3+} and Eu^{3+}. Those with a prolonged crystal growth cycle are partially reduced to Sm^{2+} and Eu^{2+}. Reduction is suppressed in glassy carbon due to its low porosity. However, a drawback is the difficulty of forming feed channels and cavities for those cases where suppression of the block nature in single crystals is necessary or when the melts are prone to appreciable supercooling.

The supercooling effects, which are distinctly observed for RF_3 melts, are rather substantial [36] whereas for the MF_2–RF_3 systems up to concentrations of 50 mole % RF_3 they are insignificant.

3. CONCLUSIONS

Figure 1 shows graphically the principal results of many years of work on the synthesis of single crystals of grossly* nonstoichiometric $M_{1-x}R_xF_{2+x}$ fluorite phases by the Bridgman–Stockbarger method. The compositions of melts, from which single crystals were grown at the Institute of Crystallography, Academy of Sciences of the USSR, are denoted by arrows. Crystals grown by other authors are shown by circles.

Figure 1 is complemented by Table 2, which contains data on RF_3 concentration limits in the synthesized $M_{1-x}R_xF_{2+x}$ single crystals. References to the main literature sources are also given there. These sources describe preparation of the crystals. Compositions of several fluorite phases with 0.5 mole % RF_3 are included in Table 2 as exceptions. The method was developed and the k_0 coefficients for the Tiller region of solid solutions were determined for these.

About 300 single crystals of grossly nonstoichiometric (concentrated) fluorite solid solutions of various chemical composition had been synthesized at the start of 1987. Of these, more than 80% were prepared through the present synthetic, physicochemical, and crystal chemical program for study of grossly nonstoichiometric fluorides [1, 2].

At various stages in the preparation of the crystals, V. E. Bozhevol'nov, O. E. Izotova, M. Z. Kazakevich, V. V. Kas'yanov, G. V. Molev, Yu. N. Orlov, Ya. Rustamov, K. B. Seiranyan, V. S. Sidorov, N. L. Tkachenko, and T. V. Uvarova also participated along with the authors of the present review.

The skewed distribution of synthesized crystals in the alkaline earth series (Fig. 1) need not be mentioned. The largest quantity of compositions were prepared for $Sr_{1-x}R_xF_{2+x}$. In each family of fluorite phases there are highly as well as insufficiently assimilated combinations of M^{2+} and R^{3+}. For the $Ca_{1-x}R_xF_{2+x}$ family, data on crystallization in the systems with R = Pr, Nd, Sm, Tb, Dy, and Lu are scarce whereas for $Ba_{1-x}R_xF_{2+x}$ the paucity is in systems with R = Sm, Dy, Pr, Tm, and Lu. The inadequacy of data on crystallization of solid solutions near the saturation limits at eutectic temperatures is common.

In spite of the wide gaps in compositions of $M_{1-x}R_xF_{2+x}$ crystals, grossly nonstoichiometric fluorite phases at present are one of few families of inorganic materials with grossly perturbed stochiometry whose properties are widely studied for single-crystalline specimens. This plays an important role in the maturing of $M_{1-x}R_xF_{2+x}$ phases as models and as objects for study of fundamental problems of gross nonstoichiometry in general. The most systematic studies have been related to atomic structure, ion transport, and stability to radiation.

Study of spectrally generated characteristics of $M_{1-x}R_xF_{2+x}$ comprised the first program. This revealed the unusual partially disordered nature of the nonstoichiometric fluorite phases which is in many respects favorable for resolution of quantum electronic problems [3, 40].

Later [41], the defect structure in the $Ca_{0.61}Ce_{0.39}F_{2.39}$ phase was found by direct x-ray structural studies. It is the first nonstoichiometric fluoride in which the structural defects of the anionic motif was localized. Then, a series of theoretical studies of possible variants of cluster defects in similar phases followed. Recently, a switch to neutron diffraction (single crystal) methods [42-45] has provided a new driving force for a program of $M_{1-x}R_xF_{2+x}$ atomic structure solution.

Studies on the atomic structure of fluoride solid solutions of cerium and yttrium in CaF_2, lanthanum and lutetium in SrF_2, and cerium, praseodymium, and holmium in BaF_2 have been completed on our crystals.

---

*The term "grossly nonstoichiometric" was introduced by J. Anderson [39] in order to distinguish crystals in which structural defects cannot be described as points. For $M_{1-x}R_xF_{2+x}$, these composition regions from various data and for various systems lie between compositions from fractions to several mole % RF_3.

Studies of fluoride ion transport in $M_{1-x}R_xF_{2+x}$ crystals as a result of disturbed atomic structure (anionic motif) is one of the most complete programs which describe grossly nonstoichiometric fluoride materials [46-56], where the effect of structural defectiveness of the phases on the ionic conduction has been shown to be exceptionally strong. It increases by 8-10 orders of magnitude relative to the starting stoichiometric MF_2. These efforts led to the discovery of a family of medium-temperature fluoride conductors with ionic conduction and activation energy values that are controlled over wide limits. Measurements of ion transport values for crystals of fluorite phases grown for all systems are shown in Fig. 1, except for CaF_2-SmF_3, SrF_2-SmF_3, SrF_2-LuF_3, and BaF_2 with SmF_3, DyF_3, TmF_3, YbF_3, and LuF_3.

A series of studies on determination of the changes that occur in $M_{1-x}R_xF_{2+x}$ crystals due to various types of radiation was carried out. The nature of radiation stability of the prepared crystals, except the CaF_2-YF_3, CaF_2-LuF_3, BaF_2-EuF_3, and BaF_2-LuF_3 systems, was investigated [57-62]. The dynamics of the crystal lattice were studied by ^{19}F NMR [63-65]. Several separate publications on other physical properties of $M_{1-x}R_xF_{2+x}$ phases have appeared. The electrical and dielectrical properties of concentrated $M_{1-x}R_xF_{2+x}$ phases are especially interesting [66-74]. These studies are the most complete for grossly nonstoichiometric (concentrated) $M_{1-x}R_xF_{2+x}$ fluorite solid solutions. The investigation of their properties continues and is determined mainly by progress in preparation of single crystals.

REFERENCES

1. B. P. Sobolev, "Nonstoichiometry in MF_m-RF_n systems and new multicomponent fluoride materials," in: Abstracts of Papers of the VIIth All-Union Symp. on the Chemistry of Inorganic Fluorides [in Russian], 9-11 Oct., 1984, Leninabad, Nauka, Moscow (1984), p. 112.

2. B. P. Sobolev, "Nonstoichiometry in systems of alkaline earth and rare earth fluorides," Author's Abstract of Doctoral Dissertation in Chemical Sciences, Moscow (1978).

3. A. A. Kaminskii, V. V. Osiko, A. M. Prochorov, and Yu. K. Voronko, "Spectral investigation of the stimulated emission of Nd^{3+} in CaF_2-YF_3," Phys. Lett., 22, No. 4, 419-421 (1966).

4. B. Joukoff and J. Primot, "Crystal growth and structural particularities of $(BaF)_{1-x}(Y, LuF_3)_x$ solid solutions," Mater. Res. Bull., 11, No. 10, 1201-1208 (1976).

5. V. V. Kas'yanov and V. V. Sinitsyn, "Growth of crystals based on barium fluoride and REE fluorides," Nauchn. Tr. Gos. Nauchno-Issled. Proektn. Inst. Redkomet. Promst., Moscow (1979), Vol. 91, pp. 105-107.

6. R. C. Pastor, A. K. Pastor, and K. T. Miller, "Solid solution of RF_3 in CaF_2," Mater. Res. Bull., 9, No. 9, 1247-1250 (1974).

7. B. P. Sobolev and P. P. Fedorov, "Phase diagrams of the $CaF_2-(Y, Ln)F_3$ systems. 1. Experimental," J. Less-Common Met., 60, No. 1, 33-46 (1978).

8. B. P. Sobolev and N. L. Tkachenko, "Phase diagrams of the $BaF_2-(Y, Ln)F_3$ systems," J. Less-Common Met., 85, No. 2, 155-170 (1982).

9. B. P. Sobolev and K. B. Seiranian, "Phase diagrams of the systems $SrF_2-(Y, Ln)F_3$. Pt. 2. Fusibility of systems and thermal behavior of phases," J. Solid State Chem., 39, No. 2, 17-24 (1981).

10. O. S. Karapetyan, M. A. Arutyunyan, B. E. Adamyan, et al., "Magnetic susceptibility and x-ray phase analysis of the SrF_2-EuF_3 system as functions of temperature," Izv. Akad. Nauk Arm. SSR, Fiz., 9, No. 4, 338-340 (1974).

11. V. V. Karelin, "Physicochemical bases for preparation of single-crystalline materials in solid solutions of alkaline earth and rare earth fluorides," Author's Abstract of Doctoral Dissertation in Chemical Sciences, Moscow (1985).

12. P. P. Fedorov and B. P. Sobolev, "On conditions of formation of maxima on fusibility curves of solid solutions in salt systems," Zh. Neorg. Khim., 24, No. 4, 1038-1040 (1979).

13. Yu. K. Voron'ko, A. A. Kaminskii, and V. V. Osiko, "Analysis of optical spectra of crystals," Zh. Éksp. Teor. Fiz., 49, No. 2(8), 420-428 (1965).

14. V. V. Karelin, M. Z. Kazakevich, A. F. Red'kin, et al., "Distribution coefficients of rare earths in single-crystalline CaF_2," Kristallografiya, 20, No. 4, 758-762 (1975).

15. M. Z. Kazakevich, "Study of dopant distribution of some lanthanides in single-crystalline alkaline earth fluorides using radionuclides," Author's Abstract of Candidate's Dissertation in Chemical Sciences, Moscow (1979).

16. M. Z. Kazakevich, S. N. Neretin, and V. V. Karelin, "Equilibrium distribution coefficients of Tm^{3+}, Tb^{3+}, and Lu^{3+} in single-crystalline SrF_2 and BaF_2," All-Union Institute of Scientific and Technical Information (VINITI), deposited paper No. 1623 (1976).

17. M. Z. Kazakevich and V. V. Karelin, "Distribution of Dy, Ho, and Er dopants upon directed crystallization from SrF_2 and BaF_2 melts," All-Union Institute of Scientific and Technical Information (VINITI), deposited paper No. 255 (1977).

18. V. V. Karelin, "Physicochemical bases for preparation of new single-crystalline materials in systems formed from alkaline earth and rare earth elements," Vestn. Mosk. Univ., Ser. 2: Khim., 25, No. 4, 331-348 (1984).

19. F. Delbove and S. Lallemand-Chatain, "Determination cryometrique a la limite de dilution infinie, des coefficients de distribution entre solution solide et solution ignee fondue, des ions (3+) des terres rares dissous dans les fluorures alcalinoterreux," *C. R. Acad. Sci., Ser. C*, **270**, No. 11, 964-966 (1970).

20. K. Nassau, "Application of Czochralski method to divalent metal fluorides," *J. Appl. Phys.*, **32**, No. 10, 1820-1821 (1961).

21. W. A. Tiller, K. A. Jackson, J. W. Rutter, and B. Chalmers, "The redistribution of solute atoms during the solidification of metals," *Acta Metall.*, **1**, No. 4, 428-437 (1953).

22. Yu. P. Grigorash, V. V. Karelin, and V. S. Urusov, "Abrupt change of distribution coefficient in the CaF_2–BaF_2 system," *Dokl. Akad. Nauk SSSR*, **258**, No. 3, 652-655 (1983).

23. V. S. Urusov and I. F. Kravchuk, "Effect of microdopant capture by crystal lattice defects and its geochemical meaning," *Geokhimiya*, No. 7, 963-978 (1978).

24. E. A. Krivandina, A. I. Fedyushkin, and V. Ya. Khaimov-Malkov, "On the mechanism of melt stirring during growth of single crystals by vertical directed crystallization," *J. Cryst. Growth*, **52**, Pt. 1, 455-458 (1981).

25. J. A. Burton, R. C. Prim, and W. P. Slichter, "The distribution of solute in crystals growth from the melt. Pt. I. Theoretical," *J. Chem. Phys.*, **21**, No. 11, 1987-1991 (1953).

26. K. B. Seiranyan, "Study of phase diagrams of the SrF_2–(Y, Ln)F_3 systems and preparation of single crystals based on them," Author's Abstract of Candidate's Dissertation in Chemical Sciences, Erevan (1975).

27. N. L. Tkachenko, "Study of the BaF_2–(Y, Ln)F_3 systems and preparation of single crystals based on them," Author's Abstract of Candidate's Dissertation in Geological–Mineralogical Sciences, Moscow (1974).

28. V. A. Meleshina, E. A. Krivandina, E. V. Yakovenko, and B. P. Sobolev, "Variation of composition of $Ca_{1-x}Ho_xF_{2+x}$ and $La_{1-y}Sr_yF_{3-y}$ crystals relative to development of their cellular structure," in: Abstracts of Papers of the IVth All-Union Conf. on Growth of Crystals, Vol. 1, *Growth of Crystals from Melts* [in Russian], 26-30 Sept., 1985, Tsakhkadzor, Izd. Akad. Nauk Arm. SSR, Erevan (1985), pp. 239-240.

29. T. M. Turkina, P. P. Fedorov, and B. P. Sobolev, "Stability of planar crystallization front during growth of single crystals of solid solutions of $M_{1-x}R_xF_{2+x}$ (where M = Ca, Sr, Ba; R = rare earth element) from the melt," *Kristallografiya*, **31**, No. 1, 146-152 (1986).

30. W. Bontinck, "The hydrolyses of solid CaF_2," *Physica*, **22**, No. 8, 650-658 (1958).

31. N. V. Baryshnikov, Yu. A. Karpov, and T. V. Gushchina, "On sources of oxygen in rare earth fluorides," *Izv. Akad. Nauk SSSR, Noerg. Mater.*, **4**, No. 4, 532-536 (1968).

32. Yu. K. Voron'ko, V. V. Osiko, V. T. Udovenchik, and M. M. Fursikov, "Optical properties of CaF_2–Dy^{3+} crystals," *Fiz. Tverd. Tela*, **7**, No. 1, 267-273 (1965).

33. R. C. Pastor and M. Robinson, "Crystal growth of alkaline earth fluorides in a reactive atmosphere: Pt. 3," *Mater. Res. Bull.*, **11**, No. 10, 1327-1334 (1976); R. C. Pastor and K. Arita, "Crystal growth of alkaline earth fluorides in a reactive atmosphere, Pt. 2," *ibid.*, No. 8, 1037-1042 (1976).

34. L. M. Shamovskii, P. M. Stepanukha, and A. D. Shushakov, "Growth of fluorite single crystals activated by rare earth elements," in: *Spectroscopy of Crystals* [in Russian], Nauka, Moscow (1970), pp. 160-164.

35. H. Guggenheim, "Growth of highly perfect fluoride single crystals for masers," *J. Appl. Phys.*, **34**, No. 8, 2482-2485 (1963).

36. B. P. Sobolev, P. P. Fedorov, D. B. Shteynberg, et al., "On the problem of polymorphism and fusion of lanthanide trifluoride. Pt. 1," *J. Solid State Chem.*, **17**, No. 2, 191-199 (1976); "On the problem of polymorphism and fusion of lanthanide trifluoride, Pt. 2," *ibid.*, 201-212 (1976).

37. V. A. Gorbulev, P. P. Fedorev, and B. P. Sobolev, "Interaction of oxyfluorides of rare earth elements with fluorides of fluorite structure," *J. Less-Common Met.*, **76**, No. 1/2, 55-62 (1981).

38. L. S. Garashina, E. G. Ippolitov, B. M. Zhigarnovskii, and B. P. Sobolev, "Study of phase composition of the CaF_2–YF_3 system," in: *Study of Natural and Technical Mineral Formation* [in Russian], Nauka, Moscow (1966), pp. 289-294.

39. J. S. Anderson, "Thermodynamics and theory of nonstoichiometric compounds," in: *Problems of Nonstoichiometry* [Russian translation], Metallurgiya, Moscow (1975), pp. 11-96.

40. A. A. Kaminskii, *Laser Crystals* [in Russian], Nauka, Moscow (1975).

41. V. B. Aleksandrov and L. S. Garashina, "New data on the structure of CaF_2–TRF_3 solid solutions," *Dokl. Akad. Nauk SSSR*, **189**, No. 2, 307-310 (1969).

42. V. B. Aleksandrov, L. P. Otroshchenko, L. E. Fykin, et al., "New variety of defect structure of a nonstoichiometric phase of the fluorite type $Ba_{0.73}Pr_{0.27}F_{2.27}$," *Kristallografiya*, **29**, No. 2, 381-383 (1984).

43. V. B. Aleksandrov, B. A. Maksimov, V. I. Simonov, and B. P. Sobolev, "Study of the structures of nonstoichiometric phases and binary compounds in systems formed by alkaline earth and rare earth fluorides," in: Abstracts of Papers of the VIIth All-Union Symp. on the Chemistry of Inorganic Fluorides [in Russian], 9-11 Oct., 1984, Leninabad, Nauka, Moscow (1984), p. 17.

44. L. A. Muradyan, B. A. Maksimov, B. F. Mamin, et al., "Atomic structure of the nonstoichiometric phase $Sr_{0.69}La_{0.31}F_{2.31}$," *Kristallografiya*, **31**, No. 2, 248-251 (1986).

45. V. B. Aleksandrov, L. P. Otroshchenko, L. E. Fykin, et al., "Neutron diffraction study of the defect structure of the fluorite solid solution $Ba_{0.5}Ce_{0.5}F_{2.5}$ with partial population of octahedral vacancies of the cationic motif," in: Abstracts of Papers of the IVth All-Union Conf. on Crystal Chemistry of Inorganic and Coordination Compounds [in Russian], Dec. 1986, Bukhara, Nauka, Moscow (1985), p. 203.

46. P. P. Fedorov, "Study of phase diagrams of the CaF_2–(Y, Ln)F_3 systems and polymorphism of rare earth trifluorides," Author's Abstract of Candidate's Dissertation, Chemical Sciences, Moscow (1977).

47.   P. P. Fedorov, T. M. Turkina, V. P. Sobolev, et al., "Ionic conductivity in single crystals of nonstoichiometric fluorite phases $M_{1-x}R_xR_{2+x}$ (M = Ca, Sr, Ba; R = Y, La–Lu)," *Solid State Ionics*, 6, No. 4, 331-335 (1982).

48.   M. Svantner, E. Mariani, P. P. Fedorov, and B. P. Sobolev, "A study of electrical conductivity in concentrated solid solutions $Ca_{1-x}Ln_xF_{2+x}$ (Ln = La, Gd, Er, Y)," *Cryst. Res. Technol.*, 16, No. 5, 617-622 (1981).

49.   M. Svantner, E. Mariani, P. P. Fedorov, and B. P. Sobolev, "Solid solution with fluorite structure in the CaF_2–LaF_3 system," *Krist. Tech.*, 14, No. 3, 365-369.

50.   E. Mariani, M. Svantner, P. P. Fedorov, and B. P. Sobolev, "Influence of high yttrium concentration on the electrical properties of CaF_2 crystals," *Acta Phys. Slovaca*, 27, No. 4, 260-265 (1977).

51.   A. K. Ivanov-Shits, N. I. Sorokin, P. P. Fedorov, and B. P. Sobolev, "Conductivity of solid solutions $Sr_{1-x}La_xF_{2+x}$ ($0.03 \leq x \leq 0.40$)," *Fiz. Tverd. Tela*, 25, No. 6, 1748-1753 (1983).

52.   E. Mariani, V. Trnovceva, B. P. Sobolev, and P. P. Fedorov, "Superionic conduction in heavily doped alkaline earth fluorides," in: Abstracts of Papers of the Int. Conf. "Defects in Dielectric Crystals" [in Russian], 18-23 May, 1981, Riga, Zinatne, Riga (1981), pp. 495-496.

53.   P. P. Fedorov and B. P. Sobolev, "Relation of high ionic conduction to maxima on fusibility curves of heterovalent solid solutions," in: Abstracts of Papers of the VIIIth All-Union Conf. on Physical Chemistry and Electrochemistry of Melts and Solid Electrolytes [in Russian], 11-13 Oct., 1983, Leningrad, Vol. 3, p. 141.

54.   P. P. Fedorov, B. P. Sobolev, A. K. Ivanov-Shits, and N. I. Sorokin, "Ionic transfer in fluorides with fluorite structure," in: Abstracts of Papers of the VIIIth All-Union Conf. on Physical Chemistry and Electrochemistry of Melts and Solid Electrolytes [in Russian], 11-13 Oct., 1983, Leningrad, Vol. 3, p. 140.

55.   A. K. Ivanov-Shitz, N. L. Sorokin, B. P. Sobolev, and P. P. Fedorov, "Ionic transport in systems MF_2–RF_3 (M = Ca, Sr, Ba; R = La–Lu)," Abstracts of the Int. Symp. Syst. with Fast Ionic Transport, Bratislava (1985), pp. 99-103.

56.   É. R. Khairetdinov, A. K. Ivanov-Shits, N. I. Sorokin, et al., "Electric conductivity of the solid solution $Ba_{1-x}Ho_xF_{2+x}$ with fluorite structure," *Fiz. Tverd. Tela*, 28, No. 8, 2546-2548 (1986).

57.   Ya. Rustamov, G. A. Tavshunskii, P. K. Khabibulaev, et al., "Radiative coloration of single crystals of nonstoichiometric phases $M_{1-x}R_xF_{2+x}$ with defect fluorite structure," *Zh. Tekh. Fiz.*, 55, No. 6, 1150-1158 (1985).

58.   Ya. Rustamov, "Study of radiative-optical properties of nonstoichiometric phases $M_{1-x}R_xF_{2+x}$ with fluorite structure," Author's Abstract of Candidate's Dissertation in Physical-Mathematical Sciences, Tashkent (1984).

59.   G. A. Tavshunskii, Ya. Rustamov, T. S. Bessonova, et al., "Luminescence and absorption by single crystals of $Ca_{1-x}(Y, Ln)_xF_{2+x}$ subjected to nuclear radiation," Abstracts of Papers of the All-Union Seminar "Radiative Effects in Wide-Band Optical Materials" [in Russian], 23-25 May, 1979, Samarkand, Fan, Tashkent (1979), p. 168.

60.   Sh. A. Vakhidov, G. A. Tavshunskii, Ya. Rustamov, et al., "Effect of nuclear radiation on crystals of mixed fluorides," *Zh. Tekh. Fiz.*, 49, No. 9, 1943-1949 (1979).

61.   Sh. A. Vakhidov, G. A. Tavshunskii, Ya. Rustamov, et al., "Luminescence of solid solutions $M_{1-x}R_xF_{2+x}$ under the influence of ionizing radiation," in: Abstracts of Papers of the XXVIIth All-Union Conf. on Luminescence (Crystal Phosphors) [in Russian], 13-16 May, 1980, Ézernieki, Zinatne, Riga (1980), p. 110.

62.   Sh. A. Vakhidov, G. A. Tavshunskii, Ya. Rustamov, et al., "Luminescence and absorption of crystals of mixed fluorides subjected to nuclear irradiation," *Izv. Akad. Nauk Uzb. SSR, Ser. Fiz.-Mat. Nauk*, No. 1, 79-82 (1981).

63.   A. I. Livshits, V. M. Buznik, P. P. Fedorov, and B. P. Sobolev, "NMR study of anionic mobility in defect phases of fluorite and tysonite structures in the CaF_2–LaF_3 system," *Izv. Akad. Nauk SSSR, Neorg. Mater.*, 18, No. 1, 135-139 (1982).

64.   V. M. Buznik, V. A. Vopilov, A. I. Livshits, et al., "Fluorine-containing solid electrolytes studied by NMR," in: Abstracts of Papers of the VIth All-Union Symp. on Chemistry of Inorganic Fluorides [in Russian], 21-23 July, 1981, Novosibirsk, Nauka, Moscow (1981), p. 22.

65.   A. I. Livshits, V. M. Buznik, V. A. Vopilov, et al., "Study of ionic mobility in fluorides with defect fluorite and tysonite structure," Preprint Inst. Fiz. Sib. Otd. Akad. Nauk SSSR, No. 244F, Krasnoyarsk (1983).

66.   H. W. den Hartog, "Defect structure and defect–defect interaction in solid solution of AF_2 and RF_3 doped with Gd^{3+} probes," *Phys. Rev. B: Condens. Matter*, 27, No. 1, 20-26 (1983).

67.   J. Meuldijk and H. W. den Hartog, "Charge transport in $Sr_{1-x}La_xF_{2+x}$ solid solution. An ionic thermocurrent study," *Phys. Rev. B: Condens. Matter*, 28, No. 2, 1036-1047 (1983).

68.   J. Meuldijk, R. van der Meulen, and H. W. den Hartog, "Dielectric-relaxation experiments on cubic solid solution SrF_2 and CeF_3 or PrF_3," *Phys. Rev. B: Condens. Matter*, 29, No. 4, 2153-2159 (1984).

69.   J. Meuldijk and H. W. den Hartog, "Depolarization and conduction experiments on solid solutions of the type $M_{1-x}R_xF_{2+x}$" in: Proceedings of Solid State Chemistry 1982: IInd Europ. Conf., 7-9 June, 1982, Netherlands, in: Studies in Inorganic Chemistry, H. J. M. Heijigers and J. Schooman (eds.), Elsevier, Amsterdam (1983), Vol. 3, pp. 169-172.

70.   J. Meuldijk, H. H. Mulder, and H. W. den Hartog, "Depolarization experiments on space charges in concentrated solid solutions of NdF_3 in SrF_2," *Phys. Rev. B: Condens. Matter*, 25, No. 8, 5204-5213 (1982).

71.   J. Meuldijk, G. Kiers, and H. W. den Hartog, "Effect of clustering on the space–charge relaxation phenomena in fluorite-type solid solutions ($Sr_{1-x}Dy_xF_{2+x}$) and ($Sr_{1-x}Er_xF_{2+x}$)," Phys. Rev. A, 28, No. 10, 6022-6030 (1983).

72.   J. Meuldijk and H. W. den Hartog, "Ionic thermocurrents and ionic conductivity of solid solutions of SrF_2 and YbF_3," *Phys. Rev. B: Condens. Matter*, 27, No. 10, 6376-6384 (1983).

73.   H. W. den Hartog, K. F. Pen, and J. Meuldijk, "Defect structure and charge transport in solid solutions $Ba_{1-x}La_xF_{2+x}$" *Phys. Rev. B: Condens. Matter*, 28, No. 10, 6031-6040 (1983).

74. H. W. den Hartog and J. C. Langevoort, "Ionic thermal current of concentrated cubic solid solutions of $SrF_2:LaF_3$ and $BaF_2:LaF_3$," *Phys. Rev. B: Condens. Matter*, **24**, No. 5, 3547-3554 (1981).

75. A. A. Kaminskii, B. P. Sobolev, Kh. S. Bagdasarov, et al., "Investigation of stimulated emission of the $^4F_{3/2} \to {}^4I_{3/2}$ transmission of Nd^{3+} ions in crystals," *Phys. Status Solidi A*, **26**, No. 1, K63-K65 (1974).

76. B. P. Sobolev, P. P. Fedorov, I. L. Tkachenko, et al., "High-melting fluoride materials in the systems $(Ca, Sr, Ba)F_2-LnF_3$ (Ln = La-Nd)," in: *Study of Materials for Optical Coatings* [in Russian], Tr. Vses. Nauch.-Issled. Inst. Luminoforov, Stavropol' (1977), No. 15, pp. 73-78.

77. A. Z. Arakelyan, K. B. Seiranyan, P. P. Fedorov, and B. P. Sobolev, "Crystallization of nonstoichiometric fluorite phases in the systems MF_2-RF_3," in: Abstracts of Papers of the Vth All-Union Conf. on Growth of Crystals, Vol. 2: Growth of Crystals and Their Structure [in Russian], Tbilisi (1977), pp. 135-136.

78. C. R. A. Catlow, A. V. Chadwick, and J. Corish, "The defect structure of anion-excess CaF_2," *J. Solid State Chem.*, **48**, No. 1, 65-76 (1983).

79. K. Recker and F. Wallrafen, "Untersuchung der Zweistoffsysteme CaF_2-YF_3 und CaF_2-NdF_3," *Ber. Dtsch. Keram. Ges.*, **50**, No. 3, 68-73 (1973).

80. V. A. Arkhangel'skaya, V. M. Reiterov, and P. L. Smolyanskii, "Redox processes during growth of activated fluorite crystals," *Izv. Akad. Nauk SSSR, Neorg. Mater.*, **12**, No. 9, 1560-1564 (1976).

81. A. K. Ivanov-Shits, N. I. Sorokin, S. R. Arutyunyan, et al., "Thermal conductivity of ionic conductors: solid solutions with fluorite structure," *Fiz. Tverd. Tela*, **28**, No. 4, 1235-1237 (1986).

82. R. C. Pastor, A. C. Pastor, and K. T. Miller, "Solid solution of RF_3 in CaF_2," *Mater. Res. Bull.*, **9**, No. 9, 1247-1250 (1974).

83. A. Z. Arakelyan and K. B. Seiranyan, "Distribution of components in single crystals of solid solutions $Ca_{1-x}Ho_xF_{2+x}$" in: Expanded Abstracts of the VIth Int. Conf. on Growth of Crystals, 10-16 Sept., 1980, Vol. 3: Growth from Melts and High-Temperature Solutions [in Russian], Moscow (1980), pp. 163-164.

84. A. A. Kaminskii, B. P. Sobolev, Z. I. Zhmurova, and S. É. Sarkisov, "Generation of stimulated emission of Nd^{3+} ions in disordered crystals of a solid solution of the system CaF_2-LuF_3," *Izv. Akad. Nauk SSSR, Neorg. Mater.*, **20**, No. 5, 869-870 (1984).

85. K. B. Seiranyan, R. O. Sharkhatunyan, L. S. Garashina, and B. P. Sobolev, "Solubility of rare earth trifluorides in SrF_2 and preparation of single crystals of $Sr_{1-x}Ln_xF_{2+x}$" in: Abstracts of Papers of the IVth All-Union Conf. on Growth of Crystals, Part 1: Growth of Crystals and Their Structure [in Russian], Izd. Akad. Nauk Arm. SSR, Erevan (1972), pp. 127-129.

86. K. B. Seiranyan, R. O. Sharkhatunyan, and B. P. Sobolev, "On the question of maxima in the fusibility curve of solid solutions of $Sr_{1-x}Ln_xF_{2+x}$" in: *Growth of Crystals* [in Russian], Izd. Erevan Univ., Erevan (1977), Vol. 12, pp. 150-152.

87. A. A. Kaminskii, S. É. Sarkisov, K. B. Seiranyan, and B. P. Sobolev, "Study of stimulated emission of crystals of $Sr_{0.9}Gd_{0.1}F_{2.1}$ with Nd^{3+} ions," *Izv. Akad. Nauk SSSR, Neorg. Mater.*, **9**, No. 2, 340 (1973).

88. M. R. Brown, H. Thomas, J. S. S. Whiting, and W. A. Shand, "Experiments on Er^{3+} in SrF_2. Pts. 1, 2," *J. Chem. Phys.*, **50**, No. 2, 881-899 (1969).

89. L. S. Garashina, A. A. Kaminskii, L. Li, and B. P. Sobolev, "Lasers based on cubic $SrF_2-YF_3-Nd^{3+}$," *Kristallografiya*, **14**, No. 5, 925 (1969).

90. V. S. Sidorov, "Study of the phase diagrams and preparation of single crystals of solid solutions in systems of some rare earth fluorides and barium," Author's Abstract of Candidate's Dissertation in Chemical Sciences, Dushanbe (1979).

91. A. A. Kaminskii, "On the possibility of studying Stark structure of spectra of TR^{3+} ions in disordered fluorides of crystalline systems," *Zh. Éksp. Teor. Fiz.*, **58**, No. 2, 407-419 (1970).

92. C. G. Andeen, J. J. Fontanella, M. C. Wintersgill, et al., "Clustering in rare-earth-doped alkaline earth fluorides," *J. Phys. C*, **14**, No. 24, 3557-3574 (1981).

93. A. A. Kaminskii, B. P. Sobolev, Kh. S. Bagdasarov, et al., "Investigation of stimulated emission from crystals with Nd^{3+} ions," *Phys. Status Solidi A*, **23**, K135-K136 (1974).

94. A. M. Golubev, N. I. Sorokin, P. P. Fedorov, and B. P. Sobolev, "Atomic structure and electric conductivity of solid solutions of $Ba_{1-x}Ho_xF_{2+x}$ $(0.05 \le x \le 0.28)$," in: Abstracts of Papers of the XIVth All-Union Conf. on Application of X-Rays to Materials Science [in Russian], Kishinev (1985), pp. 97-99.

95. A. M. Golubev, B. P. Sobolev, and V. I. Simonov, "Structure of the solid solution $Ba_{0.625}Er_{0.375}F_{2.375}$ with a tetragonally distorted fluorite-like crystal lattice," *Kristallografiya*, **30**, No. 2, 314-319 (1985).

96. S. Kh. Batygov and A. A. Kaminskii, "On the nature of aging of fluorite and yttrofluorite type crystals activated by Nd^{3+} with induced emission," *Zh. Éksp. Teor. Fiz.*, **53**, No. 3, 839-852 (1967).

HYDROTHERMAL CHEMISTRY
AND GROWTH OF HEXAGONAL GERMANIUM DIOXIDE

T. B. Kosova and L. N. Dem'yanets

Hexagonal germanium dioxide GeO_2 is stable at $T > 1035°C$. This modification of GeO_2 has theoretical and practical interest as a structural analog of α-quartz.

Three polymorphic modifications of crystalline GeO_2 are known: a low-temperature tetragonal modification with the rutile structure $GeO_{2(tetr)}$, a high-temperature hexagonal form with the α-quartz structure $GeO_{2(hex)}$, and a tetragonal modification with the low-temperature structure of cristobalite. The stability region of the last form has not been determined. The transition temperature $GeO_{2(tetr)} \rightarrow GeO_{2(hex)}$ is 1035°C at 0.1 MPa [1-3]. Two modifications exist under normal conditions: the tetragonal as a stable phase and the hexagonal as a metastable phase.

Two modifications can be prepared under hydrothermal conditions: GeO_2 with the rutile and α-quartz structures.

$GeO_{2(hex)}$ has a 32 symmetry form and parameters which are similar to α-quartz:

$$\alpha\text{-quartz: } a = 4.903 \text{ Å}, c = 5.393 \text{ Å, space group } C3_12 \ (C3_22) \ [4];$$
$$GeO_{2(hex)}: a = 4.987 \text{ Å}, c = 5.652 \text{ Å, space group } C3_12 \ (C3_22) \ [5].$$

However, α-quartz is a stable phase under normal conditions up to the α–β transition temperature of 573°C whereas the analogous GeO_2 modification is a metastable phase. The difference in chemical behavior of SiO_2-quartz and of the GeO_2 quartz modification in water and aqueous solutions at room and elevated temperature has its origins here. This causes a significant difference for their growth conditions both with respect to temperature and solution compositions which can be used for single crystal growth.

1. BEHAVIOR OF GeO_2 IN AQUEOUS SOLUTIONS AT 25°C

1.1. Solubility of GeO_2 in Water at 25°C

A large number of publications are devoted to study of solubility and behavior of GeO_2 in water and aqueous solutions of acids and bases at room temperature. However, data on the state of GeO_2 in aqueous media are to a significant degree contradictory. Critical analysis of literature data and results of our investigations are examined in detail in [6, 7].

Results of experimental study of $GeO_{2(hex)}$ solubility [4-20] in water and aqueous solutions at 25°C are converted by us to a single scale, $\log m_{Ge}$, and given in Fig. 1 (curve 1). It should be mentioned that the

81

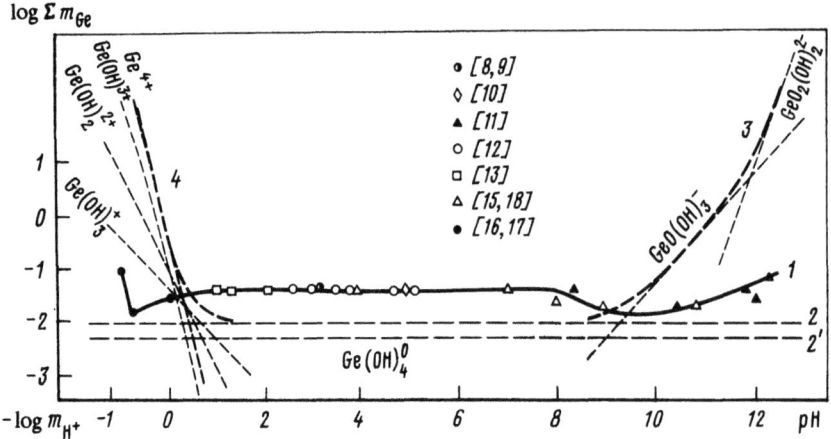

Fig. 1. Solubility of $GeO_{2(hex)}$ at 25°C. 1) Experimental curve; 2) concentration of mononuclear Ge forms [24, 25]; 2') concentration of mononuclear Ge forms [14]; 3) theoretical curve of solubility under formation conditions of anionic forms of germanium in solution; 4) theoretical curve of solubility under formation conditions of cationic forms of germanium in solution.

equilibria attained during study of $GeO_{2(hex)}$ solubility will be metastable since the stable solid phase is the rutile modification of GeO_2. $GeO_{2(tetr)}$ besides $GeO_{2(hex)}$ is observed as the solid phase in our experiments [7] even at 100°C with prolonged storage (20-30 days) in equilibrium with solution. The solubility of $GeO_{2(hex)}$ in water at 25°C varies from $3.9 \cdot 10^{-2}$ to $4.8 \cdot 10^{-2}$ m. We note that the solubility of $GeO_{2(hex)}$ is two orders of magnitude higher than the solubility of α-quartz ($0.8 \cdot 10^{-4}$ m) under analogous conditions and comparable to the solubility of amorphous silica [21].

The path of curve 1 in the pH 1-8 portion shows that the GeO_2 solubility does not depend on solution pH. This indicates that Ge transfers into solution as neutral particles* (monomeric and to a small degree polymeric):

$$GeO_2 + 2\,H_2O = Ge(OH)_4^0,$$
$$n\,GeO_2 + (2n + m)\,H_2O = \{Ge(OH)_4^0\}_n \cdot mH_2O.$$

Solubility determinations show that $GeO_{2(hex)}$ solubility in both strongly acidic (pH < 1) and basic (pH 8-9) regions (Fig. 1, curve 1) is decreased. The GeO_2 solubility falls to a minimal value of 0.004 m (pH 9.1) with increased base concentration and then increases with further increase of pH.

A series of studies [9, 10, 15, 16] has shown that polygermanates corresponding to the cations $M_2Ge_5O_{11}$, $M_2Ge_5O_{11} \cdot nH_2O$, $M_2Ge_4O_9$, and others and not GeO_2 appear as the solid phase in equilibrium with the solution at pH above 8.5-9. The composition of these changes with increasing base concentration. Thus, curve 1 (Fig. 1) in the range pH $\gtrsim$ 9 corresponds to the solubility of the corresponding polygermanates and not of GeO_2.

An analogous change of $GeO_{2(hex)}$ solubility is observed in the strongly acidic region (pH < 1) with increasing acid concentration. Experimental studies in solutions of hydrohalic acids [16-20, 22, 23] showed that

*The composition of these Ge-containing particles, depending on the coordination number of germanium in solution which by analogy with the solid phase can be equal to 4 or 6, can be described with a variable amount of water molecules, for example $Ge(OH)_4^0 \cdot 2H_2O$ or $GeO(OH)_3^- \cdot 2H_2O$. However, the formulas for the hydroxy-complexes given above will be used for further thermodynamic calculations since the free energy change of the hydration reaction in the standard state is equal to zero.

the $GeO_{2(hex)}$ solubility curve is characterized by a minimum at an acid concentration of 4-6 m. The $GeO_{2(hex)}$ solubility increases with further increase of acid concentration. This enhanced solubility is related in [18, 22] to formation of complexes such as $Ge(OH)_xCl_{6-x}^{2-}$ and $Ge(OH)_xCl_{5-x}^-$. A new solid phase, $GeCl_4$ [17], was discovered at HCl > 8 m in solubility experiments. Consequently, we do not have a "pure" GeO_2–H_2O system even in the strongly acidic region.

1.2. Forms of Germanium in Solution

Predominance of this or that form of germanium in solution in the absence of a complexing agent, as for a number of other cations such as Si, Al, and Sn, is determined by its total concentration Σm_{Ge}, the coordination number of germanium in solution, the concentration of H^+ (OH^-) ions, and the solution temperature.

A number of authors showed that monomeric germanium-containing particles dominate in solution at low total germanium concentration ($\Sigma m_{Ge} < 0.005\ m$ [24] or $\Sigma m_{Ge} < 0.01\ m$ [14, 24]) and in the absence of anion-complexing agents such as F^-, Cl^-, etc. These are monomers such as $Ge(OH)_4^0$, $GeO(OH)_3^-$, and $GeO_2(OH)_2^{2-}$ [24, 25] in neutral and basic solutions (pH 7-13). Polynuclear particles, for example $\{[Ge(OH)_4]_8(OH)_3\}^{3-}$ [24], appear in solutions with germanium concentrations which exceed $0.005\ m$ ($0.01\ m$). The fraction of these increases with increased germanium concentration in solution. The logarithm of the formation reaction equilibrium constant of the $\{[Ge(OH)_4]_8(OH)_3\}^{3-}$ polyion is 29.14 ± 0.5 [24, 25]. Thus, it follows that the concentration of polyions with 3− charge at pH $\approx$ 4.3 and a total Ge concentration in solution above $0.005\ m$ is 8 orders of magnitude lower than the monomer concentration, i.e., the concentration of charged polynuclear particles is negligible. The degree of polymerization of germanium particles depends on germanium concentration and solution pH.

The maximal Ge concentration in solution when all of the germanium is found as uncharged monomeric particles corresponds to curve 2 (or 2') (Fig. 1) according to [14, 24]. The experimentally determined curve lies above curve 2 (or 2'). This shows that uncharged polynuclear germanium particles besides the monomeric particles are also present in solution. The lowering of $GeO_{2(hex)}$ solubility in the basic region up to the minimal value at pH 9.1 (NaOH concentration $0.004\ m$) attests to the disintegration of uncharged polynuclear particles to neutral $Ge(OH)_4^0$ particles. Decreased $GeO_{2(hex)}$ solubility in the acidic region up to the concentration of the $Ge(OH)_4^0$ monomer (Fig. 1), as in the case of basic solutions, can be explained by depolymerization of Ge-containing particles.

Curves 3 and 4 (Fig. 1) correspond to the calculated change of GeO_2 solubility on pH if the GeO_2 dissolution occurs with formation of cationic and anionic forms according to the equations

$$GeO_2 + 2H_2O = Ge(OH)_4^0, \tag{1}$$

$$Ge(OH)_4^0 + (OH)^- = GeO(OH)_3^- + H_2O, \tag{2}$$

$$GeO(OH)_3^- + (OH)^- = GeO_2(OH)_2^{2-} + H_2O, \tag{3}$$

$$Ge^{4+} + H_2O = Ge(OH)^{3+} + H^+, \tag{4}$$

$$Ge(OH)^{3+} + H_2O = Ge(OH)_2^{2+} + H^+, \tag{5}$$

$$Ge(OH)_2^{2+} + H_2O = Ge(OH)_3^+ + H^+, \tag{6}$$

The calculated curves 3 and 4 (Fig. 1) do not concur with the experimental GeO_2 solubility curve (curve 1). Since curve 1 in the pH > 9 region corresponds to polygermanate solubility and not to hexagonal GeO_2, the right branch of the curve (Fig. 1, curve 3), which is constructed based on the formation reaction equilibrium constants of the $GeO(OH)_3^-$ and $GeO_2(OH)_2^{2-}$ complexes, pertains to metastable equilibrium in the GeO_2–H_2O system. If it is accepted that germanium in the acidic region is present as the cations Ge^{4+}, $Ge(OH)_2^{2+}$, $Ge(OH)^{3+}$, and $Ge(OH)_3^+$, in agreement with [26], then the solubility at pH < 1 should change according to

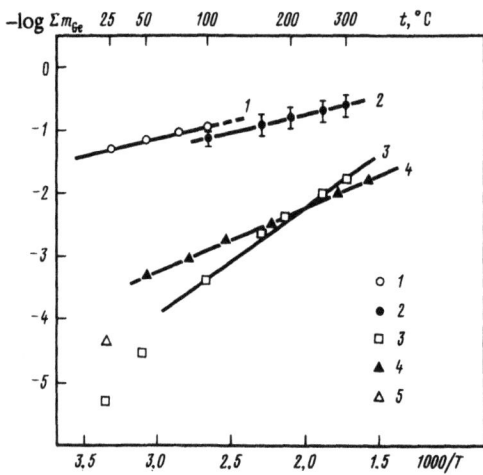

Fig. 2. Solubility of polymorphic modifications of GeO_2 in water as a function of temperature. 1) $GeO_{2(hex)}$ (experimental solubility curve); 2) $GeO_{2(hex)}$ (calculated curve); 3) $GeO_{2(tetr)}$ (experimental solubility curve); 4) α-quartz from [21]; and 5) $GeO_{2(tetr)}$ from [20].

curve 4 (Fig. 1). The path of the $GeO_{2(hex)}$ solubility curve in this pH region shows that the data of [22] on the absence of cationic forms of germanium in solution should be considered most reliable.

Thus, cationic forms are absent both in the case of silicon (the SiO_2–H_2O system) and for the GeO_2–H_2O system. The difference between the germanium system and the SiO_2–H_2O system, where silicon in the pH 1-9 range exists in the form of $Si(OH)_4$ particles [21], is that neutral forms of germanium in solution can be both mononuclear and polynuclear. The increased solubility of $GeO_{2(hex)}$ in comparison with quartz by two orders of magnitude (as a metastable phase) also explains the presence in solution of polynuclear forms of germanium.

2. BEHAVIOR OF GeO_2 AT ELEVATED TEMPERATURES

2.1. Solubility of GeO_2 in Water at Elevated Temperatures

The solubility of $GeO_{2(hex)}$ in water (Fig. 2) between 25-100°C is described by the equation [7]

$$\log m_{GeO_2} = -\frac{767}{T} + 1.207. \tag{7}$$

Experiments on the determination of $GeO_{2(hex)}$ solubility in H_2O at $T \gtrsim 120$°C showed that the obtained values of Ge concentration cannot be values of $GeO_{2(hex)}$ solubility since the stable tetragonal GeO_2 phase was identified in experiments of more than 30 days duration. The quantity of $GeO_{2(tetr)}$ increases with increased duration of the experiment.

We studied *in situ* the transition $GeO_{2(hex)} \rightarrow GeO_{2(tetr)}$ in water at 183°C using an energy-dispersive analysis which was described in [27]. The results are shown in Figs. 3 and 4. Lines of the $GeO_{2(tetr)}$ spectrum are distinctly revealed at $T = 183$°C already 15 min after the initiation of heating. The starting $GeO_{2(hex)}$ is completely converted into the tetragonal modification after 60 min. This fact suggests that the 185 ± 5°C temperature, proposed in the literature as the transition temperature for $GeO_{2(hex)} \rightarrow GeO_{2(tetr)}$ in the presence

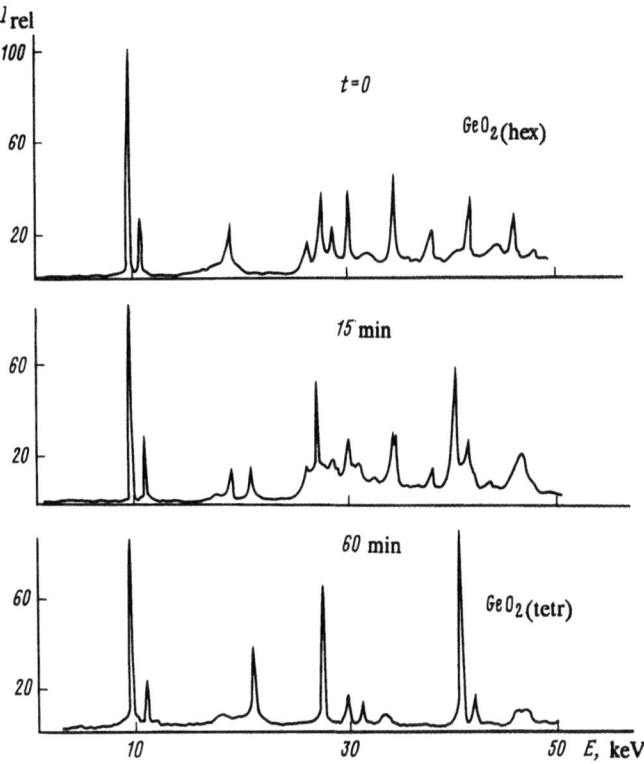

Fig. 3. Results of energy-dispersive analysis of the solid phase in the GeO_2–H_2O system at 183°C.

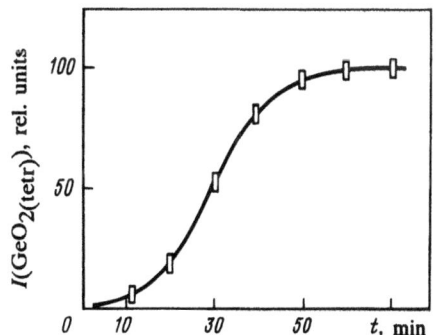

Fig. 4. Kinetics of the $GeO_{2(hex)} \rightarrow Geo_{2(tetr)}$ transition at 183°C from energy-dispersive analysis.

of water [28], is rather arbitrary. The transition of $GeO_{2(hex)}$ into the stable $GeO_{2(tetr)}$ modification is decidedly broadened and the reaction rate quickly increases with increasing temperature.

Thus, it seems impossible to determine experimentally the $GeO_{2(hex)}$ solubility in water and aqueous solutions at $T > 120°C$ (at lower temperatures the transition rate is sufficiently slow). It can be estimated from data on $GeO_{2(tetr)}$ solubility and ΔG^0 as a function of temperature for the transition reaction $GeO_{2(tetr)} \rightarrow GeO_{2(hex)}$.

Solubility of the polymorphic GeO_2 modification with the rutile structure was studied by us [6, 7] between 25-300°C at pressures from the saturated vapor pressure up to 140 MPa. The results are incorporated into Fig. 2.

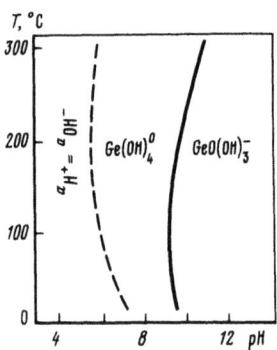

Fig. 5. Relation of hydroxo-complexes of germanium in aqueous solutions in neutral and basic regions as a function of temperature.

The experimentally determined value of $GeO_{2(tetr)}$ solubility at 25°C is $4 \cdot 10^{-6}$ m [7]. This is approximately an order of magnitude lower than the value given in [19]. The $GeO_{2(tetr)}$ solubility increases from $3.8 \cdot 10^{-4}$ to $1.7 \cdot 10^{-2}$ m with increasing temperature from 100 to 300°C. These values are substantially lower than the metastable solubility of $GeO_{2(hex)}$.

Between 25-300°C, $GeO_{2(tetr)}$ solubility is described by the equation

$$\log m_{Ge(OH)_4^0} = \frac{-4132}{T} + 11.969 - 0.011559\,T, \tag{8}$$

and between 100-300°C, by the linear equation

$$\log m_{Ge(OH)_4^0} = -\frac{1774}{T} + 1.403. \tag{9}$$

Pressure in this temperature range practically does not affect $GeO_{2(tetr)}$ solubility. Even at 300°C, when a change of the filling coefficient entails a sharp change of pressure (from the saturated vapor pressure up to 130-140 MPa), the obtained $GeO_{2(tetr)}$ solubility values are identical within experimental error. The solubility does not depend on pH (as for $GeO_{2(hex)}$ also) between pH 2-9 at constant temperature.

The solubility values of the stable α-quartz modification in water at these same temperatures [21] are shown in Fig. 2 for comparison. The order of magnitude of solubility of the stable α-SiO_2 modification is seen to be comparable to the solubility of the stable tetragonal modification of GeO_2 and not the isostructural but metastable $GeO_{2(hex)}$.

The behavior of germanium hydroxo-complexes at various temperatures and solution pH values must be known in order to describe the equation of GeO_2 solubility under hydrothermal conditions. Estimation of reaction (3) equilibrium constants as functions of temperature allowed the behavior of germanium hydroxo-complexes at various temperatures and solution pH values to be determined [6]. Figure 5 shows that the field of dominance by germanium hydroxo-complexes depends weakly on temperature and that the mononuclear $Ge(OH)_4^0$ particles dominate at elevated solution temperatures.

The $GeO_{2(tetr)}$ solubility values are independent of pH, as shown by the fact that germanium transfers into solution under high-temperature conditions primarily as uncharged particles (Fig. 5). The low absolute value of solubility at 100-300°C (<0.01 m) and the effect of temperature, which facilitates depolymerization processes, allow the $GeO_{2(tetr)}$ dissolution reaction under the studied conditions to be described as

$$GeO_{2(tetr)} + H_2O_l = Ge(OH)_{4d}^0. \tag{10}$$

The $GeO_{2(hex)}$ dissolution reaction, with a single mechanism for $GeO_{2(tetr)}$ and $GeO_{2(hex)}$ dissolution, can also be written as

$$GeO_{2(hex)} + 2H_2O_l = Ge(OH)_{4d}^0. \tag{11}$$

The solubility curve of $GeO_{2(hex)}$ at $T > 100°C$ (Fig. 2) is obtained from data on the $GeO_{2(tetr)}$ solubility and the Gibbs free energy change for the transition $GeO_{2(tetr)} \rightarrow GeO_{2(hex)}$ as functions of temperature, calculated by us in [6]. The calculated curve, obtained accounting only for mononuclear forms of germanium in solution, lies lower than the experimental curve 1 (Fig. 2), i.e., part of the germanium in solution at $T > 25°C$ is found as polynuclear particles.

2.2. Stability of GeO_2 in the Presence of Mono- and Divalent Cations

Growth of GeO_2 crystals will be limited by the kinetics of the $GeO_{2(hex)} \rightarrow GeO_{2(tetr)}$ transition, as follows from the arguments above. Besides this, the stability of GeO_2 depends substantially on the cationic composition and pH of the medium. Use of solvent mineralizers leads to replacement of GeO_2 by germanates. This was demonstrated above for pH > 9 and $T = 25°C$. The stability of GeO_2 in the presence of mono- and divalent cations can be estimated using thermodynamic calculations. Equilibrium constants of the GeO_2 replacement reactions in aqueous solutions by the corresponding germanates must be considered for this. The germanate compositions were chosen in agreement with phase diagrams for GeO_2–$MO(M_2O)$:

$$M_d^{2+} + 0.5\,GeO_2 + H_2O_l = 0.5\,M_2GeO_{4cr} + 2H_d^+, \tag{12}$$

where M^{2+} = Ca, Zn, Fe, Co, Ni

$$Li_d + 3.5\,GeO_2 + 0.5\,H_2O_l = 0.5\,Li_2Ge_7O_{15} + H_d^+, \tag{13}$$

$$Na_d^+ + 2.25\,GeO_2 + 0.5\,H_2O = 0.25\,Na_4Ge_9O_{20} + H_d^+, \tag{14}$$

$$K_d^+ + 2\,GeO_2 + 0.5\,H_2O = 0.5\,K_2Ge_4O_9 + H_d^+. \tag{15}$$

Calculations were carried out for both $GeO_{2(hex)}$ and $GeO_{2(tetr)}$. The equilibrium constants for reactions (12)-(15) at various temperatures were calculated by us from the equations $\Delta G_T^0 = -RT \ln K$. The values of the Gibbs free energies ΔG_T^0 were found by summation of the free energies of the necessary reactions for which thermodynamic data are known

$$\begin{array}{lll}
MO_{cr} + 0.5\,GeO_{2cr} = 0.5\,M_2GeO_{4cr} & \Delta G_T^0 \text{ reaction} & (16) \\
+ & + & \\
M_d^{2+} + H_2O_l = MO_{cr} + 2H_d^+ & \Delta G_T^0 \text{ reaction} & (17) \\
\hline
M_d^{2+} + 0.5\,GeO_{2cr} + H_2O_l = 0.5\,M_2GeO_{4cr} + 2H_d^+ & \Delta G_T^0 \text{ substitution reaction (12)} &
\end{array}$$

Equilibrium constants of reactions (13)-(15) for the corresponding alkaline germanates can be obtained analogously.

Standard formation enthalpies $\Delta H_{298.5}$, standard entropies $S_{298.5}^0$, and heat capacities as functions of temperature for the germanates examined must be known for calculating ΔG_T^0 of reaction (16).

Calorimetric data of [29, 30] were used for calculating the thermal change of reaction (16). These studies provide enthalpies of dissolution for the germanates Fe_2GeO_4, Co_2GeO_4, Ni_2GeO_4, Zn_2GeO_4, and Ca_2GeO_4 of the corresponding oxides of divalent metals and of $GeO_{2(hex)}$ and $GeO_{2(tetr)}$ in a melt of $2PbO$–B_2O_3 at 966 K. We calculated changes of enthalpy ΔH_{966} of reaction (16) from these measurements. The thermal effects of these reactions at 298.15 K were calculated with the assumption that the ΔC_p of reaction (16) as a function of temperature is the same as for the formation reaction from the oxides of the corresponding silicates. The silicate heat capacities as functions of temperature were taken from [31].

Calculation of the entropy change at 298.15 K in (16) necessitates availability of standard entropy values of the corresponding germanates. Values of the standard entropies of Li_2GeO_3, Na_2GeO_3, K_2GeO_3, $CaGeO_3$, and $BaGeO_3$ are given in [32]. Comparing these data with the standard entropies of silicates of analogous

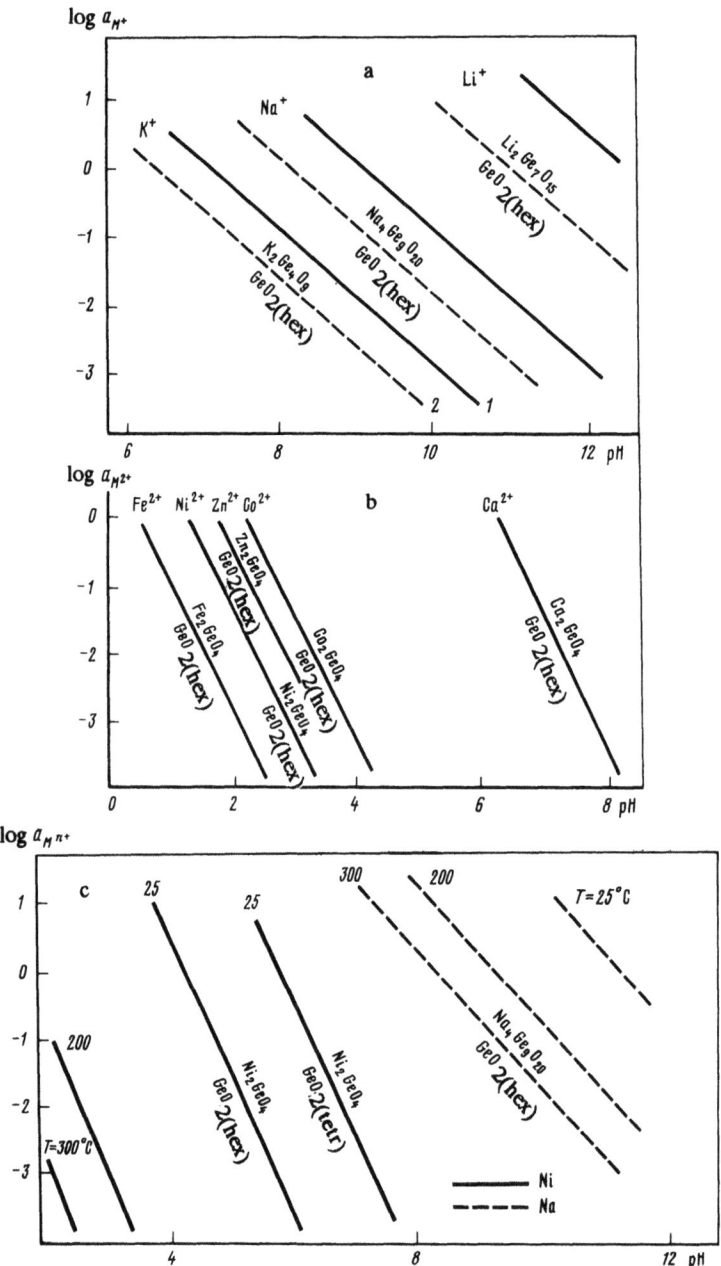

Fig. 6. Phase equilibria in systems based on GeO_2. a) The Me_2O–GeO_2–H_2O (Me = Li, Na, and K) systems at 200 (1) and 300°C (2); b) the MeO–GeO_2–H_2O (Me = Ca, Fe, Ni, Zn, and Co) systems at 200°C; c) effect of temperature on phase equilibria in the Na_2O–GeO_2–H_2O, NiO–$GeO_{2(hex)}$–H_2O, and NiO–$GeO_{2(tetr)}$–H_2O systems.

composition, we found that the relation between them can be expressed by the equation

$$S^0_{298,15}(\text{germanate}) = 5{,}071 + 0.915 S^0_{298.15}(\text{silicate}) \cdot \qquad (18)$$

Coefficients of this equation were found by a least squares method for silicates and germanates of Li, Na, K, Ca, and Ba accounting for experimental error of the data. Equation (18) was used for estimation of the standard entropies of germanates $S^0_{298.15}$. The values obtained were 33.8 (Zn_2GeO_4), 36.8 (Fe_2GeO_4), 36.2 (Co_2GeO_4), and 29.4 (Ni_2GeO_4).

Free energy changes in the reactions (17) between 25-300°C were calculated according to [31].

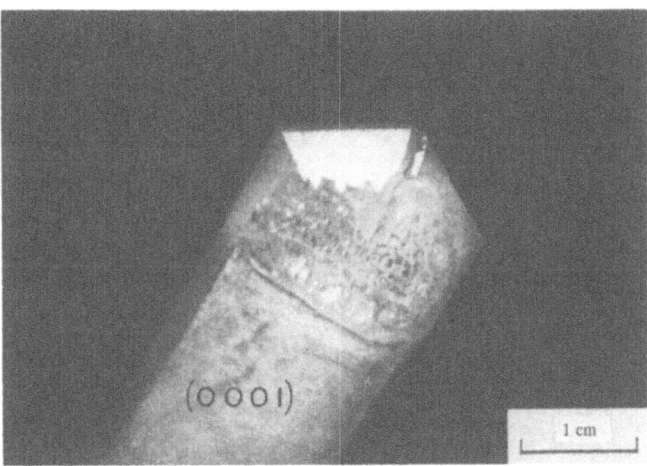

Fig. 7. Single crystal $GeO_{2(hex)}$ grown on an α-quartz seed.

Results of calculations of phase equilibria in $M_2O(MO)-GeO_2-H_2O$ systems between 25-300°C are given in Fig. 6a-c. The indicated tendencies hold for all the systems studied. $GeO_{2(hex)}$ in the presence of alkali cations can be formed at temperatures below 200°C between pH 1-11 (at low cation activity). The stable field of GeO_2 is shortened with increasing ionic radius of the M^+ cation. The presence of K^+ ions leads to formation of germanates already in neutral and even weakly acidic solutions.

Introduction of divalent cations (Fig. 6b) leads to further shifting of the GeO_2 stability limits to more acidic solutions. The effect of hydrolytic processes on GeO_2 stability was evaluated for systems with divalent cations. Hydrolysis is important at low values of metal activities in solution ($\log a_{M^{2+}} \leq -2$ for Zn^{2+} and less than -4 and -5 for other divalent cations). The concentration of cations is usually substantially higher, $a_{M^{n+}} > 0$, than n moles in actual hydrothermal experiments on crystal growth. Therefore, hydrolytic processes can be neglected in this case. The M_2GeO_4/GeO_2 boundary accounting for hydrolysis is shown for Zn^{2+} in Fig. 6b.

The stability region of the stable GeO_2 tetragonal phase is wider than that of $GeO_{2(hex)}$ in all cases in the presence of cations. This is demonstrated in Fig. 6c for the example of the system with Ni and Na. The increase of temperature also leads to a decrease of the stability field for GeO_2.

The data obtained indicate that the region of possible $GeO_{2(hex)}$ crystal growth is severely limited both in temperature and in solution pH value, as well as by the cationic composition of the medium. Growth of $GeO_{2(hex)}$ is possible in weakly basic, neutral, or acidic solutions depending on the solvent. Stable germanates of basic divalent metals are formed in strongly basic solutions.

3. GROWTH OF $GeO_{2(hex)}$ CRYSTALS
UNDER HYDROTHERMAL CONDITIONS

We chose weakly acidic media — aqueous solutions of alkali metal fluorides and ammonium fluoride [33] — for growth of $GeO_{2(hex)}$ crystals, in agreement with the examined features of $GeO_{2(hex)}$ behavior in aqueous media at elevated temperatures. Preliminary experiments showed that $GeO_{2(hex)}$ solubility in fluoride solutions practically does not differ from $GeO_{2(hex)}$ solubility in pure water. An insignificant rise in solubility with increased concentration is observed in fluoride solutions, for example NH_4F from 0 to 1.5 mass %. With further increase of C_{NH_4F}, the solubility value is practically unchanged. The F^- ions which are added to the growth system should facilitate improved transport of germanium into the growth zone in comparison to pure water.

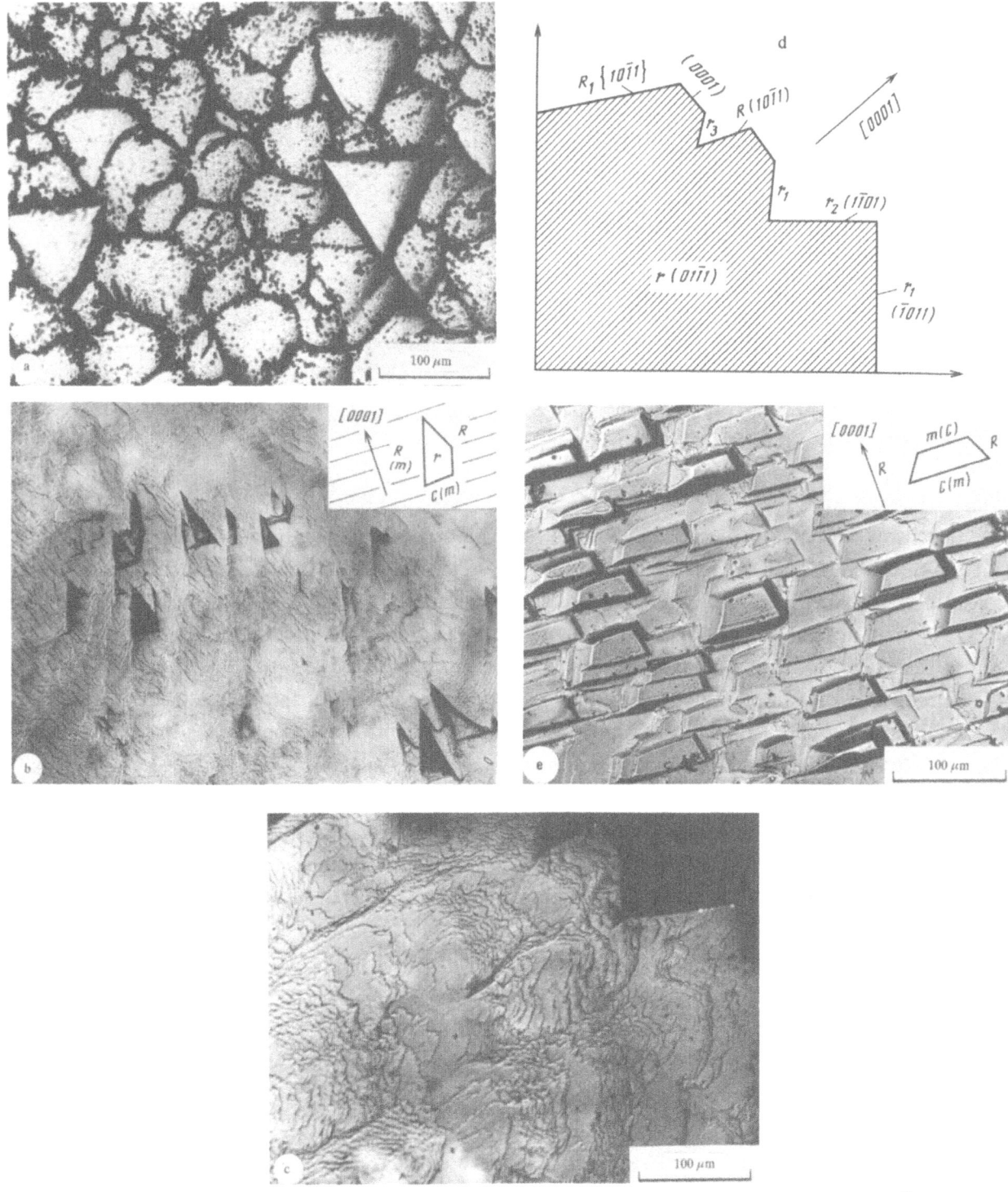

Fig. 8. Surface growth of GeO_2 crystal faces at small saturations. a) (0001) pinacoid face, V 0.04 mm/day; b) (01$\bar{1}$1) small rhombohedron r face, initial growth stage; c) (0$\bar{1}$1) small rhombohedron r face, cross section of growing pinacoid surface; d) scheme of limited mound growth on (0001) pinacoid face; e) (10$\bar{1}$1) large rhombohedron R face, V 0.01 mm/day.

The possible preparation of crystalline GeO_2 outgrowths on α-quartz in pure water was noted in [34]. The authors demonstrated a clearly elevated $GeO_{2(hex)} \rightarrow GeO_{2(tetr)}$ transition temperature in the presence of water at 225°C. The layer that grew up to 2 mm in thickness was uneven and the quality was low.

The $GeO_{2(hex)}$ was grown in 1-liter autoclaves fitted with floating fluoroplastic inserts. The temperature of the dissolution zone was 130-150°C. The temperature drop inside the autoclave was 1.5-7°C. The relatively high temperatures were determined by the kinetics of the $GeO_{2(hex)}$ crystallization process on a seed. They in turn determine the duration of the experiment, not more than 30 days, in order to avoid complete inversion of $GeO_{(hex)}$ into $GeO_{2(tetr)}$.

Growth of $GeO_{2(hex)}$ under metastable conditions was initiated by introduction of quartz substrate. Quartz plates which were cut parallel to the pinacoid faces (C-cut, {0001}), the {10$\bar{1}$1} large R, and the {01$\bar{1}$1} small r faces of the rhombohedra were used as seeds. Distilled water and solutions of ammonium, potassium, rubidium, and lithium fluorides were used as solvents.

Growth of GeO_2 was observed on quartz seeds of all these orientations. Significant difference in growth rates in various solvents was not observed. The crystallographic direction of growth and solvent type substantially affect adhesion of the grown layer. The maximal thickness of the grown layer was obtained on seeds perpendicular to the threefold axis (up to 5 mm, Fig. 7). The poorest growth was observed on seeds parallel to the {10$\bar{1}$1} large rhombohedron. In this case, growth was practically not observed in NH_4F, H_2O, and KF solutions. A heteroepitaxial layer, which due to the heterometry and weak adhesion was easily separated mechanically from the substrate, was formed on the seed surface in LiF solutions. In contrast to quartz, as in [35], appreciable growth of the {10$\bar{1}$0} hexagonal prism faces is observed. The growth rate of the prism faces exceeded the growth rate of the large rhombohedron faces. The relation of growth rates of $GeO_{2(hex)}$ crystals in the solutions studied was: V{0001} > V{01$\bar{1}$1} > V{10$\bar{1}$0}, V{10$\bar{1}$1}.

The growth rate of {0001} pinacoid faces reached 0.2 mm/day with ΔT inside the autoclave of about 7°. V{0001} was ~0.04 and V{10$\bar{1}$1} was ~0.01 mm/day for $\Delta T \approx 2°$.

The morphology of the growing faces of GeO_2 crystals is different for crystals grown under low ($\sigma \approx 0.1$) and high ($\sigma \approx 0.5$) saturation conditions.

At low saturations ($\Delta T \approx 2°$, $\sigma \approx 0.1$) and in the initial growth stages, the macrorelief of the growth layer surface is unique for each of the crystallographic substrate orientations studied. Figure 8a-e shows the surface growth for faces of GeO_2 crystals grown in lithium fluoride solutions on the C {0001} pinacoid faces, the {01$\bar{1}$1} small r, and the {10$\bar{1}$1} large R faces of the rhombohedra. The orientational relations of the growth shapes on the various surfaces were determined by construction of the corresponding stereographic projections onto the faces.

Cellular growth, characterized by a specific "cobblestone" relief built primarily by rounded blocks, is observed on the pinacoid faces. Triangular blocks can be seen in a number of cases. The sides of the triangles correspond to cross sections of the planes by a particular type of rhombohedral face. The dark points in Fig. 8a are surface etching pits at the end of the growth experiment. The morphology of the pinacoid surface attests to the normal growth of the {0001} faces.

Figure 8b shows the initial growth stage of the {01$\bar{1}$1} small rhombohedron face. Growth occurs by layered growth. Traces of shifting of the pinacoid face (light hatching) are visible on the face. The growth front is perpendicular to [0001]. Stagnant portions which have a clear crystallographic orientation remain during shifting of layers. The sides of the negative growth shapes correspond to cross-sectional lines of the {01$\bar{1}$1} R face and the faces of the pinacoid C (of the prism m)* and the {10$\bar{1}$1} large rhombohedron R.

The portion of the face of the same crystal that is completely covered by the growing layer is shown in Fig. 8c. Expansion of the growth layers from the central portions of the face to the periphery is visible. The edges of the crystals are bounded by the planes of r- and R-rhombohedra in various combinations (Fig. 8c).

The surface of the large rhombohedron is covered by characteristic vicinal trapezoidal shapes that are regularly oriented on the face (Fig. 8d) and are formed by planes of the large and small rhombohedra and the pinacoid (of the prism m). The GeO_2 film peels in portions from the {10$\bar{1}$1} R face.

---

*The face C(m) belongs to the ⟨$\bar{1}2\bar{1}0$⟩ zone, the face R(m), to the ⟨$\bar{1}\bar{1}23$⟩ zone.

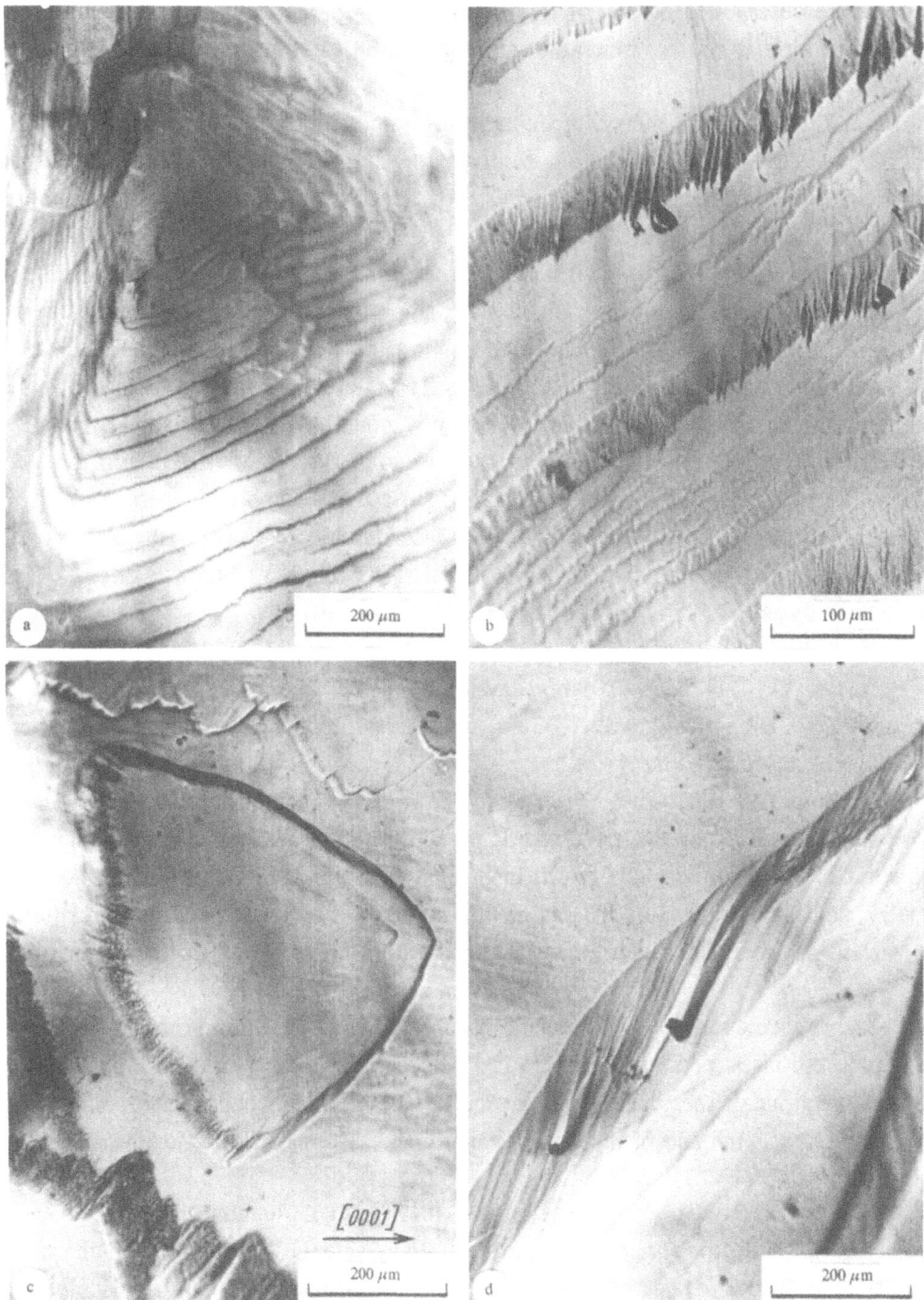

Fig. 9. Growth surface of GeO$_2$ crystal faces at large saturations. Spiral growth on the (01$\bar{1}$1) rhombohedron face (a), characteristic view of growth steps on faces of rhombohedra and prisms (b), isolated step on the (01$\bar{1}$1) rhombohedral face (c), and repulsion of mechanical impurities by the shifting face (d).

    The morphology of the growth surfaces of GeO$_2$ crystals that are grown at large saturations often does not carry information on the crystallographic orientation of the grown surface. The morphology of the surfaces of the rhombohedra and prism is identical in general features. It is stepped and has rounded vicinal mounds (with macrospirals and without). The vicinal mounds with macrospirals (Fig. 9a) are usually located in the upper part of the crystal (the seed plates are positioned vertically). The density of steps is higher in the upper

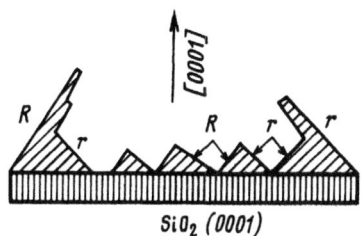

Fig. 10. Scheme of skeletal growth of
GeO_2 on (0001) faces of SiO_2.

part of the crystals, to which the current of most saturated solution flows, and gradually diminishes in the middle and lower portions of the face. The steps, which are curved in the middle portions of the face, bulging toward the lower side relative to the position in the autoclave, have assorted heights (Fig. 9b). The distance between high steps is on the order of ~100 μm. A series of steps of smaller width, 10-20 μm, can be seen on the surface of these steps. Sometimes isolated islets are observed on the step surface. The surface of the step can be covered by characteristic rounded protruding mounds of growth (Fig. 9c). Repulsion of impurities during step advance can be seen in Fig. 9d. The GeO_2 growth mechanism on a quartz seed for the small rhombohedron and prism faces at low saturations can be considered layered. At large saturations, it is spirally layered.

Multinucleated growth is maintained on the pinacoid face. However, the relief of the face is not "cobblestone" but pointed triangular mounds that are bounded by rhombohedral r and R faces. Skeletal growth is often encountered at growth rates above 0.2 mm/day (Fig. 10).

4. CONCLUSIONS

1. The solubility of polymorphic modifications of GeO_2 in water at 25-300°C is studied. The theoretical path of the solubility curve of the metastable modification $GeO_{2(hex)}$ is calculated as a function of temperature. Dissolution of $GeO_{2(hex)}$ in water and aqueous solutions at pH 1-8 occurs with formation of neutral, primarily mononclear particles. Cationic forms of germanium are absent at pH < 1. The stable solid phases at pH > 9 are germanates of alkali metals and ammonium.

2. The field of predominance of germanium hydroxo-complexes depends weakly on temperature. Uncharged germanium hydroxo-complexes predominate at elevated temperatures in neutral and weakly basic solutions.

3. Thermodynamic analysis of the stability of GeO_2 polymorphic modifications in ternary systems which contain alkali, alkaline earth, and transition elements showed that the stability of GeO_2 is also limited by the cationic composition of the medium. GeO_2 can be formed over a wide range of solution pH values (1-11). The field of its stability is shortened with increasing ionic radius of alkali metal (Li—Na—K) added to the system and with increasing temperature. Introduction of divalent cations leads to further shifting of the boundaries for GeO_2 formation to more acidic solutions. Stable alkali metal germanates crystallize from strongly basic solutions.

4. Growth of $GeO_{2(hex)}$ under hydrothermal conditions in aqueous fluoride solutions on α-quartz seeds is studied. Rates of GeO_2 growth as a function of growth conditions are evaluated for various crystallographic directions. The morphology of the growing surfaces of GeO_2 crystals is studied.

REFERENCES

1.   A. W. Laubengauer and D. S. Morton, "Germanium. XXXIX. Polymorphism of germanium dioxide," *J. Am. Ceram. Soc.*, **54**, No. 6, 2303-2320 (1932).

2.   V. G. Hill and L. L. Y. Chang, "Hydrothermal investigation of GeO₂," *Am. Miner.*, **53**, 1744-1748 (1968).

3.   H. Bohm, "The cristobalite modification of GeO₂," *Naturwissenschaften*, **55**, No. 12, 648-649 (1968).

4.   J. D. Dana and C. Frondel, *The System of Mineralogy. Vol. III. Silica Minerals*, Wiley, New York (1962)

5.   G. S. Smith and P. B. Isaaks, "The crystal structure of quartz-like GeO₂," *Acta Crystallogr.*, **17**, No. 4, 842-846 (1964).

6.   T. B. Kosova and L. N. Demianets, "Hydrothermal solubility of GeO₂ polymorphs. Thermodynamics and experiment," in: Proc. Ist Int. Symp. Hydrothermal Reactions, Japan, 1982, Tokyo (1983), pp. 588-611.

7.   T. B. Kosova, L. N. Dem'yanets, and T. G. Uvarova, "Study of solubility of germanium dioxide in water at 25-300°C," *Zh. Neorg. Khim.*, **32**, No. 3, 620-627 (1987).

8.   V. A. Vekhov, V. S. Vitukhnovskaya, and R. F. Doronkina, "Change of solubility of germanium dioxide in water at elevated temperature from 0 to 100°C," *Izv. Vyssh. Uchebn. Zaved., Khim. Khim. Tekhnol.*, **7**, No. 6, 1018-1019 (1964).

9.   V. A. Vekhov, V. S. Vitukhnovskaya, and R. F. Doronkina, "Change of state and solubility of germanium dioxide in ammoniacal aqueous solutions," *Zh. Neorg. Khim.*, **11**, No. 2, 237-252 (1966).

10.   O. D. Lyakh, I. A. Shcheka, and A. I. Perfil'ev, "On the reaction of germanium dioxide with ammonia and urotropine in aqueous solutions," *Zh. Neorg. Khim.*, **10**, No. 8, 1822-1826 (1965).

11.   K. H. Gauer and O. T. Zajicek, "The solubility of germanium (IV) oxide in aqueous NaOH solution at 25°C," *J. Inorg. Nucl. Chem.*, **26**, 951-954 (1964).

12.   V. A. Leitsin, "On the question of germanium dioxide solubility," *Izv. Vyssh. Uchebn. Zaved., Khim. Khim. Tekhnol.*, **5**, No. 3, 679-681 (1962).

13.   L. V. Denisova, A. M. Andrianov, and N. A. Vasyutinskii, "Effect of aging germanium dioxide on its solubility in water and solutions of hydrochloric acid," *Izv. Akad. Nauk SSSR, Neorg. Mater.*, **5**, No. 1, 63-66 (1969).

14.   A. K. Babko and G. I. Gridchina, "Polyions of germanium in basic solutions," *Ukr. Khim. Zh.*, **34**, No. 9, 948-952 (1968).

15.   E. A. Knyazev and S. V. Borisova, "The Na₂O–GeO₂–H₂O system at 25°C," *Zh. Neorg. Khim.*, **12**, No. 10, 2785-2788 (1967).

16.   G. Brauer and H. Muller, "Losungen von Germanium(IV)–Oxyd in anorganischen Sauren," *Z. Anorg. Allg. Chem.*, **87**, No. 1/2, 71-86 (1956).

17.   L. N. Shigina and V. M. Andreev, "Solubility of germanium dioxide in hydrochloric acid," *Zh. Neorg. Khim.*, **11**, No. 4, 870-877 (1966).

18.   E. A. Knyazev, "Solubility in the GeO₂–H₂O–HCl–GeCl₄ system," *Zh. Neorg. Khim.*, **8**, No. 10, 2384-2388 (1963).

19.   W. Pugh, "Germanium. Part IV. The solubility of germanium dioxide in acids and alkalis," *J. Chem. Soc.*, 1537-1549 (1929).

20.   F. Muller, "Further studies on allotropy of germanic oxide," *Proc. Am. Philos. Soc.*, **65**, No. 3, 183-192 (1926).

21.   A. G. Volosov, I. L. Khodakovskii, and B. N. Ryzhenko, "Equilibria in the SiO₂–H₂O system at elevated temperatures along the lower triphasic curve," *Geokhimiya*, No. 5, 575-591 (1972).

22.   D. A. Everest and J. C. Harrison, "Studies in the chemistry of quadrivalent germanium. Part IV. The chemical nature of solution of quadrivalent germanium in hydrochloric or hydrobromic acid," *J. Chem. Soc.*, 1820-1823 (1957).

23.   D. A. Everest and J. E. Salmon, "Studies of quadrivalent germanium: ion-exchange studies of solutions of germanates," *J. Chem. Soc.*, 2438-2443 (1954).

24.   N. Ingri, "Equilibrium studies of polyanions. 12. Polygermanates in Na(Cl) medium," *Acta Chem. Scand.*, **17**, No. 3, 567-616 (1963).

25.   N. Ingri and G. Schorch, "On the determination of the formation of $GeO_2(OH)_2^{2-}$ using a hydrogen-electrode measurement in 3 M Na(Cl) medium," *Acta Chem. Scand.*, **17**, No. 3, 590-596 (1963).

26.   V. A. Nazarenko and S. V. Flyantikova, "Cationic hydroxo-complexes of quadrivalent germanium in solutions with ionic strength 0-1," *Zh. Neorg. Khim.*, **13**, No. 7, 1855-1860 (1967).

27.   Yu. N. Dem'yanets, Yu. E. Gorbatyi, and A. V. Chichagov, "X-ray diffraction study of phase equilibria and kinetics of mineral formation at high pressures and high temperatures," in: Expanded Abstracts of the VIth Int. Conf. on Crystal Growth [in Russian], Moscow (1980), Vol. 4, pp. 377-378.

28.   R. Schwartz and E. Huf, "Beitrage zur Chemie des Germanium. 8. Mitt. Uber das Germanium Dioxid," *Z. Anorg. Allg. Chem.*, **203**, 188-197 (1931).

29.   V. W. Kother and F. Muller, "Thermochemische Untersuchung zur Stabilitat von Orthosilicaten und -Germanaten," *Z. Anorg. Allg. Chem.*, **444**, No. 1, 77-90 (1978).

30.   A. Navrotsky, "Enthalpies of transformation among the tetragonal hexagonal and glassy modifications of GeO₂," *Z. Anorg. Nucl. Chem.*, **33**, No. 12, 1119-1124 (1971).

31.   G. B. Naumov, B. N. Ryzhenko, and I. L. Khodakovskii, *Handbook of Thermodynamic Values* [in Russian], Atomizdat, Moscow (1971).

32.   *Thermodynamic Constants of Substances* [in Russian], Vol. 9, Vses. Inst. Nauch. Tekh. Inform., Moscow (1980).

33.   T. B. Kosova and L. N. Dem'yanets, "Hydrothermal chemistry and growth of hexagonal germanium dioxide," Abstracts of Papers of the VIth All-Union Conf. on Crystal Growth [in Russian], Tsakhkadzor, 1985, Izd. Akad. Nauk Arm. SSR, Erevan (1985), Vol. 2, pp. 45-46.

34.   R. Roy and S. Theokritoff, "Crystal growth of metastable phases," *J. Cryst. Growth*, **12**, No. 1, 69-72 (1972).

35.   I. B. Makhina, "Features of morphology of α-GeO₂ crystals grown on quartz substrates," Abstracts of Papers of the VIth All-Union Conf. on Crystal Growth [in Russian], Tsakhadzor, 1985, Izd. Akad. Nauk Arm. SSR, Erevan (1985), Vol. 2, pp. 146-147.

Part III

CRYSTALLIZATION OF FILMS

LIQUID-PHASE ELECTROEPITAXY

F. A. Kuznetsov and V. N. Demin

This article discusses a comparatively new method for preparation of epitaxial layers, liquid-phase electroepitaxy (LEE). This method is based on crystallization under the influence of current which is passed through saturated or nearly saturated solutions of the crystallized substance. In contrast to processes of electro-crystallization, where the crystallizing substance is the product of an electrode reaction, crystallization in processes which are used in LEE is a secondary effect which results from changes of temperature and concentration of the crystallizing substance as a result of the passage of current.

For the past ten years, sufficient material which attesting to the fact that the LEE method possesses enormous potential has been accumulated. The efforts on LEE, the main part of which are dedicated to the study of crystallization of epitaxial layers of $A^{III}B$ compounds, began in the 1970s at the A. F. Ioffe Physicotechnical Institute of the Academy of Sciences of the USSR [1] and at Massachusetts Institute of Technology [2]. These works concerned crystallization of layers of GaSb and InSb, respectively, under the influence of current. However, study of the effects which form the basis of LEE have a more lengthy history.

In 1956, A. F. Ioffe noticed that the Peltier effect at the interface of semiconducting crystals with their melts can be so large that crystallization occurs upon passage of direct current [3]. In 1957, Pfann, Benson, and Wernik observed experimentally shifts of the liquid zone (a solution of germanium in Au, Pd, Pt, Ni, and Al) on the germanium surface under the influence of current [4]. In 1961, Angus, Radone, and Hucke [5] studied the role of electromigration in crystallization processes caused by current passage through a multicomponent liquid phase. In 1963, Tiller [6], and in 1964, Hurle, Mullin, and Pike [7] examined questions on the motion of the liquid zone through a crystal under the influence of direct current. Processes of diffusion and electromigration, the Peltier effect, and the effect of crystallization kinetics were treated in the first of these reports. The second considered the Peltier and Thompson effects and the electromigration effect.

In 1960, results of experiments on leveling of the dislocation density in germanium crystals grown by the Czochralski method by passage of direct current of 20 A/cm^2 through the crystallization front were described by E. Yu. Kokorish [8]. In 1969, V. A. Mikhailov, V. N. Vertoprakhov, and M. V. Kornievich [9] reported experimentally observed crystallization of single crystal bismuth from solution in a gallium melt.

The change of dopant distribution coefficient upon application of electric current in a crystallizing system was already examined in 1969 by Pfann and Wagner [10]. Observations of change in the dopant segregation coefficients upon current passage through the crystal—melt interface during crystal growth by the Czochralski method were reported in a series of works by Gatos and co-workers [11, 12].

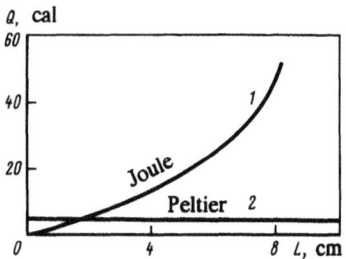

Fig. 1. Comparison of heats at the crystal–melt interface of germanium, caused by Joule and Peltier effects, as functions of crystal length. $j = 18$ A/cm^2 [15].

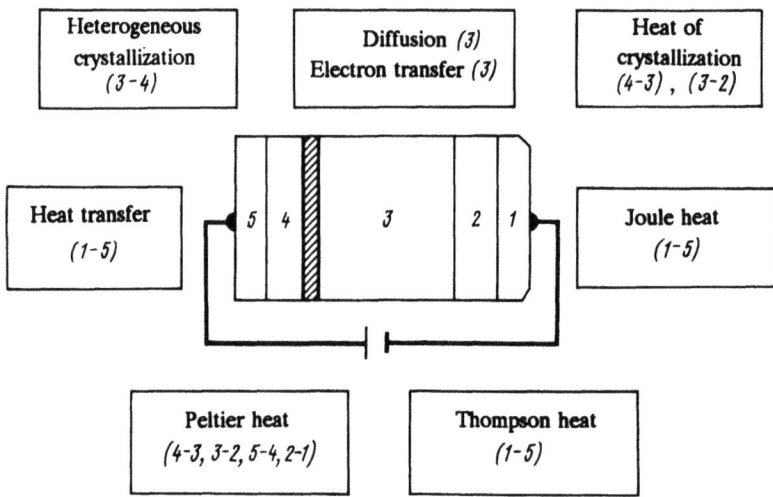

Fig. 2. Effects which arise in the LEE cell with current.

In the early works [9-11], besides the use of electric current in the zone refining process, application of electric current directly during growth of single crystal semiconductors by directed crystallization and the Czochralski method appeared promising. Calculations of O'Connor [13, 14], which consider the different temperature of the ends of the crystallizing bar in contrast to the estimates of Ioffe [3], showed the nature of the current effect on directed crystallization for such important semiconductors as Si, Ge, GaSb, and InSb.

It should be noted that the Joule heat increases as the single crystal grows if the Peltier heat and the heat of crystallization remain constant during the whole process. Calculation of the values of these heats as a function of growth conditions showed [15] that the Joule heat at the crystal—melt interface prevails over the other heats during growth of single crystals, beginning at a certain bar length. The dependence of the heat values on crystal thickness (Fig. 1) is the reason that the thickness of the structure formed (substrate with epitaxial layer) reaches an order of magnitude of 1 mm under typical conditions of LEE with direct current.

1. EFFECTS WHICH CONTRIBUTE TO LEE PROCESSES

In the first studies devoted directly to processes of epitaxial layer growth from solutions in a melt under the influence of electric current, the authors attempted to explain the observed features (primarily the growth rate as a function of current density) by some type of singular mechanism. It is now clear that a whole series

Table 1. Relations of Effects Which Comprise the LEE Process

Effect	Relation	Symbols
Peltier effect	$q_p = -\Delta \alpha T j$	q_p, Peltier heat; $\Delta \alpha$, difference of thermo-emf coefficients of substances which contact this interface
Electron transfer	$v = uE = uj\rho$	u, electric mobility of the component in solution–melt; ρ, specific resistivity of solution–melt
Joule heat	$q_J = j^2 R$	R, electric resistivity; q_J, Joule heat
Thompson effect	$g_T = \tau_T j \dfrac{dT}{dx}$	τ_T, Thompson coefficient
Diffusion	$j_{diff} = -D_i \operatorname{grad} C_i$	j_{diff}, current of atoms of the substance; D_i, diffusion coefficient
Heat transfer	$q = -\lambda \operatorname{grad} T$	λ, heat transfer coefficient
Crystallization	$v = k(T) f(T, Z) Z =$ $= \dfrac{C - C^p}{C^p}$	

of effects must be considered in the spectrum of conditions used for realization of LEE processes. These effects arise during current passage and lead to changes of temperature at the phase interface or of concentration of crystallizing substance near this interface.

Figure 2 shows a schematic cell in which the LEE process is carried out and the several effects that have the greatest significance for this process. Numbers in the cell scheme denote: 1, 5) electric leads; 2) source; 3) solution–melt zone; and 4) substrate. Numbers under the names of effects refer to zones or interfaces where the respective effects are encountered. The relations which describe these effects are given in Table 1.

Thus, several mechanisms of evolution and absorption of heat exist in general in LEE systems. Electron transfer, crystallization, and dissolution of solids lead to emergence of concentration gradients of the crystallizing substance in the liquid zone. Diffusion leads to a leveling of these gradients.

The first question which must be answered for analysis of the crystallization processes is that of how the driving forces of crystallization depend on the conditions of the process. These driving forces are determined by the change of temperature and concentration of the crystallizing substance at the crystallization front relative to the equilibrium values of these parameters. A number of typical situations, for which the values of these driving forces are estimated, are examined below.

2. THE STEADY-STATE HEAT PROBLEM FOR LEE PROCESSES

Steady-state heat problems for LEE processes are examined in a number of works [16-21]. The one-dimensional problem [16-18, 20, 21] or a cell with cylindrical symmetry [19] is treated in the majority of cases. The contribution of the Thompson effect is negligibly small in comparison with the remaining heat sources which arise in the LEE cell upon current passage (Table 1). As the estimates show, the heat of crystallization can be comparable to the other heat sources at large concentrations of crystallizing component in the solution–melt. However, this value is intimately related to the crystallization rate and can be evaluated when solving the heat problem for the LEE process.

Much has already been said about the negative role of the Joule effect in the formation of single crystals of great length. However, the contribution of Joule heat to the total crystallization rate can be significant at large current densities for LEE also since the ratio of Joule heat to Peltier heat grows with increasing current density j (Table 1).

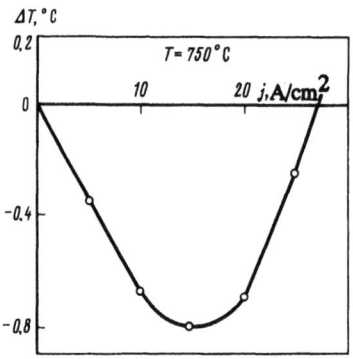

Fig. 3. Change of gallium arsenide crystallization front temperature at various current densities. $n = 3 \cdot 10^{18}$ cm^{-3}.

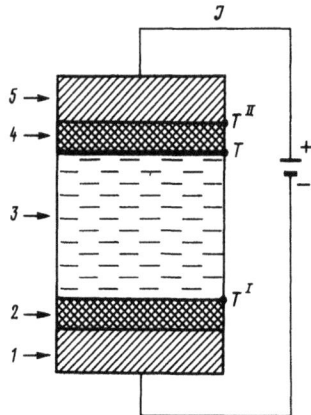

Fig. 4. Scheme of the LEE process model examined.

Attempts have been made to evaluate the Joule heat when solving the heat problem of the LEE process [19, 21]. Thus, heat sources $0.5\,j^2 R_P$ at the solution—substrate interface were introduced in [21] (R_P is the substrate resistivity).

It should be noted that the elements of the growth apparatus which have various electric and heat characteristics are constructively placed by trial-and-error. As a result of this, the majority of growth cells are non-uniform from the viewpoint of separation and dissipation of Joule heat. Therefore, working currents in which the Joule heat in the heat distribution in the cell can be neglected are preferred. Usually, the value of the limiting current does not exceed 15-20 A/cm^2 and is found from the experimental crystallization rate or crystallization temperature as a function of the current density. Thus, Fig. 3 shows the results obtained by us at which the working current density was chosen as 10 A/cm^2.

Taking the above considerations into account, we will illustrate the results of an analysis of the simultaneous effect of a number of heat sources for the case described in [18]. The process scheme is shown in Fig. 4. The zone designations are the same as in Fig. 2. The temperatures of the 2—3 and 4—5 interfaces are constant and equal to T^I and T^{II}, respectively. The temperature of the crystallization front I changes with a change of current density. The reason for the temperature change is the influence of two heat sources which are localized at the crystallization front: the evolution (absorption) of Peltier heat and the heat of crystallization.

For the one-dimensional case (absence of radial temperature gradients) and a steady-state process, the problem is formulated as follows:

$$\frac{\partial^2 T_i}{\partial x^2} = 0 \quad (i = 3, 4), \tag{1}$$

$$T_3(0) = T^{\text{I}}, \quad T_4(l_4) = T^{\text{II}}, \quad \lambda_4 \frac{\partial T_4(0)}{\partial x_4} - \lambda_3 \frac{\partial T_3(l_3)}{\partial x_3} = -qv + (\alpha_4 + \alpha_3)\, Tj, \tag{2}$$

where v is the crystallization rate, q is the heat of crystallization, α_4 and α_3 are thermo-emf coefficients of the crystallizing substance and the solution—melt, respectively, λ_i is the heat transfer of the substance of the ith zone, l_i is the length of the ith zone (taken from below to above), and j is the current density, A/cm^2. The solution of this problem has the following form. The crystallization front temperature is

$$T = \left(\frac{\lambda_3}{l_3} T^{\text{I}} + \frac{\lambda_4}{l_4} T^{\text{II}} + q^v \right) \bigg/ \left(\frac{\lambda_3}{l_3} + \frac{\lambda_4}{l_4} + \Delta\alpha j \right), \tag{3}$$

where $\Delta\alpha = \alpha_4 - \alpha_3$. The temperature difference at the interfaces of the solution zone is

$$\Delta T_1 = T^{\text{I}} - T = \left(\Delta T_\Sigma + \frac{l_4}{\lambda_4} T^{\text{II}} \Delta\alpha j \right) \bigg/ \left(1 + \frac{l_4}{l_3} \frac{\lambda_3}{\lambda_4} \right), \tag{4}$$

where $\Delta T_\Sigma = T^{\text{I}} - T^{\text{II}}$.

The examined situation (constant temperature of the 2—3 interface) is realized, for example, in the cell without the source or with the source of crystallized component as finely broken pieces which float on the solution—melt surface. Naturally, the boundary conditions will be somewhat different for the case when the electric current passes through the semiconductor—source wafer. However, a linear temperature distribution through the thickness of the liquid zone is also obtained in the final analysis. The temperature distribution in the liquid zone for all possible combinations of substrate and source conductivity types is given in [20].

3. THE CONCENTRATION PROBLEM

The question of concentration distribution of the crystallized component was most completely examined in [16-18, 23]. The distinguishing features of the process model adopted in [16, 17, 23] are the Peltier effect and electron transfer, the assumption of the absence of process limitation by an internal crystallization stage, and the absence of concentration gradients in all directions except that normal to the crystallization front.

For the one-dimensional situation, the concentration field is described in general by the equation

$$D \frac{\partial^2 C}{\partial x^2} - v \frac{\partial C}{\partial x} - uE \frac{\partial C}{\partial x} = \frac{\partial C}{\partial t}, \tag{5}$$

where D is the diffusion coefficient of the crystallized substance in the liquid zone, C is the concentration of the crystallized substance, x is the distance from the crystallization front, v is the growth rate, t is the time, u is the mobility of the dissolved substance, and E is the electric field potential.

Two cases were examined in [17] for the model described above.

1. Convection can arise in the liquid-phase layer. The convection is sufficiently strong to level the concentration of the crystallized substance at all points of the liquid phase that lie outside the diffusion layer of thickness δ that adjoins the crystallization front.

2. Convection is absent in the liquid layer. Equation (5) can be solved for the examined model using the following set of boundary conditions:

$$D \frac{\partial C}{\partial x}\bigg|_{x=l} - uEC_L = -v\,(C_L - C_S), \tag{6}$$

$C = C_0$ at $t = 0$ for all x values;
$C = C_L$ for $t > 0$ at $x = 1$ (crystallization front);
$C = C_0$ at $t > 0$ for $x > \delta$ (with convection);   (7)
$C = C_0$ at $t > 0$ for $x \to \infty$ (without convection).

Here C_0 is the initial concentration of the crystallized substance which corresponds to equilibrium at the initial temperature, C_L is the concentration at the liquidus line at the crystallization front temperature, and C_S is the concentration of the crystallized substance in the solid.

A model was developed in [18] which considers the possibility of process limitation by inhibition of the crystallization step. We can calculate the kinetics of the heterogeneous crystallization process if instead of the boundary condition (6) we use the expression

$$\left(D \frac{\partial C}{\partial x} - uEC \right)_{x=1} = -K_{(T)} f(T, Z).$$ (8)

The process that occurs in the cell, which is schematically depicted in Fig. 4, is analyzed in [18]. The beginning of the calculation here is the 2–3 boundary. The coordinate l (thickness of liquid layer) corresponds to the crystallization front position. Equilibrium is assumed to be established at the 2–3 interface, which is calculated from the boundary condition

$$C(0) = C_l(T^{II}).$$ (9)

The right part of (8) represents the kinetic function of the surface process. This equation is described in general as the product of the factors $K(T)$ of the kinetic coefficient, which depends only on temperature and has dimensionality of crystallization rate, and the functions $f(T, Z)$, which depend on both temperature and the relative saturation $Z = (C_l - C_L(T))/C_L(T)$ at the crystallization front.

It was shown in the cited work that the process rate as a function of its behavior for various types of kinetic equations can be represented uniquely. For this, the two-variable function $f(T, Z)$ is transformed to the form $F(\bar{Z})$, where $\bar{Z}$ is the corrected relative saturation which is determined from the expression $\bar{Z} = Z/Z^*$. The correction factor Z^* is different for the various forms of the kinetic equations. Thus, for the three forms of the $f(T, Z)$ functions which are usually examined, $f_1 = Z$ (normal growth), $f_2 = Z^2$ (stepwise growth and growth on screw dislocations), and $f_3 = \exp(-v(T)/Z)$ (two-dimensional nucleation), the correction factor is determined by the relations

$$Z^* = \begin{cases} 1 & \text{at} \quad f = f_1 = Z \text{ and at } f = f_2 = Z^2, \\ v(T) & \text{at} \quad f = f_3 = \exp(-v(T)/Z). \end{cases}$$ (10)

A significant conclusion of [18] is the inference of nonadditivity in general of the contributions of electron transfer and the Peltier effect to the final crystallization rate. This conclusion is illustrated in Fig. 5, which shows the ratio of the sum of crystallization rates as functions of current density, which are calculated assuming the influence of only the Peltier effect v_1 or only due to electron transfer v_2, to the rate v, which is calculated for the case of simultaneous influence of both effects. The calculations were carried out for the growth of gallium arsenide in the Ga–GaAs system at $T = 880°C$ for three different liquid zone thicknesses.

The limiting step can change with a change of process conditions. When mass transfer is the rate determining step, this can be described by the limiting transfer $K \to \infty$. The expression for deposition rate in this case acquires, in agreement with [18], the form

$$v_\infty = BC_L \, j - \frac{B_j}{1-e^{-\varphi}} \frac{dC_L}{dT} \Delta T_s \, ,$$ (11)

where $B = u\rho$.

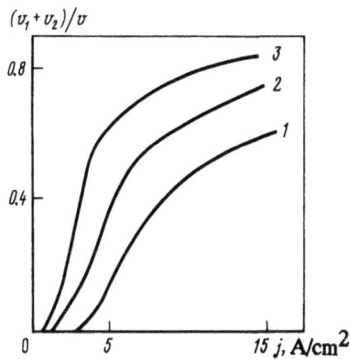

Fig. 5. The $(v_1 + v_2)/v$ value as a function of the current density for the Ga–GaAs system at 880°C. 1-3) $l = 1, 2,$ and 3 mm, respectively [18].

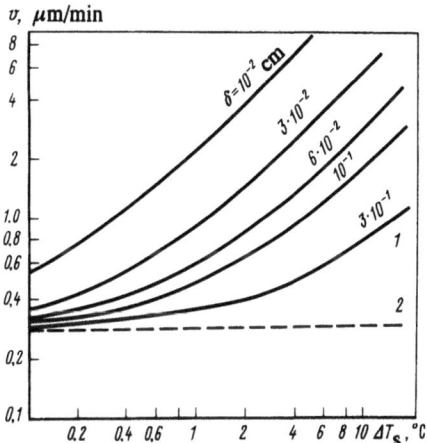

Fig. 6. Contribution of Peltier effect and electron transfer to the final LEE rate [17]. 1) Peltier + electron transfer; 2) electron transfer only.

Actually, estimates of the value of the parameter $\varphi = (B/D)lj$ for solutions of gallium arsenide in gallium and for indium antimonide in indium [18] show that the exponential functions of this parameter are approximated with good accuracy by the linear expressions. Thus, Eq. (11) with this approximation is transformed into (12):

$$v_\infty = BC_L j + \frac{D}{L} \frac{dC_L}{dT} \Delta T_s .$$ (12)

The applicability criterion of the proposed expressions for growth rates which correspond to the condition $K \to \infty$ was formulated and quantitatively determined in [18].

The process rate can be represented as the sum of the contributions of electron transfer and crystallization due to the Peltier effect (the first and second terms of (12), respectively) for the limiting cases described by (11) or the equivalent expression (12). The ratio of the contribution of these effects changes under different conditions. The nature of these changes is illustrated in Fig. 6 [17].

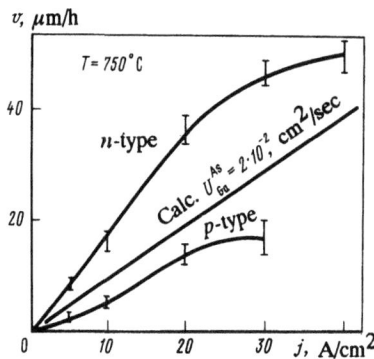

Fig. 7. Crystallization rate as a function of current density for substrates with various types of conductivity. $n = 3 \cdot 10^{18}$ cm^{-3} and $p = 2 \cdot 10^{18}$ cm^{-3}.

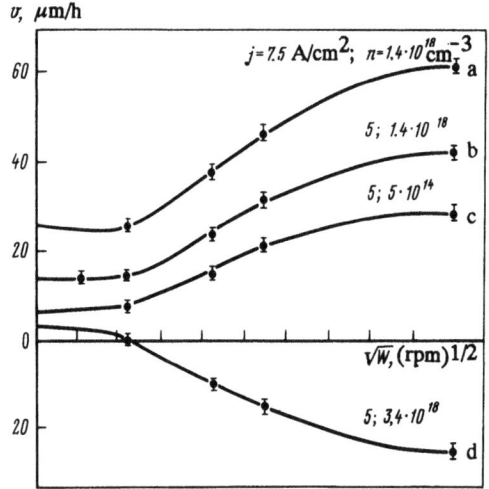

Fig. 8. Gallium arsenide crystallization rate as a function of substrate rotation rate. $T = 800°C$, orientation (111) A.

The question of the contributions of these effects was debated at the initial stage of LEE investigations [15, 24, 25]. It is now clear that it is practically impossible to neglect the contributions of both effects. This consideration can be illustrated by the experimental results on LEE of gallium arsenide [26] (Fig. 7). The upper curve in Fig. 7 corresponds to crystallization on a n-type substrate, the lower to that on a p-type substrate. The straight line gives the calculated values of the contribution of electron transfer to the final rate. The observed deviations from this calculated line agree with the sign of the Peltier effect (cooling of the crystallization front on the n-type substrates, heating with growth on the p-type substrates). This qualitative relationship of the experimentally observed functions to the examined models can be illustrated repeatedly. However, more detailed quantitative analysis of the experimental results indicates that it is not always possible to restrict the calculation of heat and concentration change at the crystallization front to electron transfer. The possible limitation of the process by the kinetics of the heterogeneous step under several conditions that are realized in the experiment can be illustrated by the results of the LEE experiments of gallium arsenide with a rotating substrate [27]. The experimentally determined growth rate (on the rotating substrate) as a function of the rotation rate approaches a constant value with an increase of rotation rate (Fig. 8). In our opinion, this attests to the transition from diffusional control to kinetic. And finally, one more consideration should be mentioned which complicates construction of models which adequately describe the LEE process. In [28, 29] it was noted that comparatively small additions of extraneous substances lead to a significant change of LEE process rate. This is illustrated by the data of [28], which are included in Fig. 8a and b. Here, the crystallization of gallium arsenide from a pure melt of GaAs—Ga and from a melt with an additive of 2.5% silicon (curves a and b, respectively) is compared. Such a significant change of rate is explained by complex formation in the melt [28]. Assuming such a possibility, it should also be noted that the presence of an dopant can affect the kinetics of the heterogeneous step, leading to a change of the final rate. The data that have been accumulated until now are insufficient for a final conclusion on the nature of the effect of small dopant concentrations on the crystallization rate. Undoubtedly, the most important problems for further study of LEE processes are the study of complex formation in the melt and the kinetics of the heterogeneous step.

4. DOPANT TRAPPING

The current state of the question of dopant trapping during growth of layers by LEE is reflected in [29-31]. The assumption that the crystallization processes of the dopant and main substance occur independently lies at the basis of the proposed model. The change of dopant concentration in the depositing substance is related to the change of its concentration in the solution at the crystallization front. The distribution of the dopant concentration in the liquid zone is described by the equation, analogous to (5), for concentration distribution of the deposited component:

$$D_i \frac{\partial^2 C_i}{\partial x^2} + (v + u_i E) \frac{\partial C_i}{\partial x} = \frac{\partial C_i}{\partial t}. \tag{13}$$

The condition of dopant balance at the crystallization front with the assumption (relative to the subsurface dopant concentration) of equilibrium trapping has the form

$$v C_i \delta - (v + u_2 E) C_i \mid_{x=0} = D \frac{\partial C_i}{\partial x}. \tag{14}$$

In this case, at the initial time point

$$C_i = C_{i0}, \; t = 0. \tag{15}$$

The solution of (13) under the conditions of (14) and (15) allows the distribution coefficient as a function of the LEE conditions to be found.

Two cases are usually examined: without convection in the liquid zone, when $C_i = C_{i0}$ for $x = \infty$, and the case of strong convection when $C_i = C_{i0}$ for $x > \delta$.

In 1962, Pfann and Wagner gave the solution of (13) for the steady-state [10]:

$$k_{\text{eff}} = \frac{C_{is}}{C_{i0}} = \frac{k_0 (1 + u_i E / v)}{k_0 + (1 + u_i E / v - k_0) \exp\{ - [(v + u_i E) \delta] / D_i\}} . \tag{16}$$

Equation (16) transforms into the well known Barton—Prim—Slichter equation for the effective distribution coefficient for the case where electric current is absent, i.e., $E = 0$. When $[(v + u_i E)\delta]/D_i \ll 1$, conditions in LEE are realized where the rate of crystallization front advance is usually several times smaller than the dopant migration rate since the solution—melts have a small concentration of deposited component

$$v \approx \frac{C_0}{d} uE \ll u_i E.$$

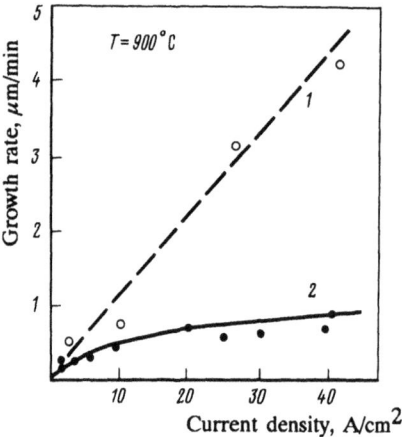

Fig. 9. Gallium arsenide crystallization rate as a function of current density for solution–melts undoped (a) and doped with 2.5 at. % silicon [28] (b).

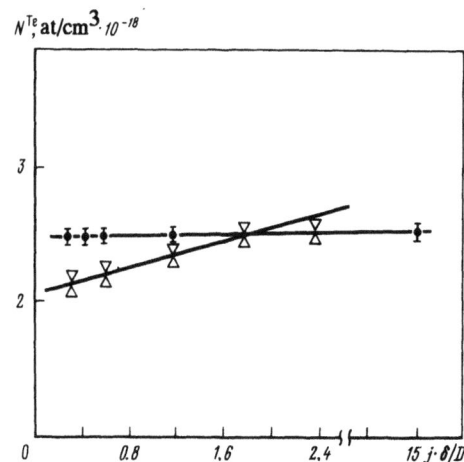

Fig. 10. Tellurium concentration in an epitaxial gallium arsenide layer as a function of the product $(j\delta/D)$ with a change of diffusion layer thickness (a) and current density (b).

In this case, (16) has the form

$$k_{\text{eff}} = k_0 \left[1 + \frac{\delta}{D_i} \; u_i E \right].$$ (17)

Recalling that $E = j\rho$ and accounting for the temperature dependence of the distribution coefficient caused by the temperature change of the crystallization front as a result of the passage of electric current, we have

$$k_{\text{eff}} = k_0 \left[1 + \frac{\delta j}{D_i} u_i \rho \right] + \left(\frac{\partial k_0}{\partial T} \right)_{T_0} \Delta T_s \; .$$ (18)

Without convection, the non-steady-state solution of (13), which was first carried out in [29], is necessary for times $t < 4l^2/(\pi^2 D)$ (l is the liquid zone thickness for determination of the effective dopant distribution coefficient). The full expression for k_{eff} has a rather cumbersome form. However, the inequality $u_i E(t)^{1/2} \ll 2(D_i)^{1/2}$ is valid in most cases of LEE. For it, the following expression for k_{eff} is obtained [29]:

$$k_{\text{eff}} = k_0 \left[1 + 2 \left(\frac{t}{\pi D_i} \right)^{\frac{1}{2}} u_i E \right] - 2 \left(\frac{t}{\pi D_i} \right)^{\frac{1}{2}} v \, (k_0 - 1). \tag{19}$$

The contribution to the effective electron transfer distribution coefficient for a dopant with $k_0 < 1$ will exceed the contribution of the crystallization rate ($v \ll u_i E$), and we finally obtain

$$k_{\text{eff}} = k_0 \left[1 + 2 \left(\frac{t}{\pi D_i} \right)^{\frac{1}{2}} u_i E \right] + \left(\frac{\partial k_0}{\partial T} \right)_{T_0} \Delta T_s. \tag{20}$$

Equation (20) shows that k_{eff} increases proportionally to $t^{1/2}$, approaching (18), which is the steady-state solution. Equation (18) shows that the change of distribution coefficient occurs by two different mechanisms. First, the change of electron transfer rate leads to a change of the surface concentration of dopant. Second, the crystallization front temperature is changed. The contributions of these two mechanisms to the change of k_{eff} are represented respectively by the first and second terms of the right side of (18). It is important to note that the first term, which refers to the contribution of electron transfer to the change of k_{eff}, depends on both the value of the current density j and on the thickness of the diffusion layer δ (Fig. 9) whereas the temperature change of the crystallization front depends only on the change of current density. Thus, k_{eff} changes in the experiment with a change of the diffusion layer thickness only with a substantial contribution of electron transfer. The tellurium concentration in the gallium arsenide epitaxial layer, grown using LEE, as a function of the product $j\delta$ is shown in Fig. 10 [22]. The tellurium distribution upon a change of diffusion layer thickness during gallium arsenide growth ($j = 4$ A/cm^2) caused by a change in the substrate rotation rate from 100 rpm ($\delta = 0.048$ mm) to 0 ($\delta = 3$ mm) is shown in Fig. 10a. In this case, the tellurium concentration in the layer and, consequently, k_{eff} do not depend on δ within the limits of accuracy of concentration change (± 5 rel. %). This indicates a small contribution of electron transfer to the change of k_{eff}. The change of tellurium concentration in the layer at fixed rotation rate ($\delta = 0.12$ mm) and various current densities (2.5-20 A/cm^2) is shown in Fig. 10b. The change of dopant concentration and, consequently, k_{eff} is approximately 17% for a change of $j\delta$ in the same range as for the case shown in Fig. 9a.

Comparison of the two functions shows the predominant contribution of the tellurium distribution coefficient as a function of temperature to the total change of k_{eff} in LEE. Estimates calculated by us using (18) for both dopants based on their known values of mobility in liquid gallium and indium [32] show that the change of k_{eff} in LEE due to electron transfer does not exceed 1-5% of its equilibrium k_0 value. Known values of the mobilities of dopants u_i in liquid gallium determined for the binary gallium–dopant system were used for these estimates.

The presence of a third component (in our case arsenic) can substantially affect the mobility of some dopants as a result of their possible reaction in the melt. Thus, the value of electric mobility of tin in gallium, which was determined by various authors, ranges from 1 to $3 \cdot 10^{-3}$ cm^2/(V · sec).

Estimation of the change of k_{eff} using this value gives $\Delta k / k_0 \ll 1\%$ for an increase of diffusion layer thickness from 0 to 5 mm. The experiments carried out by us on growth of gallium arsenide layers from a solution–melt doped with tin (3.5 at. %) at 800°C showed that the effective tin distribution coefficient changes with diffusion layer thickness (with a change of substrate rotation rate) significantly more than expected from the estimates made above (Fig. 11a).

An analogous change of k_{eff} with diffusion layer thickness in the same system was noted by Gatos [29] during growth of gallium arsenide layers at 950°C (Fig. 11b). The value of tin mobility estimated from (18) (Fig. 11) in our case was $5 \cdot 10^{-2}$ cm^2/(V · sec). This is near that of arsenic in gallium, equal to $2 \cdot 10^{-2}$ cm^2/(V · sec), and suggests formation of complicated complexes of tin and arsenic in the solution–melt.

Having examined the nature of the change of dopant distribution coefficient in a steady-state LEE process, it can be concluded that this change in an actual LEE process does not exceed 20-30% and is limited by

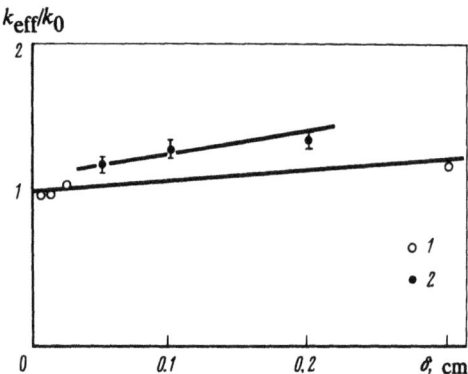

Fig. 11. The k_{eff}/k_0 ratio as a function of δ during growth of gallium arsenide layers from a solution–melt doped with tin: a) from the data of the authors; b) data of [29].

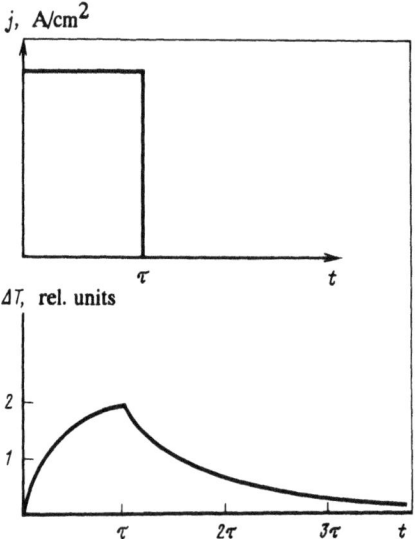

Fig. 12. Change of crystallization front temperature upon imposition of a pulse of duration τ which is less than the time for establishment of thermal equilibrium in the system.

the diffusion layer thickness and the value of electric current passage. Moreover, a drastic change of dopant concentration is difficult to produce in a steady-state LEE process since the thickness of the transition layer, which is equal approximately to $\tau_D v$, has a value on the order of a micron and more (τ_D is the time for establishment of diffusional equilibrium in the melt, usually not less than a minute, and v is the growth rate, on the order of tenths of a micron per hour). More meaningful changes of k_{eff} as a result of significant changes of crystallization front temperature can be obtained by imposing short current pulses with a duration τ, which is less than the time for establishment of thermal equilibrium in the system, on the principal current. Thus, the current density in the pulse and, therefore, the values of ΔT_p attained can be significantly larger than in steady-state processes since the effect of Joule heat on the crystallization front temperature in the non-steady-state case is limited to the region which is determined by the nature of the thermal front expansion length during a pulse of duration τ. The change of crystallization front temperature upon imposition of a square-wave electrical pulse of duration τ is shown in Fig. 12.

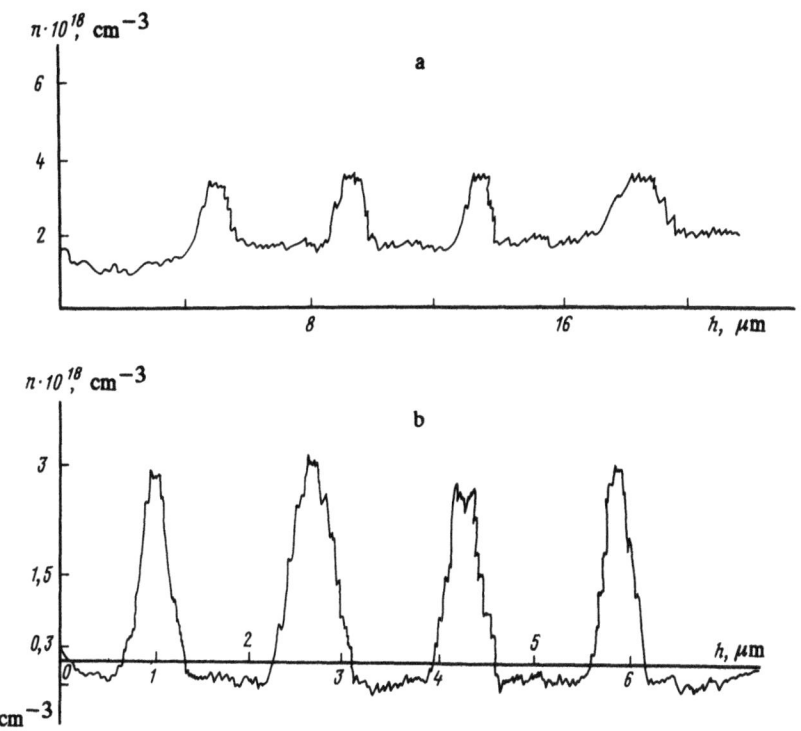

Fig. 13. Production of periodic structures in the LEE process of gallium arsenide by imposition of current pulses on the main current. Growth from a solution–melt doped with tellurium (a) and doped with zinc and tellurium (b).

The change of crystallization front temperature Δt_P, as seen in Fig. 12, quickly decreases after the current is stopped and practically differs little from 0 at $t \geq 3\tau$. These considerations were experimentally verified by us [33]. Figure 13a shows the free carrier concentration along the thickness of the film, which was grown at 800°C from a solution–melt, doped with tellurium as a function of short current pulses (1 sec) of 70 A/cm² imposed on the main current (5 A/cm²). The tellurium concentration upon pulse imposition increases, reaching approximately 200%. This is significantly larger than the possible change in the steady-state LEE process. Considering that the principal change of crystallization front temperature, and also k_{eff}, occurs within time 3τ, we can estimate the crystallization rate. Figure 13a shows that the crystallization rate for growth in the pulse regime is 0.5 μm/sec, i.e., approximately two orders of magnitude higher than rates that are usually employed in liquid epitaxy. Attempts to increase the growth rate of gallium arsenide layers by the usual method of liquid epitaxy with cooling above 200 μm/h lead to trapping of solvent and a rapid degradation of layer properties [34]. However, our studies showed that layers grown in the pulse regime at such high rates were single-crystalline and did not have solvent inclusions.

Repetition of the electric pulse after certain time intervals $\Delta t > 3\tau$ can produce semiconducting periodic structures (SPS) in one LEE process with significant concentration differences. It is interesting that the ratio between the concentrations of several dopants in the layer grown by LEE with application of current pulses will be different since the distribution coefficient as a function of temperature is different for the various dopants. Thus, for zinc and tellurium, $(\partial k_0/\partial T)_{T_0}$ differ not only in absolute value but also in sign. Since these dopants are p- and n-type dopants in gallium arsenide, a periodic change of conductivity type in the gallium arsenide layer grown by this method from the solution–melt doped with zinc and tellurium can be achieved. Separate experiments carried out by us on the LEE of gallum arsenide with double doping of a melt by zinc and tellurium demonstrated the possibility of preparation of periodic structures with distinct boundaries between regions with different conductivity types (Fig. 13b).

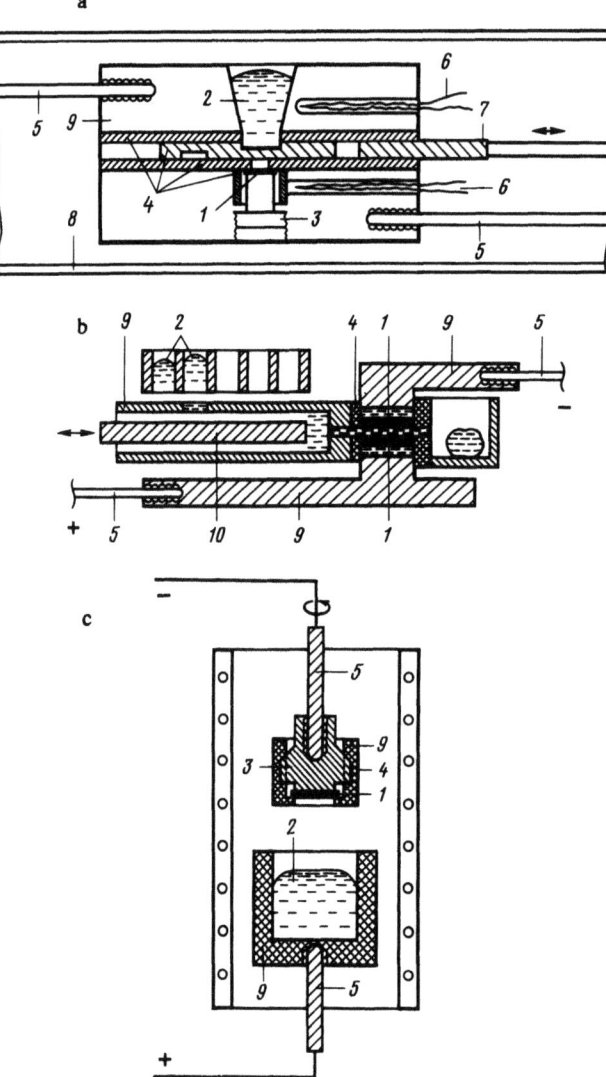

Fig. 14. Scheme of LEE cells. Penal, piston, and cylindrical cell with a rotating substrate (a-c), respectively. 1) substrate; 2) melt; 3) tightening screw-electrode; 4) insulator (BN); 5) electrical lead; 6) thermocouple; 7) insulating sleeve; 8) quartz reactor; 9) graphite body; 10) piston for supply of liquid phases into the space between substrates.

5. APPARATUS DESIGN FOR THE LEE METHOD

Investigators in the early stages of development of the LEE method used somewhat modified liquid epitaxy cells (Fig. 14a). However, subsequent physicochemical study of the LEE process showed that other requirements, caused by the specifics of the process, arise besides those for usual liquid epitaxy. These include: the necessity of guaranteed even current density on the substrate, lowering of the lead—substrate contact resistivity to a minimum, decrease of electric resistivity of the whole cell, and a provision for uniform evolution of Joule heat in the cell volume. The first two problems were successfully solved by further perfection of the barrel cell. A piston system for supply of melt as a liquid electric contact from the opposite side of the substrate (Fig. 14b) is used in the most successful construction of such a cell [35]. However, barrel type cells are not optimal for the LEE process due to possible nonuniform Joule heat evolution in the volume of the graphite cell of complex configuration and disruption of the uniformity of temperature on the substrate surface, as well as due to possible uncontrolled convection in the melt (especially at large liquid zone thicknesses). We believe that the most successful construction in which the nature of mass- and heat-transfer effects in the LEE process are considered is the cell of cylindrical symmetry with a rotating substrate [22] (Fig. 14c).

6. CONCLUSIONS

Thus, studies on growth of epitaxial layers by liquid-phase electroepitaxy show that this method has great possibilities. The rapid crystallization rate, the possibility of dopant concentration variation, and the high structural perfection of the layers make the method promising for technical applications. The method is convenient for automation since the most important technical parameter, the current density, is easily changed and regulated.

In spite of the comparative novelty of the LEE method and the small number of advocates at present, a good understanding of the principal features of the processes has been achieved and quantitative methods, which can serve as a basis for development of optimal methods for growth of pure and doped layers by this method, have been developed.

In the overwhelming majority of works on LEE, the crystallization process is assumed to occur in quasi-equilibrium: the limiting step is mass transfer in the solution volume. Results of our studies show that conditions of kinetic control of LEE processes [28] are experimentally attainable and that interesting features of LEE processes arise under kinetic control [36]. Undoubtedly, the LEE method can be used for studying surface step kinetics of the crystallization process.

In spite of the fact that the variety of effects that comprise the LEE process are known and that methods for quantitative description of the process have been developed, the many possibilities of this type of epitaxy have not been explored. Although the A^3B^5 compounds have been studied in almost all works on LEE, there is no reason to believe that other substances cannot be crystallized by this method. Evidently, these broad possibilities include use of pulsed non-steady-state processes. Some problems that are especially important in further development of the LEE method are: the mechanism and principles of surface processes, including electric properties of the interface; complex formation in the melt; and crystallization from solution—melts with nonmetallic conductivity.

REFERENCES

1. L. V. Golubev, T. V. Pakhomova, O. A. Khachaturyan, and Yu. V. Shmartsev, "Liquid epitaxy in an electric field," in: IVth All-Union Conf. on Crystal Growth [in Russian], Izd. Akad. Nauk Arm. SSR, Erevan (1972), pp. 164-167.
2. M. Kumagawa, A. F. Witt, M. Lichtensteiger, and H. C. Gatos, "Current-controlled growth and dopant modulation in liquid-phase epitaxy," *J. Electrochem. Soc.*, **120**, No. 4, 583-584 (1973).
3. A. F. Ioffe, "Two new applications of the Peltier effect," *Zh. Tekh. Fiz.*, **26**, No. 2, 478-483 (1956).
4. W. G. Pfann, K. E. Benson, and J. H. Wernik, "Some aspects of Peltier heating at liquid–solid interfaces in germanium," *J. Electron.*, **2**, 597-608 (1957).
5. J. Angus, D. V. Radone, and E. E. Hucke, "The effect of an electric field on the segregation of solute atoms at a freezing interface," *Met. Soc. Conf.*, **8**, 833-840 (1961).
6. W. A. Tiller, "Migration of a liquid zone through a solid: Pt. 2," *J. Appl. Phys.*, **34**, No. 9, 2763-2767 (1963).
7. D. T. Hurle, J. B. Mullin, and E. R. Pike, "The motion of liquid alloy zones along a bar under the influence of an electric current," *Philos. Mag.*, **9**, No. 97, 423-434 (1964).
8. E. Yu. Kokorish, "Influence of the Peltier effect on the perfection of single crystal germanium prepared by extrusion from a melt," *Kristallografiya*, **5**, 815-816 (1960).
9. V. A. Mikhailov, M. V. Kornievich, and V. N. Vertoprakhov, "Crystallization of bismuth and tin from a melt in liquid gallium in an electric field," *Fiz. Met. Metalloved.*, **28**, No. 1, 180-183 (1969).
10. W. G. Pfann and R. S. Wagner, "Principles of field freezing," *Trans. Metall. Soc. AIME*, **224**, No. 6, 1139-1146 (1962).
11. M. Lichtensteiger, A. F. Witt, and H. C. Gatos, "Modulation of dopant segregation by electric currents in Czochralski-type crystal growth," *J. Electrochem. Soc.*, **118**, No. 3, 1013-1015 (1971).
12. A. F. Witt, M. Lichtensteiger, and H. C. Gatos, "Application of interface demarcation to the study of facet growth and segregation. Germanium," *J. Electrochem. Soc.*, **121**, No. 6, 887-890 (1974).
13. J. R. O'Connor, "The use of thermoelectric effects during crystal growth," *J. Electrochem. Soc.*, **108**, No. 7, 713-715 (1961).
14. J. R. O'Connor, "Peltier coefficient at a solid–liquid interface," *J. Appl. Phys.*, **31**, No. 9, 1690-1691 (1960).
15. D. I. Levinzon, V. B. Zakutnyi, and V. A. Shershel', "Improvement of growth processes of germanium single crystals through use of Joule and Peltier effects," in: *Silicon and Germanium* [in Russian], Metallurgiya, Moscow (1969), pp. 80-84.

16. V. A. Gevorkyan, L. V. Golubev, S. G. Petrosyan, et al., "Electrical liquid epitaxy. I," *Zh. Tekh. Fiz.*, **47**, No. 6, 1306-1313 (1977).

17. L. Jastrzebski, J. Lagowski, H. C. Gatos, and A. F. Witt, "Liquid-phase electroepitaxy: growth kinetics," *J. Appl. Phys.*, **49**, No. 12, 5909-5920 (1978).

18. Ya. M. Buzhdyan, F. A. Kuznetsov, and L. N. Belyaeva, "Toward analysis and optimization of the liquid-phase electroepitaxy process," in: *Growth of Semiconducting Crystals and Films* [in Russian], Nauka, Novosibirsk (1981), pp. 52-67.

19. A. N. Barchuk and A. I. Ivashchenko, "On the role of the Joule effect in liquid-phase electroepitaxy," *Zh. Tekh. Fiz.*, **52**, No. 9, 1878-1882 (1982).

20. V. A. Govorkyan, L. V. Golubev, A. E. Khachaturyan, and Yu. V. Shmartsev, "On the question of growth kinetics in equilibrium liquid-phase electroepitaxy," *Zh. Tekh. Fiz.*, **53**, No. 3, 545-549 (1983).

21. S. A. Nikishin, "Toward an analysis of mass transfer during liquid-phase electroepitaxy of gallium arsenide," *Zh. Tekh. Fiz.*, **53**, No. 3, 538-544 (1983).

22. F. A. Kuznetsov, V. N. Demin, and Ya. M. Buzhdyan, "Liquid-phase electroepitaxy, a new method for growth of epitaxial layers," in: *Materials of Electronics Technology* [in Russian], Nauka, Novosibirsk (1983), Part 1, pp. 45-62.

23. S. G. Mil'vidskii, S. A. Nikishin, and R. P. Seisyan, "Features of mass transfer during liquid-phase electroepitaxy of gallium arsenide," *Kristallografiya*, **27**, No. 4, 742-750 (1982).

24. J. J. Daniele, "Experiments showing the absence of electromigration of As and Al in Peltier LPE of GaAs and $Ga_{1-x}Al_xAs$," *J. Electrochem. Soc.*, **124**, No. 7, 1143-1144 (1977).

25. V. N. Demin, Ya. M. Buzhdyan, and F. A. Kuznetsov, "On the role of electron transfer in liquid-phase electroepitaxy," *Zh. Tekh. Fiz.*, **48**, No. 7, 1442-1445 (1978).

26. V. N. Demin, Yu. M. Rumyantsev, F. A. Kuznetsov, and Ya. M. Buzhdyan, "Crystallization of gallium arsenide from its solution in a melt under the influence of an electric current," in: *Growth of Semiconducting Crystals and Films* [in Russian], Nauka, Novosibirsk (1981), pp. 67-73.

27. V. N. Demin and F. A. Kuznetsov, "On the applicability of diffusional approximation in models of LEE processes," *Zh. Tekh. Fiz.*, **54**, No. 5, 934-937 (1984).

28. L. Jastrzebski and H. C. Gatos, "Current-controlled growth, segregration, and amphoteric behavior of Si in GaAs from Si-doped solutions," in: Proc. Vth Conf. Cryst. Growth, IGGG-5, Boston (1977), pp. 17-22.

29. J. Lagowski, I. Jastrzebski, and H. C. Gatos, "Liquid-phase electroepitaxy: dopant segregation," *J. Appl. Phys.*, **51**, No. 1, 364-373 (1980).

30. A. N. Barchuk and A. I. Ivashchenko, "Numerical analysis of dopant segregation in liquid-phase epitaxy," *Élektron. Tekh., Ser. Mater.*, No. 12, 44-48 (1981).

31. K. M. Gambaryan, V. A. Govorkyan, L. V. Golubev, S. V. Novikov, and Yu. V. Shmartsev, "Analysis of dopant distribution during equilibrium liquid-phase electroepitaxy," *Zh. Tekh. Fiz.*, **54**, No. 10, 2011-2012 (1984).

32. V. A. Mikhailov and D. D. Bogdanova, *Electron Transfer in Liquid Metals* [in Russian], Nauka, Novosibirsk (1978).

33. V. N. Demin and F. A. Kuznetsov, "Control of doping epitaxial layers in liquid-phase electroepitaxy," in: Papers of the VIth Int. Conf. on Crystal Growth [in Russian], Nauka, Moscow (1980), Vol. 3, pp. 376-377.

34. J. Grossley and M. B. Small, "Some observations of the surface morphologies of GaAs layers grown by liquid-phase epitaxy," *J. Cryst. Growth*, **19**, 160-165 (1973).

35. S. A. Nikishin, "New solutions and applications of liquid-phase electroepitaxy. 1," *Zh. Tekh. Fiz.*, **54**, No. 5, 938-942 (1984).

36. V. N. Demin, L. N. Krasnoperov, and F. A. Kuznetsov, "On the possible application of alternating current in liquid-phase electro-epitaxy," *Zh. Tekh. Fiz.*, **55**, No. 11, 2179-2183 (1985).

MOLECULAR BEAM EPITAXY OF SEMICONDUCTOR, METAL, AND DIELECTRIC FILMS

S. I. Stenin and A. I. Toropov

1. INTRODUCTION

The most important achievement of molecular beam epitaxy (MBE) is the ability to prepare so-called modulated semiconducting structures (MSS), a special case of which are superlattices (SL) which consist of layers with a thickness of one or several monolayers of various materials. The first SL were prepared in the $Ga_xAl_{1-x}As$—GaAs system with a small misfit of the lattice constants f. A substantially broader class of MSS and SL can be created by conjugation of crystals with a large f. In this case, defect-free strained SL are formed in which the adjoining layers are found in a pseudomorphic state. The basic principles and technical equipment of MBE are reviewed in [1-4]. In the present article, the physicochemical aspects of MBE as applied to growth of semiconducting, metal, and dielectric films are examined. Also, the effects of reflection intensity oscillation during reflection high-energy electron diffraction (RHEED), the methods of automatic ellipsometry (AE) and modulated beams which allow precision control of adsorption—desorption processes, superstructure rearrangements, kinetics of surface reactions, and mechanisms of MSS growth are discussed.

Prospects for combination of MBE with other technological methods (low-energy ion implantation and organometallic synthesis) are evaluated.

2. ANALYTICAL MEANS FOR MONITORING MBE PROCESSES

At present, three methods for monitoring film growth are known. The first is RHEED, which allows the structure and morphology of the growing film to be followed. The second are methods that ensure control of the molecular beams. The third is automatic ellipsometry which allows direct evaluation of the structure, composition, and morphology of films and indirect measurement of the growing film thickness. We will examine in detail the capabilities of these methods.

RHEED has become the dominant method for monitoring surface structure during film growth by MBE. Also, it allows the morphology of the surface growth (the presence of smooth surfaces, surfaces with highly developed relief, or formation of facets) to be revealed. In addition, the emergence of faults such as microtwinning and crystalline particles of secondary phases can be observed. Finally, a transition in the texture or polycrystalline state of the subsurface film layer can be identified. Superstructure rearrangements on the surface can be followed using RHEED. These well-known capabilities of RHEED are widely applied in all studies [5].

Recently, intensities and widths of diffraction maxima in the RHEED patterns have been used to obtain information on the film growth mechanism [6-9]. Values of reflectivity, principal reflections, and fractional

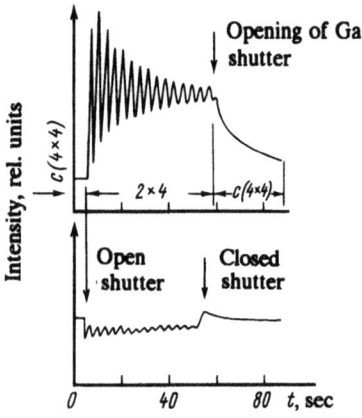

Fig. 1. Oscillation of null electron beam intensity in the initial growth stage on a GaAs(001) surface [7]. Initial substrate surface is the superstructure c (4×4) As which is converted to c (2×4) Ga during film growth (a). Initial substrate surface is the superstructure (3×1) Ga and surface reconstruction during film growth does not occur (b).

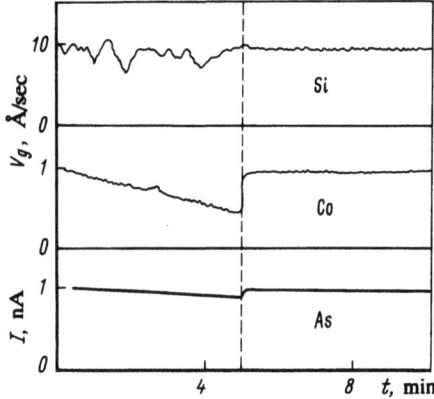

Fig. 2. Stability of growth rate V_g for silicon and cobalt and As$^+$ ion fluxes in free running mode (time t = 0-5 min) and with feedback engaged (time t = 5-10 min) as measured by atomic fluorescence modulation [12].

order reflections are studied. Figure 1 demonstrates the main effects that are observed in the initial stage of GaAs film growth [7, 8]. The intensity of the null reflection oscillates with time under certain diffraction conditions. The first such oscillations were noticed during doping of gallium arsenide with tin [6]. Later, they were also found during MBE of undoped films. The oscillation period corresponds accurately with the time for growth of one GaAs monolayer. The sharp increase of intensity at the initial growth stage in Fig. 1a is explained by superstructure rearrangement from (4×4) As to (2×4) Ga during opening of the gallium source shutter. We will discuss several details which relate to analysis of the growth mechanisms using RHEED data in Sec. 3. Suffice it to say that RHEED is an excellent tool for monitoring the structure and morphology of the film surface and has evolved into an analytical means for investigation of epitaxial growth mechanisms and a means for precise measurement of the thickness and rate of film growth.

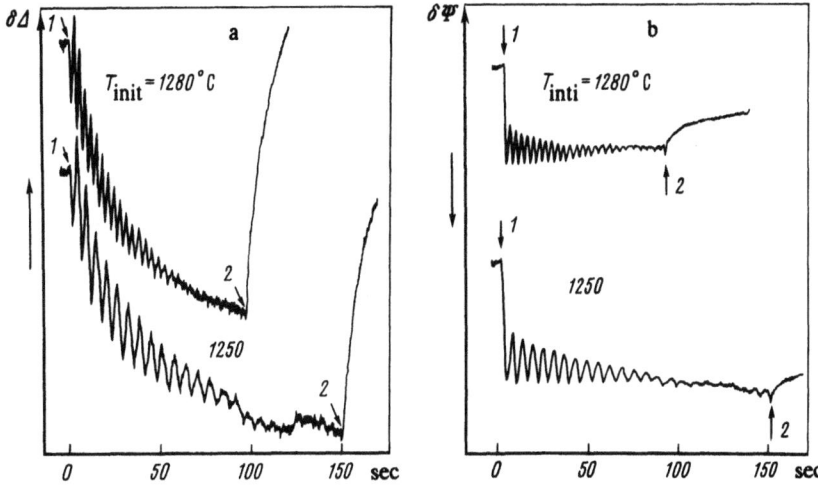

Fig. 3. Change of ellipsometry angles Δ (a) and ψ (b) as a function of germanium film growth time [18].   1) Initial, $\delta\Delta = 0.02°$, $T_g = 380°C$; 2) end of growth, $T_g = 380°C$, $\Delta\psi = 0.02°$.

Control of the molecular beams that impinge on the substrate guarantees high reproducibility during film preparation. This control was realized using quadrupole mass spectrometers [10, 11] in the first stage of MBE development. Recently, a transition to optical methods for control of the molecular beams has occurred. The most widely employed method is atomic fluorescence with optical emission excited by an electron beam. This method is successfully applied for MBE of both elementary semiconductors [12, 13] and $A_{III}B_V$ compounds [14]. Figure 2 illustrates capabilities of the atomic fluorescence method for stabilization of molecular beams and film growth rate for silicon and cobalt silicide. Also, the uniformity of thickness during doping of silicon films by arsenic ions is shown [12, 13]. Application of this method ensured short- and long-term stability of growth rate and of silicon of 3 and 1%, respectively.

The beams can be more simply controlled and stabilized during epitaxy of compounds which incorporate elements with greatly differing vapor tensions. For example, the temperature of the gallium-containing crucible is precisely stabilized during GaAs growth and the arsenic supply is controlled and stabilized by a pressure sensor placed in its beam. A similar system was successfully used during InAs synthesis in [15].

The elemental composition of the substrate surface and the film grown is evaluated by Auger electron spectroscopy (AES) and SIMS. The sensitivity of the latter for a number of elements can attain a value of 10^{15}-10^{16} cm^{-3}. However, both methods are difficult to use for monitoring *in situ*. Therefore, we note only that AES in MBE setups is the main analytical means for monitoring substrate surface purification in ultrahigh vacuum (UHV).

An MBE apparatus equipped with automatic ellipsometry was first described in [16]. The capabilities of this instrument for analysis of structural rearrangements on a clean silicon surface for the reversible transition from an ordered system with two interplanar distances ($2d_{111}$) to an ordered system with one interplanar distance (d_{111}) and for the phase transition of the superstructure Si(111) (7×7) into the structure Si(111) (1×1) were demonstrated in [17]. Recently, a new effect [18] of oscillation of optical characteristics of the surface layer during germanium epitaxy from a molecular beam (Fig. 3) was observed by reflection automatic ellipsometry. The oscillation period coincides with the time for formation of a layer of diatomic thickness. Comparison of AE and RHEED data indicates a single reason for these phenomena. Thus, the possibility of studying surface reactions that occur with formation of two-dimensional nuclei (crystal growth, dissolution, oxidation, etc.) by AE under conditions where RHEED is inapplicable (change of material under the electron beam, carrying out the process in a medium that is dense for the electron beam, and growth of a dielectric film) was shown in [18]. Apparently, AE is especially promising for investigation of the kinetics of surface rearrangements during MBE.

3. EPITAXY OF SEMICONDUCTING FILMS

Substantial differences exist both in the technical implementation of MBE and in the occurrence of physicochemical processes during epitaxy of films of monatomic semiconductors and semiconducting $A_{III}B_V$ and $A_{II}B_{VI}$ compounds. However, the problems that must be solved in setting up the epitaxial process are unique. They involve mainly preparation of a chemically pure structurally ordered substrate surface, determination of optimal conditions that guarantee a two-dimensional nucleation mechanism and film growth, study of the behavior of dopants, identification of optimal doping conditions during MBE, and analysis of the defect formation mechanisms, especially during heteroepitaxy, in order to minimize the defect density in the films [19, 20].

The problem of substrate surface purification was discussed in [19-22] and has been solved for the general case. Removal of the principal residual impurity on the semiconductor surfaces, carbon, is achieved by creation of thin films of semiconductor oxides that are "carbonophobic" [22-25] with subsequent sublimation of these films in UHV. Disruption of stoichiometry in the subsurface layer of binary $A_{III}B_V$ compounds is eliminated by annealing in vapors of the group V element. The substrate surface is bombarded by various molecular beams (for silicon, these are beams of Ga [26] and Si [23, 24], for InP, for example, this is a beam of arsenic [22]) in order to lower the oxide sublimation temperatures. The perfection of pure semiconductor surfaces (the second moment during substrate preparation) can be characterized by the presence of superstructure. Thus, the superstructures Si(111) (7 × 7) and Ge(111) (2 × 8) are formed for pure singular (111) faces of silicon and germanium at relatively low temperatures. The clean surface will contain excess group III element during purification of the surface of $A_{III}B_V$ crystals by annealing in UHV. The superstructure c (8 × 2) Ge, which is stabilized by gallium, is formed after such annealing in the case of GaAs(100). Annealing in an arsenic beam transforms this superstructure into s (2 × 8) As. This signifies stabilization of the surface by arsenic. A large number of structures that are stabilized by elements of both the III and V group are known for $A_{III}B_V$ compounds. Besides the chemical purity and structural perfection, the substrate surface should be smooth on an atomic scale. This appears in RHEED patterns as diffraction bands. Buffer layers are grown for attainment of atomically smooth substrate surfaces.

Processes on the growth surface are substantially different for monatomic semiconductors and semiconducting compounds. Therefore, we will examine them separately.

If growth is carried out during homoepitaxy of silicon and germanium at temperatures at which superstructures are stable, these superstructures are retained on the surface of the growing film [27, 28]. These superstructures correspond to the substrate orientation used. An analogous situation is important during heteroepitaxial growth of elemental semiconductors if the misfit parameter of the film and substrate lattice periods f is small and does not substantially disturb the two-dimensional growth mechanism [27]. Thus, for example, the transition from a superstructure that is characteristic for GaAs(100) to a superstructure that corresponds to Ge(100) is accomplished in 2-10 monolayers [29, 30] during growth of Ge films on GaAs ($f = 7 \cdot 10^{-4}$) for the (100) orientation. Oscillations of RHEED reflection intensities were distinctly observed in [7, 18]. Also, oscillations of the automatic ellipsometry Δ and ψ angles during Ge growth on GaAs(100) were noted. These confirm the layered growth mechanism in this case.

Successive change of superstructures on the growth surfaces during heteroepitaxy can be effected. This is due to a large number of reasons. Apparently, similar structural rearrangements were studied in the most detail for heteroepitaxial Ge films on Se ($f = 0.04$) [27, 31, 32]. In [31], a kinetic diagram of structural transformations on the growth surface was constructed using RHEED in situ (Fig. 4). Electron microscopy, AES, and SIMS were used for elucidation of the nature of various superstructures. Besides the usual superstructures Si(111) (7 × 7) and Ge(111) (2 × 8), the superstructures Si(111) (7 × 7) Ge, Ge(111) (7 × 7) Si, and Si,

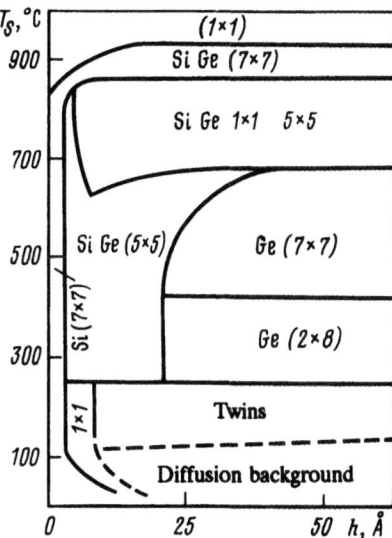

Fig. 4. Kinetic diagram of structural transformations on the surface of Ge film during growth on Si(111). Growth rate 0.1 μm/h.

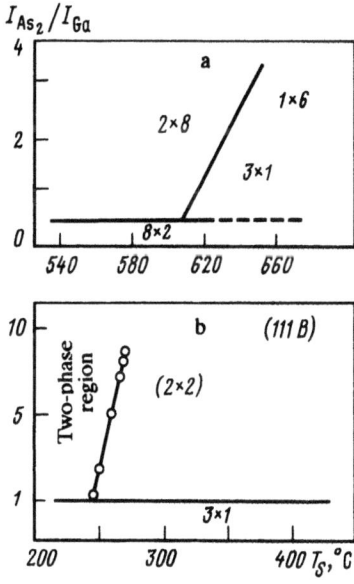

Fig. 5. Phase diagram of stability of surface superstructures for the faces GaAs(100) [37] (a) and InSb(111) B [32] (b).

Ge(111) (5 × 5) appear in the diagram. The first two of these are ordered adsorptive coatings of the element shown last in the superstructure designation. The appearance of silicon on the germanium film surface, which is accompanied by formation of the Ge (7 × 7) Si superstructure, arises from diffusion on the growth surface and from bulk diffusion through the continuous germanium film. The superstructure Si, Ge(111) (5 × 5) corresponds to the presence on the surface of a pseudomorphic germanium film. It begins to disappear upon collapse of the pseudomorphism (the moment of collapse is shown by the dashed line in Fig. 4). Thus, the change of superstructures even in the relatively simple Ge—Si heterosystem is explained by action of a number

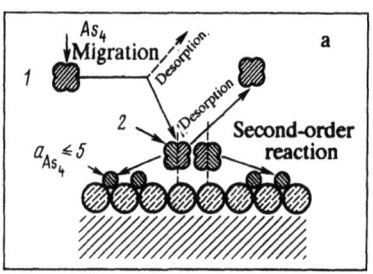

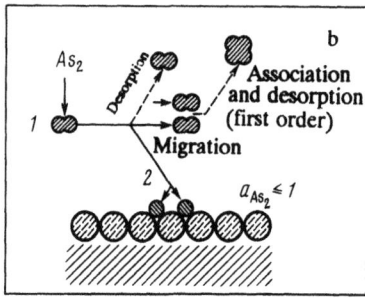

Fig. 6. Chemical reactions on the GaAs(100) surface stabilized by gallium at $T \leq 300°C$ in a beam of As_4 (a) and As_2 (b) [36-38].

of factors: possible formation of ordered adsorption coatings, a change in mobility of adatoms and diffusion rate of mass transfer in the bulk film, and the presence of elastic deformations in the films.

Exceedingly complicated superstructure processes occur during homoepitaxy of $A_{III}B_V$ compounds. The surface structure for steady-state growth depends on the substrate temperature (T_S) and the ratio of group V and III element beams (F_V/F_{III}). Figure 5 shows the phase diagrams constructed for growth of GaAs(100) (a) and InSb(111) B (b) films. As already indicated, two basic superstructures are characteristic for GaAs(100): s (8 × 2) Ga and s (2 × 8) As. The distinguishing feature of the phase diagram for InSb(111) B (Fig. 5b) is the existence of a biphasic region at low T_S. Here, epitaxial islets of InSb and Sb coexist. This effect is explained by the long lifetime of antimony adatoms at low T_S. As a result, epitaxial crystallization of antimony [33] besides the InSb formation reaction occurs.

The phase diagrams for the growth surface provide a good basis for choice of growth procedures. Numerous experiments indicate that the most perfect and smooth films are prepared near the interfaces which separate superstructures which are stabilized by atoms of the group III and V elements. Besides the effect on the surface morphology, the type of superstructure determines the inclusion of amphoteric dopants into the film. The most successful example is the doping of GaAs films by germanium [1, 34]. The presence of a surface stabilized by As leads to incorporation of Ge into Ga vacancies and, consequently, to n-type film conductivity. The Ge atoms fill arsenic vacancies on the Ge-stabilized surface, giving p-type conductivity.

Chemical reactions that occur on the growth surface during MBE of $A_{III}B_V$ compounds determine to a large degree the structure and properties of the films. The modulated molecular beam method [35] gives the most interesting information for study of these reactions. Features of the chemical reactions are discussed in [36-38] for processes on gallium arsenide (100) surfaces stabilized by gallium. To understand these features, it must be considered that the Ga molecular beam contains only monomeric Ga_1 whereas the arsenic reaches the surface basically as As_2 dimers in As_4 tetramers. Molecules of both As_4 and As_2 are physically adsorbed in the absence of gallium adatoms on the growth surface (the gallium beam is turned off) with a finite lifetime but an adhesion coefficient equal to zero. A dissociative chemosorption of As_4 and As_2 molecules occurs with introduction of gallium atoms to the surface (Fig. 6). This occurs differently for various molecules. The As_4 mo-

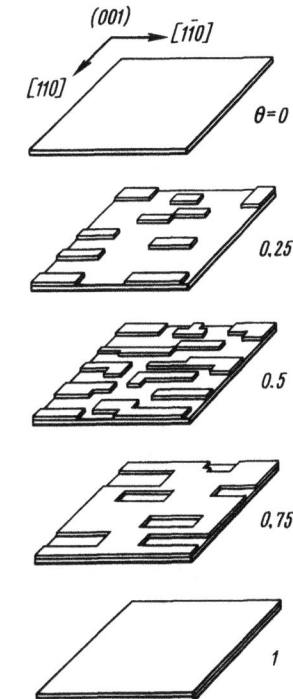

Fig. 7. Scheme of formation of one filled epitaxial layer [7]. θ is the relative surface coating.

lecules react pairwise upon encounter in neighboring Ga positions (Fig. 6a). Then, four monomers enter the growing film and one As_4 molecule is desorbed. Both chemosorption and desorption in this case are second-order reactions. The paired reaction of As_4 molecules should lead to a large concentration of point defects in films that are grown using As_4 [38]. With use of As_2 molecules, their chemosorption proceeds by a first-order reaction on one gallium atom (Fig. 6b). The excess As_2 molecules are released into the vapor. Therefore, preparation of stoichiometric high-quality layers is more probable in this case. This was confirmed experimentally in [39]. In particular, use of As_2 molecules increased the oscillation amplitudes and their numbers during measurements of the reflection intensities in RHEED patterns. This indicates a more perfect two-dimensional growth. In [40], a comparative study of GaAs films grown with various ratios of As_4 and As_2 molecules was completed. Several sharp peaks in the 1.504-1.511 eV range in the photoluminescence spectrum gradually disappear with an increase of the fraction of As_2 molecules. These are explained by emissive recombination of excitons that are related to lattice faults. The concentration of a number of deep levels and the degree of autocompensation upon doping with amphoteric germanium decrease with an increase of the As_2 fraction in the beam. However, in all the examples mentioned, it is unclear what is responsible for the improvement of film properties: a decrease of dopant concentration or native point defect concentration.

Finally, we will discuss the intrinsic MBE film growth mechanism. If the substrate surface is carefully prepared, then during homoepitaxy, as well as during heteroepitaxy with small f values, a two-dimensional mechanism of nucleation and film growth is realized. Several of its characteristics can be obtained from analysis of the oscillatory dependences of the RHEED diffraction maxima. These oscillations are observed during epitaxy of GaAs, Ge, Si, AlGaAs, and other crystals. According to the model proposed in [7], as the monolayer fills, a transition occurs from an atomically smooth surface at $\theta = 0$ and 1 (θ is the relative coating of the surface layer) to an atomically rough surface ($\theta = 0.5$) (Fig. 7). The amplitude and number of oscillations depends on the ratio of nucleation rates of the two-dimensional nuclei and their lateral expansion. In principle, conditions under which a very small density of two-dimensional nuclei of the subsequent layer will arise during the filling of one monolayer on its surface can be selected. In this case, the number of oscillations should increase substantially. A similar effect was uniquely used in [41, 42] for growth of superlattices. The number of monolayers included in each of the components of the superlattice for a total structure thickness up to 100 nm and more [42] was determined precisely from the quantity of oscillations. Thus, the beams of the group III elements can be calibrated and the magnitude of x in the ternary compound $Al_xGa_{1-x}As$ can be found:

$$x = f(AlAs)/[f(AlAs) + f(GaAs)],$$

where $f(AlAs)$ and $f(GaAs)$ are the oscillation frequencies during growth of AlAs and GaAs, respectively. The layer thickness in the superlattice determines the position of the quantum levels in the conductivity band such that a change of thickness by one monolayer shifts the level by 7 μeV.

Recently [43], intensity oscillations of RHEED reflections were also detected during GaAs sublimation in an arsenic beam. The vaporization process can be viewed as inverse growth. The difference between them consists of the parameters which control the process: during growth the oscillation period is determined by the intensity of the incident Ga beam; during vaporization, by the substrate temperature (T_S) and the partial pressure of arsenic in the chamber. The sublimation rate decreases with an increase of arsenic pressure. Thus, the sublimation rate was 18 at $P_{As} = 2 \cdot 10^{-4}$ Pa and 12 monolayers/min at $P_{As} = 3.2 \cdot 10^{-4}$ Pa for $T_S = 750°C$. The activation energy obtained, 4.7 eV, allows the conclusion to be made that the sublimation process is limited by GaAs dissociation and not by Ga vaporization.

Besides the trivial transition from a two-dimensional growth mechanism to a three-dimensional, caused by contamination of the substrate surface or the growth surface, another important case of growth mechanism change in the initial MBE stage occurs. This is the realization of the Stranski–Krastanov mechanism at large f values. This mechanism has been studied in most detail for MBE of germanium on silicon [32, 44]. Films of germanium of thickness $h \lesssim 3$ nm form a system of parallel pseudomorphic bands. At $h > 3$ nm, islets of germanium with a network of dislocation discrepancies (DD) is formed at the interface. In this case, mass transfer from the bands to the islets is observed. The islet distribution density decreased exponentially with an increase of T_S and increased with an increase of the inclination angle β from the singular (111) face.

In [45, 46], an explanation for the transition from two-dimensional growth to three-dimensional was proposed. It was based on the increase of energy of the heteroepitaxial interface upon introduction of DD. The morphological rearrangement of the pseudomorphic film into dislocated islets was shown to be possible at thicknesses which are significantly smaller than the critical thickness of the continuous film for introduction of DD. The film thickness at the moment of morphological rearrangement decreases with an increase of f and, at a very large discrepancy, growth by a three-dimensional mechanism from the initial stage itself is possible. Under actual conditions, atoms of the film can collect into an islet from a fragment of the growth surface, the radius of which is equal to the mean migration length λ of the adatoms. This parameter depends on the substrate temperature T_S and determines the islet density, with which it is related by the relation $N = (2\lambda)^{-2}$. Therefore, independent film fragments with radius λ and volume $V_\lambda = \pi\lambda^2 h$, where h is the effective film thickness, are examined during analysis of the morphological rearrangements in [46]. When V_λ reaches the limiting volume value V_L and $\lambda > R_M = [\gamma_s(1 - v)]/[G(1 + v)f^2]$, where R_M is the radius of an equilibrium islet, γ_s is the specific surface energy of the film, and v is the Poisson coefficient, the probability for transformation of fragments of continuous film into islets appears. The value of the critical discrepancy for the $Ge_{1-x}Si_x$–Si system, obtained from the condition $\lambda = R_M$, is $f_c = 4 \cdot 10^{-3}$-10^{-2} at $\lambda = 5 \cdot 10^{-2}$-50 nm, respectively. This agrees well with the experimental value $f_c = 8 \cdot 10^{-3}$ determined in [47]. Although in general the concept of the Stranski–Krastanov mechanism is clear, the problem of the effect of the f parameter on its realization has not been developed theoretically.

Concluding the discussion of MBE growth processes, we emphasize that the discovery of conditions under which a two-dimensional growth mechanism is realized is one of the most important tasks for preparation of this growth method since two-dimensional growth at low T_S allows creation of heterojunctions, p–n junctions, superlattices, and other similar structures with atomically smooth heteroboundaries, the preparation of which is impossible by other methods.

Two other problems, listed at the beginning of this section (doping features and defect formation in MBE), have been discussed in detail by us in [20, 21, 48-50] (see also [2, 13, 22, 28, 36, 38]). Therefore, we will not examine them here.

4. METAL FILM EPITAXY

Demands on the quality of metal films from the viewpoint of material purity, as a rule, are lower than those on semiconducting films. However, the quality of the semiconductor–metal interface should not be poorer than the semiconductor–semiconductor interface.

Apparently, films of Al on GaAs [51] were the first metal films on a semiconductor prepared by MBE. The deposition was carried out at ambient T_S on GaAs(100) stabilized by arsenic and gallium. The experiments showed that the Al lattice is rotated by 45° relative to the GaAs lattice in the plane of the heterojunction. The Schottky barrier height was 0.66 eV for the surface stabilized by arsenic. It increased to 0.72 eV for that stabilized by gallium. The creation by MBE of the epitaxial Schottky barrier was a step forward in comparison with the barriers based on polycrystalline metal films since a higher uniformity of heterojunction structure on the substrate should be achieved with epitaxy.

The heteroepitaxial silicon–silicide systems have been most thoroughly studied recently. Most attention has been focused on disilicides of nickel and cobalt ($NiSi_2$ and $CoSi_2$) which have the CaF_2 cubic lattice with small f ($f_{NiSi_2} = 10^{-2}$ and $f_{CoSi_2} = 2 \cdot 10^{-2}$). Apparently, these were first prepared by MBE on silicon in [52]. The $CoSi_2$ lattice is found in a twinned position relative to the silicon lattice at film thicknesses $h > 20$ nm. The twinning axis (111) coincides with the normal to the silicon surface [53]. The $NiSi_2$ at these same h values contains approximately equal quantities of segments in twinned (B-type) and normal (A-type) positions. The places of abutment of various types of segments are antiphase interfaces (API).

An interesting possibility for the study of the effect of interface structure on the properties of the Schottky barrier was discovered in [54]. Single-domain films of both the B- and A-types were prepared. Production of only the B-type or only the A-type is possible by deposition at room temperature of Ni films of thickness 0.1-0.7 and 1.6-2 nm, respectively, with subsequent annealing at 450-550°C for 3-5 min. If Ni is again deposited in small portions on the thin single-domain films of A- or B-type that are formed and the system is annealed at ~450°C, then exceptionally thick single-domain layers can be prepared. An even simpler method of growth of thick single-domain films consists in deposition of Ni or Ni and Si at $T_S \gtrsim 650$°C onto the thin single-domain coatings, i.e., MBE is carried out.

The reasons for formation of single-domain thin films are not yet clear. Comin et $al.$ [55], using the SEXAFS method, showed that an amorphous film with short-range order that corresponds to $NiSi_2$ is formed upon Ni deposition on Si at $T_S = 20$°C. Formation of B-type segments with very small coverages is explained in [54] by the lower free energy of the twinned interface. The growth of the single-domain A-type film at $h_{Ni} = 1.6$-2 nm is related to the kinetic features of epitaxial silicide formation. An important result of [54] is that two epitaxial Schottky barriers that differ only in the atomic structure of the heterojunction can be prepared.

In [56], the properties of the Schottky barriers in the single- and multidomain $NiSi_2$ films were studied. The barrier height in structures with the $NiSi_2$ A-segments is ~0.14 eV lower than in structures with the B-segments. The values of the barrier in the presence of a mixture of A- and B-segments, determined from I–V and C–V measurements, differ from each other. Thus, the possibility for control of the Schottky barrier height

Table 1. Fluoride–Semiconductor Epitaxial Systems

Fluoride	Semiconductor	Substrate orientation	f, %
GaF$_2$	Si	(100), (110), (111)	+0.61
	GaAs	(100), (110),	−3.4
	Ge	(111)	−3.5
	InP	(100)	−6.9
SrF$_2$	Si	(111)	+14.2
	GaAs	(111)	+9.6
	InP	(100), (111)	+5.5
	CdTe	(100)	−4.5
Ca$_{0.43}$Sr$_{0.57}$F$_2$	GaAs	(100), (110)	0
Ca$_{0.44}$Sr$_{0.56}$R$_2$	Si	(111)	+4.2
Ba$_{0.2}$Ca$_{0.8}$F$_2$	InP	(100)	0
Ba$_{0.07}$Sr$_{0.83}$F$_2$	InP	(100)	0

by changing the atomic structure of the metal–semiconductor heterojunction was demonstrated. It is also important that the multidomain system with API leads to a significant scatter of barrier values.

In conclusion, we note that work on MBE of multilayered metal–semiconductor structures (see, for example, [52]) is proceeding successfully besides that on monolayered metal (silicide)–semiconductor (silicon) structures. Similar studies, according to [13], form the foundation for creation of bulk integrated circuits.

5. DIELECTRIC FILM EPITAXY

Preparation of metal–dielectric–semiconductor (MDS) structures in a single epitaxial process is enticingly promising. However, epitaxy of a dielectric on a semiconducting substrate must be accomplished for this. Of course, preparation of MDS structures on a fresh semiconductor growth surface using an amorphous dielectric and a polycrystalline metal under UHV conditions is also an interesting possibility [57]. However, it is more interesting to create the epitaxial MDS structure, and the DS structure before this. At present, two directions are being pursued. The first consists in preparation of wide-band epitaxial films of A$_{III}$B$_V$ compounds by so-called reactive MBE [58], i.e., RMBE. The essence of this method consists of epitaxial synthesis of the required material on an oriented substrate under UHV conditions using various chemical compounds as beams. For example, epitaxial films of AlN (forbidden band width $E = 6.2$ eV) on silicon and sapphire substrates was grown through chemical reactions between the Al beam and an NH$_3$ stream. Use of a N$_2$ stream activated by ultrahigh frequency discharge together with a Ga molecular beam allowed epitaxy of GaN ($E = 3.4$ eV) on GaAs and sapphire. The ternary compound AlGaN was synthesized by RMBE on Si and sapphire substrates over the whole composition range. This ensured a range of $E = 3.4\text{-}6.2$ eV.

Another unique possibility for preparation of epitaxial dielectrics on monatomic semiconductor Ge, Si, and A$_{III}$B$_V$ compounds arises through use of alkaline earth fluorides [59-61]. These compounds have several features which favor their application as dielectric epitaxial layers in MBE technology. The fluorides of calcium, barium, and strontium have a face-centered cubic lattice which is similar to silicon and A$_{III}$B$_V$. CaF$_2$, BaF$_2$, and SrF$_2$ sublime at temperatures from 1200-1400°C without dissociation into alkaline earth metal and fluoride ions so that films of stoichiometric composition are easily prepared. Vegard's law is fulfilled for ternary fluoride compounds of the Ca$_x$Ba$_{1-x}$F$_2$ type. Therefore, the cell constant of the dielectrics can be varied over wide limits (from 0.546 for CaF$_2$ to 0.62 nm for BaF$_2$). The lattice constants of the majority of A$_{III}$B$_V$ compounds, germanium, silicon, and a number of A$_{II}$B$_{VI}$ compounds fall within this range. Table 1 lists systems in which epitaxy has been achieved [62]. Obviously, selection of the dielectric composition can assure that $f = 0$.

If the defectiveness of the fluoride films is evaluated, it is still rather high, in spite of the possible fit of its lattice constants with the semiconducting substrate. As in the heterosystems silicon–silicide, A- and B-type segments with API at the sites of segment abutment are observed. The problem of elimination of API for the fluoride—semiconductor heteropair is still not resolved, although it is possible that substrates with the {hkk} orientation should be used, in analogy with heteroepitaxy of $A_{III}B_V$ on Si and Ge [63, 64]. Another possible reason for the high defectiveness of the fluoride films may be the 5-7 times difference of thermal expansion coefficients of the fluoride films and the semiconductors.

In a review [69], several properties of the fluoride—semiconductor structures were presented. The field intensity for breakdown is usually $5 \cdot 10^5$ V·cm^{-1}. The best values are $3 \cdot 10^6$ and $4 \cdot 10^6$ V·cm^{-1} at 300 and 77 K, respectively (breakdown on SiO_2 usually begins at $6 \cdot 10^6$ V·cm^{-1}). The minimal density of surface states attained for the CaF_2—Si(100) system was $7 \cdot 10^{10}$ eV^{-1}·cm^{-2}. More typical values are $5 \cdot 10^{11}$-$1 \cdot 10^{12}$ eV^{-1}· cm^{-2}. For comparison, this value in the SiO_2—Si system is less than $5 \cdot 10^{10}$ eV^{-1}·cm^{-2}.

The data given indicate the possible application of dielectric fluoride films as gate dielectrics in the MDS structures.

Yet another possible application of epitaxial fluoride films is the preparation of multilayered semiconductor—dielectric— semiconductor structures for creation of three-dimensional integrated circuits [60, 65, 66]. The best results for epitaxy were obtained for a good fit of the semiconductor and fluoride film lattice constants. Thus, the following parameters were attained in the upper epitaxial layer of the GaAs/(Ca, Sr)F_2/GaAs structure [67]: electron mobility at room temperature 1300 cm^2·V^{-1}·sec^{-1} and concentration $9 \cdot 10^{16}$ cm^{-3}. Epitaxy of GaAs at these same growth conditions directly on a GaAs substrate gave 4000 cm^2·V^{-1}·sec^{-1} and $2 \cdot 10^{17}$ cm^{-3}, respectively.

6. CONCLUSIONS

New types of sources are being introduced into MBE practice with explanation of the chemical processes on the growth surface and their role in the formation of properties of the films that are grown. Thus, arsine and phosphine [68] are being used for the creation of streams of As_2 and P_2 molecules. Growth of quaternary InGaAsP compounds using the usual crystalline sources of arsenic and phosphorus is also problematical due to difficulty in controlling the P/As ratio. A novel epitaxial technique, in which one of the readily volatile components (for example, phosphorus) enters as a separate beam of low-energy ions is being developed in order to avoid this difficulty. Further, more complicated chemical reactions are being accomplished by supplying streams that do not react with each other to the substrate, hence, the emergence of RMBE. Very recently, a hybrid of MBE and organometallic synthesis [69] was introduced. Thus, evidently, gradual convergence of the MBE and GPE methods is occurring. This convergence takes advantage of the merits of both approaches and minimizes their shortcomings.

REFERENCES

1. A. Y. Cho and J. R. Arthur, "Molecular beam epitaxy," *Prog. Solid State Chem.*, **10**, Pt. 3, 157-191 (1975).
2. J. C. Bean, "Growth of doped silicon layers by molecular beam epitaxy," in: *Impurity Doping*, North-Holland, Amsterdam (1981), pp. 177-215.
3. A. G. Denisov, N. A. Kuznetsov, and V. A. Makarenko, "Equipment for molecular beam epitaxy," *Obz. Élektron. Tekh.*, Ser. 7, No. 17 (828), 1-52 (1981).
4. V. V. Anashin, G. G. Emelin, B. Z. Kanter, V. P. Migal', et al., *Analytical Apparatus and Technological Equipment for Molecular Beam Epitaxy* [in Russian], Preprint No. 1, Inst. Fiz. Khim., Novosibirsk (1985).
5. O. P. Pchelyakov, "High-energy electron diffraction applied to study of structure formation processes during molecular beam epitaxy," in: *Physics and Technical Basics of Molecular Beam Epitaxy*, Eberswalde (DDR) (1979), p. 114-140.
6. J. J. Harris, B. A. Joyce, and P. J. Bodson, "Oscillations in the surface structure of Sn-doped GaAs during growth by MBE," *Surf. Sci.*, **103**, L90-L96 (1981).
7. J. H. Neave, B. A. Joyce, F. J. Dobson, and N. Norton, "Dynamics of film growth of GaAs by MBE from RHEED observation," *Appl. Phys. A*, **31**, No. 1, 1-8 (1983).
8. J. M. Van Hove, C. S. Lent, P. R. Pukite, and P. T. Cohen, "Damped oscillations in reflection high energy electron diffraction during GaAs MBE," *J. Vac. Sci. Technol. B*, **1**, No. 3, 741-746 (1983).

9. J. M. Van Hove, P. R. Pukite, and P. I. Cohen, "RHEED oscillations as functions of MBE growth parameters," MBE Conf., San Francisco (1984), pp. 18-40.

10. L. Esaki, "Computer-controlled molecular beam epitaxy," *Jpn. J. Appl. Phys.*, **13**, No. 1, 821-828 (1974).

11. P. E. Lusher, "Crystal growth by molecular beam epitaxy," *Solid State Technol.*, No. 12, 43-52 (1977).

12. J. C. Bean and S. A. Sadowski, "Silicon MBE apparatus for uniform high-rate deposition on standard format wafers," *J. Vac. Sci. Technol.*, **20**, No. 2, 137-142 (1982).

13. J. C. Bean, "Silicon molecular beam epitaxy as a VLST processing technique," Int. Electron Devices Meeting (1981), pp. 6-13.

14. P. M. Petroff, A. C. Gossard, W. Wiegmann, and A. Savage, "Crystal growth kinetics in $(GaAs)_m-(AlAs)_m$ superlattices deposited by molecular beam epitaxy," *J. Cryst. Growth*, **44**, 5-13 (1978).

15. W. J. Schaffer, M. D. Ling, S. P. Kowalczyk, and R. W. Grant, "Nucleation and strain relaxation at the InAs/GaAs(100) heterojunction," *J. Vac. Sci. Technol. B*, **1**, No. 3, 688-695 (1983).

16. A. V. Arkhipenko, Yu. A. Blyumkina, M. A. Lamin, et al., "Apparatus for molecular beam epitaxy with automatic ellipsometry," *Poverkhnost*, No. 1, 93-96 (1985).

17. O. P. Pchelyakov, Yu. A. Blyumkina, S. I. Stenin, et al., "Study of structural rearrangements of an atomically clean silicon surface by automatic ellipsometry and reflection electron diffraction," *Poverkhnost*, No. 1, 147-149 (1982).

18. L. V. Sokolov, M. A. Lamin, V. A. Markov, et al., "Oscillations of optical properties of Ge film growth surface during molecular beam epitaxy," *Pis'ma Zh. Éksp. Teor. Fiz.*, **44**, No. 6, 278-280 (1986).

19. A. V. Rzhanov, S. I. Stenin, and B. Z. Ol'shanetskii, "Methods of monitoring surface state and problems of molecular beam epitaxy," *Mikroélektronika*, **9**, No. 4, 292-301 (1980).

20. A. V. Rzhanov and S. I. Stenin, "Molecular epitaxy: state of the question, problem, and prospects of development," in: *Growth of Semiconducting Crystals and Films* [in Russian], Nauka, Novosibirsk (1984), Part 1, pp. 5-34.

21. S. I. Stenin, "Molecular beam epitaxy of semiconductor, dielectric, and metal films," *Vacuum*, **36**, No. 7/8, 419-426 (1986).

22. A. Y. Cho, "Growth of III-V semiconductor by molecular beam epitaxy and their properties," *Thin Solid Films*, **100**, No. 4, 291-317 (1983).

23. M. Tabe, K. Arai, and H. Nakamura, "Effect of growth temperature of Si MBE films," *Jpn. J. Appl. Phys.*, **20**, No. 4, 703-708 (1981).

24. N. Tabe, "Etching of SiO_2 films by Si in ultrahigh vacuum," *Jpn. J. Appl. Phys.*, **21**, No. 3, 534-538 (1982).

25. A. Munoz-Yaque, J. Piqueras, and N. Fabre, "Preparation of carbon-free GaAs surfaces: AES and RHEED analysis," *J. Electrochem. Soc.*, **128**, No. 1, 149-153 (1981).

26. S. Wright and H. Kroemer, "Reduction of oxides on silicon by heating in a gallium molecular beam at 800°C," *Appl. Phys. Lett.*, **36**, No. 3, 210-211 (1980).

27. A. V. Rzhanov, S. I. Stenin, O. P. Pchelyakov, and B. Z. Kanter, "Molecular beam epitaxial growth of germanium and silicon films: surface structure, film defects, and properties," *Thin Solid Films*, **139**, No. 1, 169-175 (1986).

28. Y. Ota, "Silicon molecular beam epitaxy," *Thin Solid Films*, **106**, No. 1/2, 3-136 (1983).

29. B. J. Mrstik, "LEED and AES studies of the initial growth of GE epilayers on GaAs (100)," *Surf. Sci.*, **124**, 253-266 (1983).

30. C.-A. Chang, "Interface morphology of epitaxial growth of Ge on GaAs and GaAs on Ge by molecular beam epitaxy," *J. Appl. Phys.*, **53**, No. 2, 1253-1255 (1982).

31. L. V. Sokolov, M. A. Lamin, O. P. Pchelyakov, et al., "Surface rearrangements during epitaxy of germanium on silicon," *Poverkhnost*, No. 9, 75-80 (1985).

32. A. I. Toropov, L. V. Sokolov, O. P. Pchelyakov, and S. I. Stenin, "Epitaxy of germanium from a molecular beam on a vicinal surface of silicon near (111)," *Kristallografiya*, **27**, No. 4, 751-756 (1982).

33. A. J. Noreika, M. H. Francombe, and C. E. C. Wood, "Growth of Sb and InSb by molecular beam epitaxy," *J. Appl. Phys.*, **52**, No. 12, 7416-7420 (1981).

34. R. Z. Bachrach, "MBE-molecular beam epitaxial evaporative growth," in: *Crystal Growth*, Pergamon, New York (1980), pp. 1-52.

35. C. T. Foxon, M. R. Boudry, and B. A. Joyce, "Evaluation of surface kinetic data by the transform analysis of modulated molecular beam measurement," *Surf. Sci.*, **44**, No. 1, 69-92 (1974).

36. B. A. Joyce, "Present status and future directions for MBE," *Surf. Sci.*, **86**, No. 1, 92-101 (1979).

37. B. A. Joyce, C. T. Foxon, and J. H. Neave, "Fundamentals of molecular beam epitaxy," *J. Jpn. Assoc. Cryst. Growth*, **5**, 185-197 (1978).

38. C. T. Foxon, "MBE growth of GaAs and III-V alloys," *J. Vac. Sci. Technol. B*, **1**, No. 2, 293-297 (1983).

39. J. H. Neave, P. K. Larsen, J. F. van der Veen, et al., "Effect of arsenic species (As_2 or As_4) on the crystallographic and electronic structure of MBE-grown GaAs (001) reconstructed surfaces," *Surf. Sci.*, **133**, 267-278 (1983).

40. H. Kunzel, J. Knecht, H. Jung, et al., "The effect of arsenic vapor species on electrical and optical properties of GaAs grown by molecular beam epitaxy," *Appl. Phys. A*, **28**, 167-173 (1982).

41. T. Sakamoto, H. Funabashi, K. Ohta, et al., "Well-defined superlattice structures made by phase-locked epitaxy using RHEED intensity," in: Proc. Int. Conf. Superlattice, Illinois (1984), pp. 1021-1029.

42. T. Sakamoto, H. Funabashi, K. Ohta, et al., "Phase-locked epitaxy using RHEED intensity oscillation," *Jpn. J. Appl. Phys.*, **23**, No. 9, L657-L659 (1984).

43. T. Kajima, N. J. Kawai, T. Nakagawa, et al., "Layer-by-layer sublimation observed by reflection high-energy electron diffraction intensity oscillation in a molecular beam epitaxy system," *Appl. Phys. Lett.*, **47**, No. 3, 286-288 (1985).

44. A. I. Toropov, B. Z. Kanter, V. Yu. Karasev, et al., "Molecular epitaxy of germanium on silicon: surface morphology, structure, diffusion," *Poverkhnost*, No. 1, 110-115 (1982).

45. J. W. Matthews, D. S. Jackson, and A. Chambers, "Effect of coherency strain at misfit dislocations on the mode of growth of thin films," *Thin Solid Films*, **26**, No. 1, 129-134 (1975).

46. S. M. Pintus, S. I. Stenin, A. I. Toropov, and E. M. Trukhanov, "Morphological stability and mechanism of growth of heteroepitaxial films," Inst. F. P., Sib. Otd. Akad. Nauk SSSR, Novosibirsk (1986).

47. E. Kasper and W. Pabst, "Profiling of SiGe superlattices by He backscattering," *Thin Solid Films*, **37**, No. 1, L5-L7 (1976).

48. S. I. Stenin, "Problems of the formation of dislocation structure on heteroepitaxial layers," *Phys. Status Solidi A*, **55**, 519-527 (1979).

49. S. I. Stenin, "Processes of defect formation in epitaxial semiconductor films," *Thin Solid Films*, **116**, No. 1, 99-110 (1984).

50. S. I. Stenin and A. K. Gutakovskii, "Electron-microscopic studies of the mechanisms of formation of faults in heteroepitaxial systems," in: *Contemporary Electron Microscopy for Study of Substances* [in Russian], Nauka, Moscow (1982), pp. 139-147.

51. A. Y. Cho and P. D. Dernier, "Single-crystal-aluminum Schottky-barrier diodes prepared by molecular beam epitaxy (MBE)," *J. Appl. Phys.*, **49**, No. 6, 3328-3333 (1978).

52. J. C. Bean and J. M. Poate, "Silicon/metal silicide heterostructures grown by molecular beam epitaxy," *Appl. Phys. Lett.*, **37**, No. 7, 643-646 (1980).

53. J. M. Gibson, J. C. Bean, J. M. Poate, and R. T. Tung, "Direct determination of atomic structure at the epitaxial cobalt disilicide on (111) Si interface by ultrahigh resolution electron microscopy," *Appl. Phys. Lett.*, **49**, No. 9, 818-820 (1982).

54. R. T. Tung, J. M. Gibson, and J. M. Poate, "Formation of ultrathin single-crystal silicide films on Si: surface and interfacial stabilization of Si–NiSi$_2$ epitaxial structures," *Phys. Rev. Lett.*, **50**, No. 6, 429-432 (1983).

55. F. Comin, J. E. Rowe, and P. H. Citrin, "Structure and nucleation mechanism of nickel silicide on Si (111) derived from surface extended-X-ray-absorption fine structure," *Phys. Rev. Lett.*, **51**, No. 26, 2402-2405 (1983).

56. R. T. Tung, "Schottky barrier heights of single crystal silicides on Si(111)," *J. Vac. Sci. Technol. B*, **2**, No. 3, 465-470 (1984).

57. A. Y. Cho, "Recent developments in molecular beam epitaxy (MBE)," *J. Vac. Sci. Technol.*, **16**, No. 2, 275-284 (1979).

58. S. Yoshida, "Reactive molecular beam epitaxy," in: *CRC Crit. Rev. Solid State Mater. Sci.*, **11**, No. 4, 287-316 (1983).

59. C. W. Tu, T. T. Sheng, M. H. Read, et al., "Growth of single-crystalline epitaxial group II fluoride films on InP(001) by molecular-beam epitaxy," *J. Electrochem. Soc.*, **130**, No. 10, 2081-2087 (1983).

60. C. W. Tu, S. R. Forrest, and W. D. Johnson, Jr., "Epitaxial InP/fluoride/InP(100) double heterostructure grown by molecular beam epitaxy," *Appl. Phys. Lett.*, **43**, No. 6, 569-571 (1983).

61. J. M. Phillips, L. C. Feldman, J. M. Gibson, and M. L. McDonald, "Epitaxial growth of alkaline earth fluorides on semiconductors," *Thin Solid Films*, **107**, 217-226 (1983).

62. J. M. Phillips and J. M. Gibson, "The growth and characterization of epitaxial fluoride films on semiconductors," *Mater. Res. Soc. Symp.*, **25**, 381-391 (1984).

63. S. L. Wright, M. Inada, and H. Kroemer, "Polar-on-nonpolar epitaxy: sublattice ordering in the nucleation and growth of GaP on Si(211)," *Sci. Technol.*, **21**, 534 (1982).

64. S. L. Wright, H. Kroemer, and M. Inada, "Molecular beam epitaxial growth of GaP on Si," *J. Appl. Phys.*, **55**, No. 8, 2916-2927 (1984).

65. P. W. Sullivan, J. E. Bower, and G. M. Metre, "Growth of semiconductor/insulator structures-GaAs/fluoride/GaAs(001)," *J. Vac. Sci. Technol.*, **3**, No. 2, 500-507 (1984).

66. T. Asano, H. Tsiwara, H. Lee, et al., "Formation of GaAs on insulation structures on Si substrates by heteroepitaxial growth of CaF$_2$ and GaAs," *Jpn. J. Appl. Phys.*, **25**, No. 2, L139-L141 (1986).

67. S. Siskos, C. Fontaine, and A. Munoz-Vaque, "GaAs/(Ca, Sr)F$_2$/(001)GaAs lattice-matched structures grown by molecular beam epitaxy," *Appl. Phys. Lett.*, **44**, No. 12, 1146-1148 (1984).

68. H. Ogata, S. Maruno, Y. Morishita, and T. Isu, "MBE of InP using low energy P$^+$ ion beam," in: IVth Int. Conf. MBE, 7-10 Sept., Univ. of York (England), York (1986), G-10.

69. W. T. Tsang, "Chemical beam epitaxy of InP GaAs," *Appl. Phys. Lett.*, **45**, No. 2, 1234-1236 (1984).

RELATION BETWEEN GROWTH CONDITIONS, STRUCTURE, AND CATHODOLUMINESCENCE OF EPITAXIAL LAYERS OF ZnSe ON GaAs

V. G. Galstyan, M. I. Deigen, V. I. Muratova, S. A. Semiletov,*
V. Ch. Stankevich, and A. A. Tikhonova

Zinc selenide possesses a number of valuable physical properties: high photosensitivity, electrooptical effect, piezo effect, and high refractive index [1]. Due to a wide forbidden band (2.8 eV at 4.2 K), ZnSe emits in the blue spectral region which makes it a promising material for use in light-emitting devices. However, the physical properties of ZnSe are exceedingly sensitive to variations of chemical composition and to the actual crystal structure of the specimens. The type of faults which arise in the crystal and their concentration and distribution are determined by the growth mechanisms, incorporation and diffusion of dopants, and by the temperature. Therefore, they depend intricately on the method and conditions of crystal growth [2].

We studied epitaxial films of ZnSe, prepared by condensation of the vapor on GaAs substrates in a quasisealed container with hot walls. Gallium arsenide was chosen as substrate because of its well-developed production technology and the similarity of its chemical bonding and lattice periods (5.667 Å for ZnSe and 5.654 Å for GaAs). With respect to the method of film preparation, the advantage of physical deposition from the vapor phase in comparison with other methods is the possibility of keeping the substrate at a relatively low temperature [3]. This consideration is very important during epitaxy of A^2B^6 compounds because of their propensity toward formation of native and dopant point defects and their various complexes at high temperatures.

The growth kinetics of ZnSe layers in a quasisealed container with hot walls on GaAs (001), (110), (111) A, and (111) B substrates have been studied earlier [4] over a broad range of deposition parameter changes. The growth rate as a function of substrate temperature $v(T_S)$ at various incident flux densities was measured. A difference in growth mechanisms for layers at various orientations was shown. A conclusion on the kinetic limitation of the growth rate was reached.

The present work studies the effect of formation conditions of the layers on their actual structure and cathodoluminescence.

---

*Deceased.

1. EXPERIMENTAL

Growth of epitaxial layers of ZnSe was carried out in a quasisealed container (QSC) with hot walls. The cell in which the growth occurred was placed inside a vacuum chamber with a residual gas pressure of 10^{-6} mm. The ZnSe layers were grown on single-crystalline GaAs substrates oriented parallel to the (001), (110), and (111) planes. The substrates were polished chemically with subsequent annealing in vacuum before deposition.

The region of deposition conditions in which investigations were conducted included substrate temperatures T_S = 250-760°C, vaporization temperatures T_V = 600-850°C, layer thicknesses 0.1-200 μm, and growth rates v = 0.1-300 Å/sec. The ZnSe deposited in the cubic sphalerite modification over the whole studied region. More detailed characteristics of the quasisealed container and the sample preparation method are described in [4].

The principal method for study of the actual structure of the grown layers was reflection high energy electron diffraction (RHEED). Supplementary methods included x-ray diffractometry, transmission electron microscopy, and electron probe microanalysis. The overall structural perfection of the layers, which were prepared under various conditions, was evaluated from the width of the oscillation curve taken on a two-crystal x-ray diffractometer. Electron microscopic studies of ZnSe films that were separated from the substrate were performed on a JEM-100C electron microscope.

Investigation of the luminescence properties of the layers was carried out on a JSM-2 scanning electron microscope at room temperature and at liquid nitrogen temperature. Cathodoluminescence (CL) spectra of layers grown under various deposition conditions and the distribution of spectra and CL intensities along the thickness and surface of the layers were studied.

2. RESULTS

2.1. Structure of ZnSe Layers on GaAs (001)

The most characteristic faults in layers grown on substrates with (001) orientation were twinned layers and packing faults. The RHEED method revealed that the degree of twinning in such films is determined by the relationship between T_S and the deposition rate [5]. At low growth rates (v < 60 Å/sec), ideal layers were prepared only at relatively low substrate temperatures (T_S < 300°C). The electron diffraction patterns of these for various recording axes did not contain additional reflections. With an increase of incident flux density (growth rate), the temperature range at which ideal films grew was broadened. At v > 100 Å/sec, layers that do not contain twins (with accuracy up to the resolution of the electron diffraction method) can be prepared over the whole T_S range studied.

Twinning occurs in the sphalerite crystal lattice over the four dense-packed {111} planes which consist of two sublayers of different atoms with a collective shift of the atoms of one sublayer into the twinned position. Twinned reflections should appear in the two mutually perpendicular ⟨110⟩ recording axes in the RHEED patterns, as follows from the arrangement of the reciprocal lattice for the FCC crystals with twins [6]. However, twinned reflections are systematically observed only for one [110] axis in the studied films. They are absent from the other. This indicates that twinning in the (001) films occurs primarily in two {111} planes, in each case of film thickness greater than 0.5 μm.

In order to determine the types of atoms whose shift leads to twinning, selective etching was used. In view of the excessively high defect density, the GaAs substrates, and not the ZnSe films deposited on them, underwent selective etching. Initially, the direction, parallel to which the twinned layers were laid, was determined by electron diffraction on the films that were grown. Then, the film was etched in hydrochloric acid and the substrate surface was etched in the selective etchant H_2O_2:HF:H_2O = 1:1:4. The etching rate in this mixture is much faster for the (111) B face than for (111) A [7]. Therefore, the etching was bounded by the

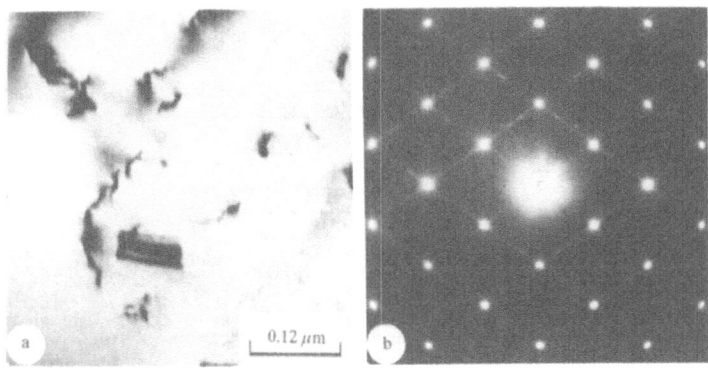

Fig. 1. Photomicrographs of ZnSe film with (001) orientation at T_S = 370°C and v > 100 Å/sec (a) and electron diffraction pattern obtained from the same film [reciprocal lattice plane (110)] (b).

(111) A faces and drawn out in the perpendicular direction. Grooves extended in the direction parallel to the stratification plane of the twinned layers in the ZnSe film were observed after etching on the substrate surface from the film side (on the opposite side of the substrate, these same etching marks were rotated by 90°). Direct resolution of atomic planes at the film–substrate interface, obtained by high-resolution electron microscopy, showed that the film planes are a direct extension of the substrate planes. This suggests that the twinned layers in the ZnSe film are laid down parallel to the {111} planes in which the upper sublayer is formed by Zn atoms and, consequently, the twinning occurs as a result of a shift of the Zn atoms into the twinned position during growth.

The density of two-dimensional faults and dislocations was estimated based on electron microscopic studies of the films with amplitude contrast transmission. Figure 1a shows photomicrographs of the film deposited at T_S = 370°C and v > 100 Å/sec. Two-dimensional faults are present in the film. These give banded contrast. Dislocations and accumulations of them are also seen. The total defect density is 10^8-10^9 cm^{-2}, whereas the two-dimensional defect density is 10^6-10^7 cm^{-2}. Figure 1b shows the transmission electron diffraction pattern from the same sample but with rotation of it relative to the beam. Besides the matrix reflections whose location corresponds to the (110) plane of the reciprocal lattice of the FCC crystal, Kikuchi lines lying in the [111] directions are present in the pattern. A slight thickening of the Kikuchi lines occurs in the twinned reflections. The formation of Kikuchi lines is caused by two-dimensional diffraction from the thin layers that lie in the {111} planes of the sample perpendicular to the diffraction plane. These are packing faults or twins (which is indicated by the thickening on the Kikuchi lines). It should be noted that twinned reflections are absent in the RHEED patterns and diffuse Kikuchi lines are observed. This indicates that the portions with twinned orientation occupy only a small part of the surface.

The overall perfection of the layers grown was estimated by x-ray diffractometry. Figure 2a shows the change of oscillation curves taken for the (004) reflection as a function of T_S for films with (001) orientation. The curves were measured using a two-crystal dispersionless setup with a Ge (004) monochromator. The films were grown at a rate v > 100 Å/sec and had a thickness ~10 μm. As Fig. 2 shows, the width of the oscillation curve depends on the substrate temperature. The curve with minimal width of 2-3' is obtained for the film grown at T_S = 350°C. An increase of deposition temperature leads to a broadening of the oscillation curve. At T_S > 600°C, their width reaches 20'. However, the electron diffraction study showed that twinned layers are absent in these films. The electron microscopy does not reveal a substantial difference in the defect density in layers grown at high and low substrate temperatures. Therefore, it is concluded that the broadening of the oscillation curves from films grown at high T_S is related to an increase in deformation of the crystal lattice of the films.

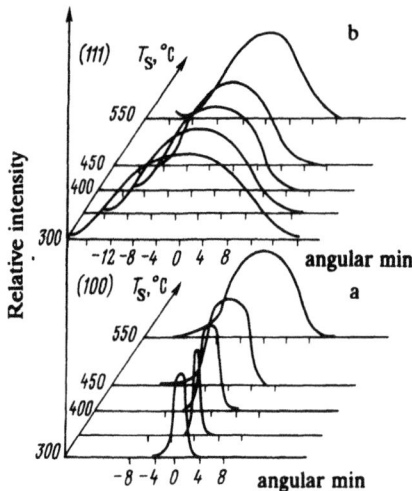

Fig. 2. X-ray oscillation curves of ZnSe layers deposited at various T_S. Cu Kα_1 radiation. a) (001) substrate orientation, (004) reflection; b) (111) A substrate orientation, (444) reflection.

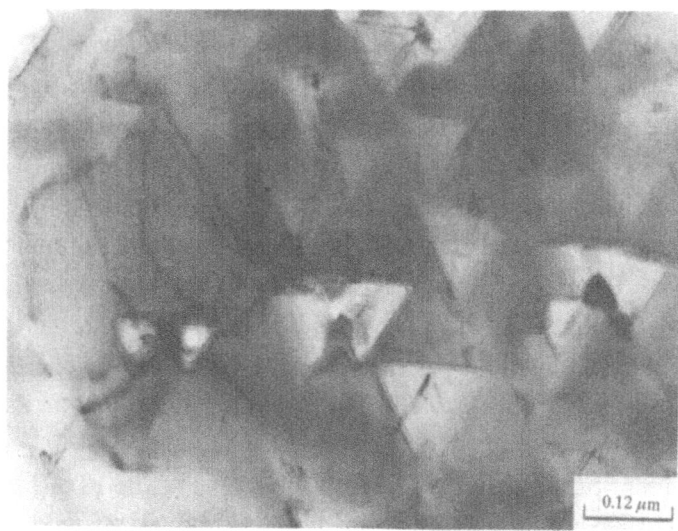

Fig. 3. Microphotographs of ZnSe films with (111) A orientation at $T_S = 500°C$.

2.2. Structure of ZnSe Layers on GaAs {111}

The structural perfection of layers with the {111} orientation is significantly lower than that of layers with the (001) orientation. Electron diffraction showed that twins are formed for all preparation conditions in the {111} films. The electron diffraction patterns from films grown at T_S below 320-350°C contain nets of reflections of equal intensity from the main orientation {111} and twins parallel to the substrate surface (surface growth). Thus, it appears that twinning under such conditions occurs readily. Furthermore, the layer structure at low T_S depends greatly on the deposition rate. Its increase leads to formation of texture along the [111] axis perpendicular to the surface instead of a mosaic crystal with (111) orientation.

The number of {111} planes along which twinning occurs increases with increase in T_S. This change correlates with the nature of the kinetic function v (T_S) [4]. Additional reflections from twins along the in-

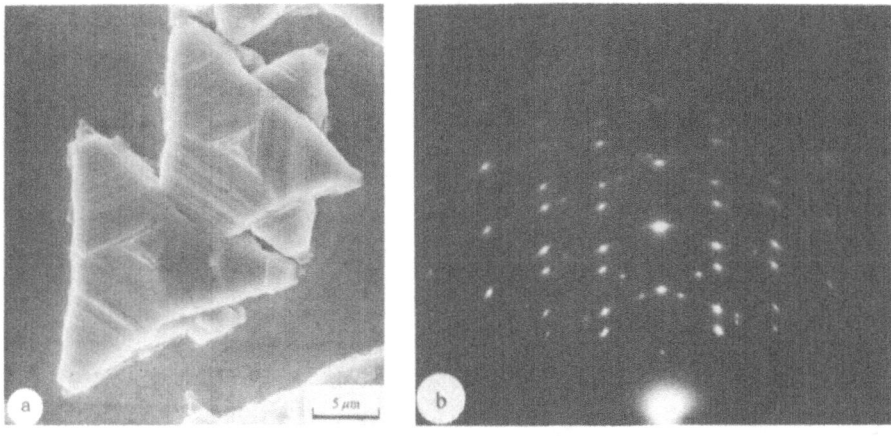

Fig. 4. Microphotographs of the growth surface (111) B at t_S = 660°C (a) and electron diffraction pattern of the same film along the [$\bar{1}$10] axis (b).

clined {111} planes appear in the electron diffraction patterns in the portion with T_S > 350°C, where the growth rate decreases with an increase of T_S and films with a highly coarse surface grow. It should be mentioned that up to T_S = 400-450°C, the structure of layers prepared on the substrates with (111) A and (111) B orientations is identical. The layer structures improved and reflections only from twins parallel to the growth plane were present in the electron diffraction patterns for films with (111) A orientation on the rising part of the $v(T_S)$ function, i.e., upon transition to a primarily layered growth mechanism. Intensities of main and twinned reflections remained identical in this case. The reflections were sharp and weak streaks appeared in the electron diffraction patterns. The existence of twins in the growth plane is confirmed by electron microscopy of such films (Fig. 3). Triangular shapes formed by packing faults or microtwinned layers that lie parallel to the film surface are observed in the microphotographs. They occupy various axial positions and their sides are elongated along the ⟨100⟩ direction. The change in contrast of the triangles relative to the background is explained by the various doping levels and the thickness of the fault planes that correspond to them. Basically, these are thin layers, but some faults extend through the whole film thickness. This is indicated by the contrast bands at the inclined interfaces that define them. The concentration of these layers is ~10^{15} cm^{-3}. Their linear dimensions are from 50 to 200 nm.

In contrast to the (111) A orientation, twins along the inclined {111} planes are formed throughout the whole range of growth conditions in the films grown on (111) B substrates. At high T_S, when separate tripyramid shapes (Fig. 4a) are resolved on the smooth portions of the growth surface, the reflections in the RHEED patterns from twins that lie parallel to the growth plane and from inclined twins have various shapes and intensities (Fig. 4b). The intensity of reflections from the parallel twins, which are arranged in rows parallel to the axis of the diffraction pattern, are almost the same as the matrix reflections. This implies a high density of such twins. In turn, reflections from inclined twins have a sharper shape. This suggests that they are formed by diffraction of the electron beam from protruding parts of the surface, i.e., the apices of the tripyramids.

The overall perfection of the structure of layers with {111} orientation, as in the case of (001), was evaluated by x-ray diffractometry. Figure 2b shows the change of the x-ray oscillation curves for layers with (111) A orientation grown at various T_S and at growth rates that correspond to a source temperature of 660°C. The curves were taken for the (444) reflection. The curves are seen from Fig. 2 to be much broader for layers with the {111} orientation than for (001). Their width exceeds 20′. The large width of the oscillation curves and their weak dependence on T_S is explained by the high concentration of faults that deform the crystal lattice. This is related to the growth mechanism of the (111) A face.

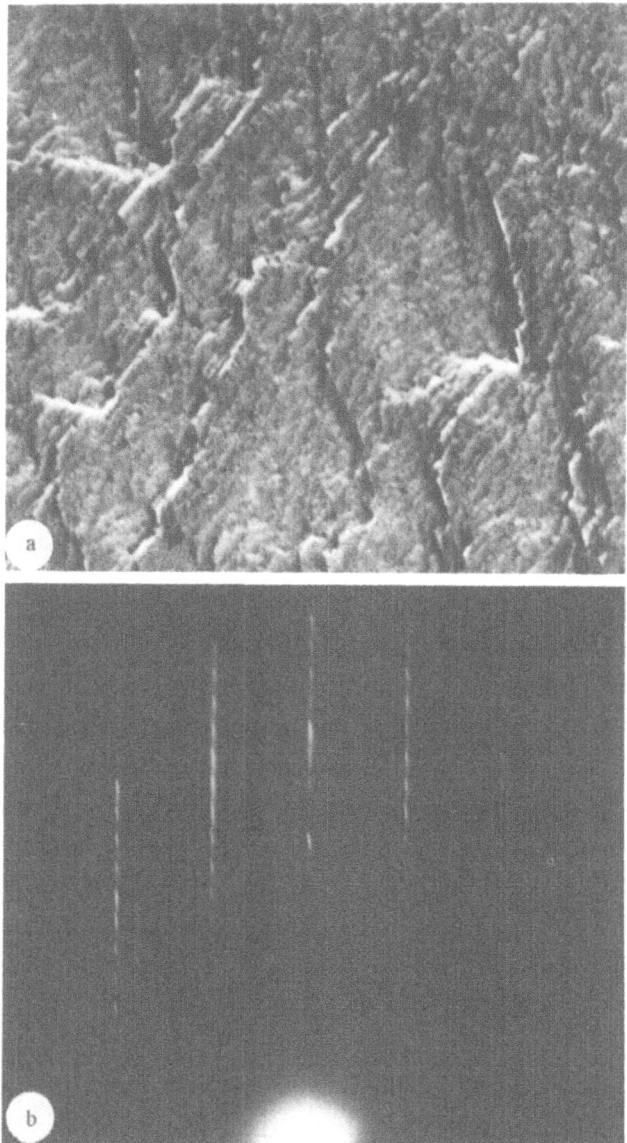

Fig. 5. Microphotograph of the growth surface of a film prepared at $T_S = 660°C$ on a substrate inclined from (111) A by 5° to the (110) side (a) and the electron diffraction pattern of the same film along the [1$\bar{1}$0] axis.

The change of actual structure of the ZnSe layers with deviation of the substrate surface from the exact (111) orientation was studied on specimens prepared by deposition on spherical substrates at $T_S = 600$-680°C. Point reflections from the matrix and twins along all {111} planes are present in the electron diffraction pattern taken in the axial (111) B direction. The density of growth features decreased with a deviation from the exact orientation by an angle of 2-3°. A net only from the matrix reflections, as well as the Kikuchi lines, remained in the patterns. The reflections were narrow and elongated normal to the (111) face due to the development of smooth portions on the surface. Further deviation from the exact orientation led only to insignificant smearing of the reflections parallel to the face. This is related to refraction of the electron beam at the ends of the macrosteps that are formed on the vicinal growth surface. Thus only substrates misaligned less than 3° along the (111) surface can be used so that the fault density in layers with an orientation near (111) B can be reduced.

The structure of layers grown on spheres with the (111) A axis is similar to that described above only in portions inclined from the sphere axis by greater than 7-8°. Fault formation near the (111) A and (111) B axes occurs differently for smaller misalignment angles. Electron diffraction patterns from film portions with the exact (111) A orientation contain few smeared, elongated reflections of identical intensity from the principal and twinned parallel orientations. The growth surface consists of a system of macrosteps (Fig. 5a) on the sphere portions inclined from the close-packed (111) A face by angles of 3-7° in the $[11\bar{2}]$ direction. Additional reflections arise in the diffraction patterns with a shift of the electron beam in the $[1\bar{1}0]$ direction (along the steps). The intensity of these reflections increases with deviation of the surface from (111) with simultaneous extinction of the matrix and twinned reflections. The maximum intensity of the additional reflections is observed for a surface deviation by 4-5°. The electron diffraction pattern in this case is a net of reflections of approximately equal intensity with a periodicity equal to 1/4(111) in the direction normal to the (111) surface and with a periodicity 1/3(224) in the [112] direction (Fig. 5b). The twinned reflections are absent and only (111), (222), (333), (442), and (600)-type reflections remain from those of the matrix. These coincide with the additional reflections. The diffraction pattern cannot be indexed as a singular twinning, as formation of a wurtzite phase, or by assuming twinned diffraction since in all these cases the lattice translation vector is 1/6-fold and not 1/4. The presence on the surface of a system of regular steps which can give superstructural reflections in the diffraction patterns should not lead to disappearance of the principal reflections. The pattern obtained can be indexed as the 4H polymorphic modification of the ZnSe lattice. It has the sphalerite structure with periodic injection of a plane. As a result, a periodic alternation of layers of the principal and twinned orientations with a four-layer repeat period occurs in the crystal: ABCBABCB... Thus, the 4H polymorphic modification is formed by a layered growth mechanism of ZnSe film on the (111) A GaAs face with a definite step configuration.

2.3. Structure of ZnSe Layers on GaAs(110)

The structural ideality of layers with the (110) orientation was determined by the substrate temperature. At $T_S < 400°C$, the growth of films was accompanied by extensive twinning. The growth rate of twinned layers is higher than the growth rate of the principal matrix. Only twinned reflections are present in the RHEED patterns taken from films of more than 10 μm thickness. The defectiveness of the layer structures with the (110) orientation decreases with an increase of T_S. At $T_S > 500°C$, only reflections from the main orientation are found in the patterns.

2.4. Cathodoluminescence from the Layer Growth Surfaces

Cathodoluminescence (CL) of the layers that were grown revealed a correlation between the structural perfection and the luminescence spectra. Figure 6a shows the CL spectra of ZnSe films grown on GaAs(001) substrates at growth rates greater than 100 Å/sec and substrate temperatures from 250 to 650°C. The spectra were obtained at room temperature and an electron beam energy of 25 keV. Only one emission band with a wavelength of 467 nm and a half-width of ~2 nm is present in the spectra of layers grown at $T_S < 450°C$. The photon energy with the same wavelength is similar to the forbidden band width of ZnSe and corresponds to edge emission. A wide band in the long-wavelength region with a maximum at 610-640 nm appears in the CL spectra with an increase of T_S whereas the intensity of the 467 nm band decreases. At $T_S = 600°C$, the maximal intensity of the red line exceeds that of the blue emission by about 3 times. The appearance of emission in the red spectral region can be assigned to self-activated luminescence (SA) caused by emissive donor—acceptor transitions [8].

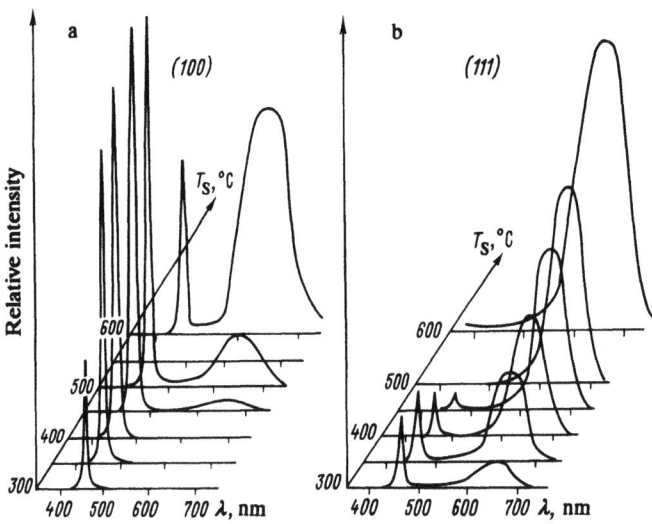

Fig. 6. CL spectra as functions of substrate temperature T_S for the orientations
(001) (a) and (111) A (b).

Figure 6b shows the CL spectra for layers grown on substrates with (111) orientation as a function of T_S. The emission in the blue spectral region is observed only at $T_S < 450°C$. The intensity is much lower than in films with the (001) orientation. At high T_S, only a broad band in the red region remains in the CL spectra.

The CL intensity is distributed unevenly on the growth surface. Local variations are observed near the separate surface relief elements. The characteristic microrelief elements of layers with the (001) orientation are traces of two-dimensional faults appearing as thin intermittent pits elongated in the [110] direction along the twinning planes. Comparison of secondary electron microphotographs and CL in films deposited at $T_S < 450°C$ shows that the sites of fault plane pitting on the growth surface correspond to portions of lowered luminescence intensity. The width of these portions is a fraction of a micron. The presence of a fault lowers the quantum yield of edge luminescence that arises in the low-temperature films. The fault density, as revealed by CL, is higher than the density of features on the growth surface but agrees with the value obtained by transmission electron microscopy (Fig. 1).

Conversely, in films prepared at high T_S that emit in the red spectral region, the portions that belong to the faults give elevated luminsecence intensity. Figure 7 shows microphotographs of the film surface in the (001) orientation that was deposited at $T_S = 600°C$, taken using reflected electrons (Fig. 7a) and CL (Fig. 7b), and the RHEED pattern obtained along the axis parallel to the direction of the elongated pits (Fig. 7c). Comparing the distribution of luminescence intensities with the surface relief, we note that the most brilliantly luminescing portions accompany the longest elongated pits. The regions of elevated luminescence are located to one side of the corresponding pits and are extended to a distance up to ~10 μm from the fault. The luminescence intensity is lowered directly at the sites of the surface faults, as for the low-temperature films. The simple geometric structure suggests that the extent of the longest pits corresponds to the cross section of the film surface with the fault planes situated in the inclined {111} planes and which have their origin in the heterojunction. Thus, faults, around which the brilliantly emitting regions are formed, arise at the interface of the film with the GaAs substrate. The local increase of CL intensity in the red spectral region at high T_S can be assumed to be a result of the local increase of dopant content, most likely Ga, which diffuses with an increased rate along the fault and in the region of the crystal adjoining the fault. An analogous picture was

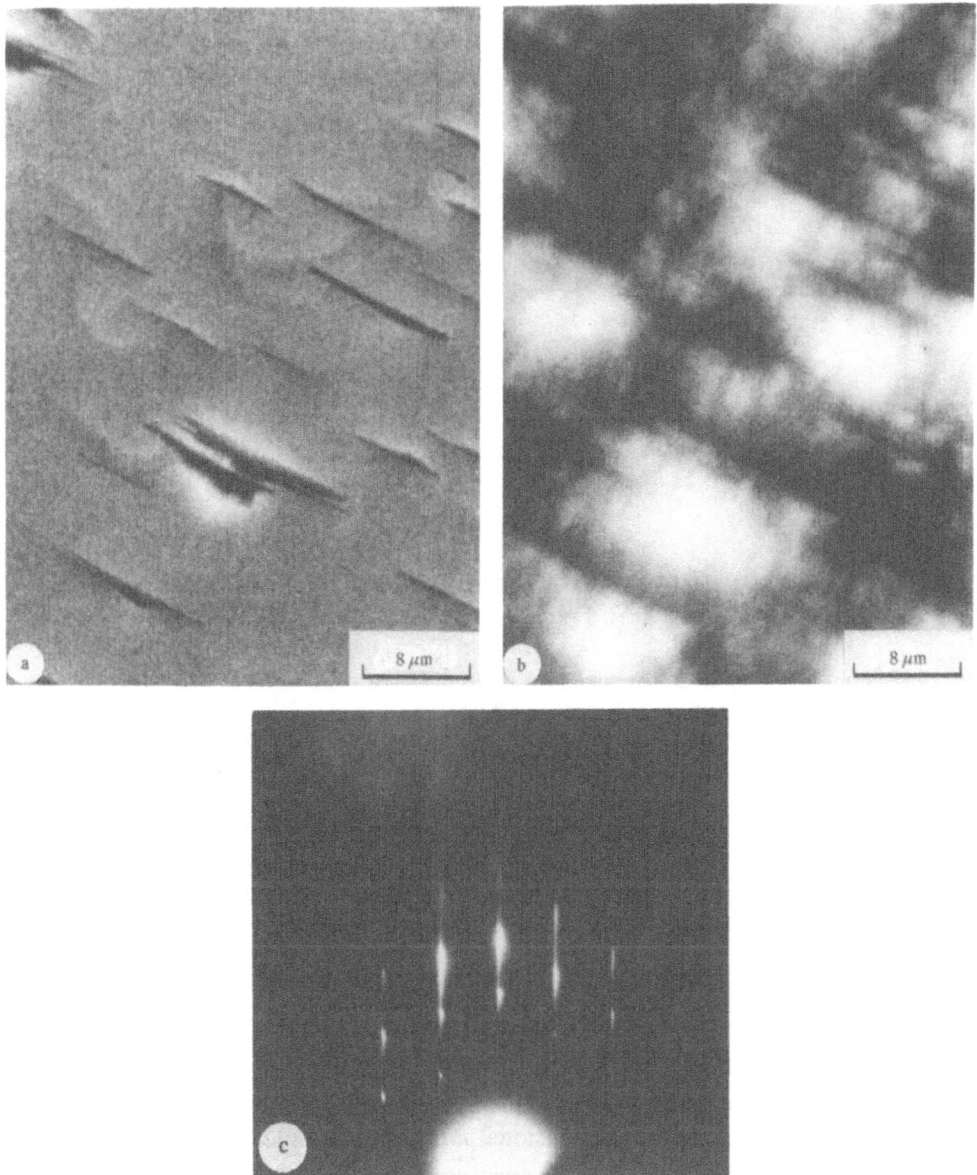

Fig. 7. Microphotographs of growth surface of a film (001) at $T_S = 600°C$ using reflected electrons (a and b) and the RHEED pattern from the same film (c).

observed in films with the (111) orientation. Figure 8 shows the micromorphology of the growth surface and the CL distribution on the film surface grown at $T_S = 600°C$ on a substrate with (111) A orientation. The maximal luminescence intensity is observed for the growth features. It significantly exceeds the CL intensity of the planar portions.

Electron probe microanalysis of the growth surface of layers was carried out for substrate elements. At high T_S, weak lines from Ga and, at a few surface points, from As were present in the spectra of samples besides the characteristic peaks from the principal elements Zn and Se. The Ga distribution on the growth surface was nonuniform. Its concentration systematically increased in the vicinity of the growth features of the type shown in Fig. 8. This confirms the assumption that the elevated CL intensity in the red spectral region is related to an increased concentration of Ga dopant.

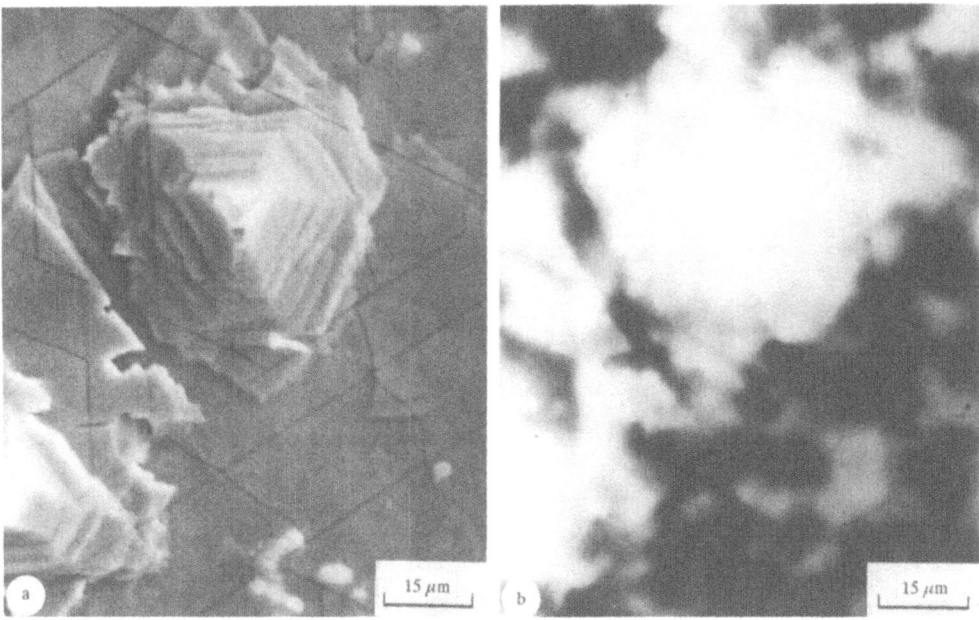

Fig. 8. Microphotographs of a pyramidal growth on the (111) A surface at T_S = 700°C using reflected electrons (a) and CL (b).

2.5. Distribution of Structural Faults and Cathodoluminescence through the Thickness of the Deposited Layer

Electron microscopic investigation on transverse sections of the film—substrate system revealed a non-uniform distribution of faults through the thickness of the film. Figure 9a shows a photomicrograph of the portion that adjoins the heterojunction in the sample with (001) orientation prepared at T_S = 370°C and a rate of ~160 Å/sec. The photograph shows a sharp boundary between the film, which contains faults, and the practically fault-free substrate. Inclined two-dimensional faults that completely traverse the film and dislocations are present. An elevated fault concentration is observed near the heterojunctions. A large part of the faults is localized in the layer of 0.1-0.2 μm thickness next to the junction. The fault concentration decreases with distance from the boundary region. It should be mentioned that the distinct break of extinction contours near the faults indicates the existence of large elastic stresses in the film regions that border the faults.

The CL distribution was also studied for the transverse sectons. Figure 9b shows the photomicrograph of the same section as in Fig. 9a but taken at lower magnification by CL. Luminescence from the section, just as from the growth surface, is concentrated in the blue spectral region. The luminescence intensity through the film thickness is nonuniformly distributed. It is lower near the film—substrate heterojunction than in the bulk film. Local regions with lower luminescence intensity are also encountered in the bulk film. Besides this, separate dark bands at an angle of ~60° to the substrate surface pass through the whole film. Slanted pits of ~1 μm or macrosteps are observed at the exit sites of these on the growth surface. The structural faults are assumed to quench the edge luminescence. The lowered intensity of blue CL near the heterojunction in the low-temperature films with (001) orientation is also caused by this.

A brilliantly emitting red region near the interface is detected in the high-temperature (T_S > 450°C) film sections with (001) orientation using CL. The spectral maximum of this is at 610 nm. The width of the elevated brilliancy region increases with an increase of T_S. It extends not only into the film but also into the substrate. This can be seen in the photomicrograph in Fig. 10, where the left part of the section is taken using

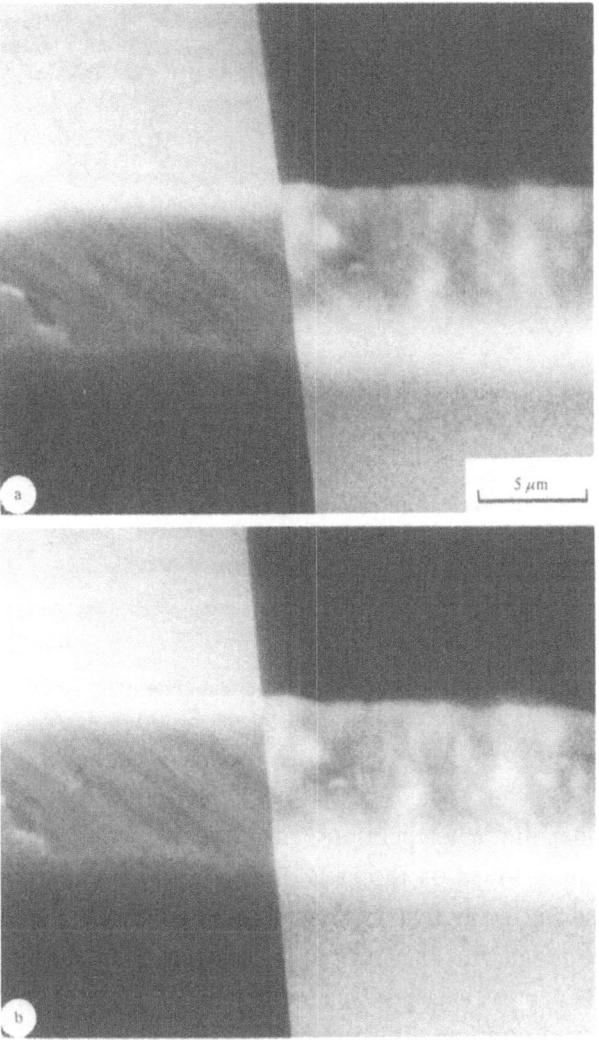

Fig. 9. Microphotographs of the transverse cleavage of film and (001) substrate at $T_S = 370°C$ using transmission (a) and CL (b).

reflected electrons and the right using CL. The position of the interface in both parts of the photomicrograph does not coincide since the region of CL intensity is shifted into the substrate.

Electron probe microanalysis of the sections established the elemental distribution of the film and substrate through the thickness. For samples prepared at low T_S, the distribution of Zn, Se, Ga, and As is characterized by sharp steps in the heterojunction region, and mutual diffusion of elements does not occur (within accuracy limits). If $T_S = 600°C$, the Ga distribution curve is drawn out in the film region to a thickness up to $2\ \mu m$. Diffusion of Zn into substrate also occurs. Comparison of microanalytical data with the CL of sections shows that the depth of penetration into the film corresponds approximately to the width of the region with increased red luminescence. Thus, the appearance of the red band in the region next to the junction is caused by diffusion of Ga atoms into the growing layer and of Zn atoms into the substrate. Apparently, these atoms enter into the composition of the CA complexes.

3. CONCLUSIONS

The comprehensive data on the actual structure and the CL characteristics of epitaxial ZnSe films as functions of the growth conditions showed that the degree of influence of the growth conditions decreases in the

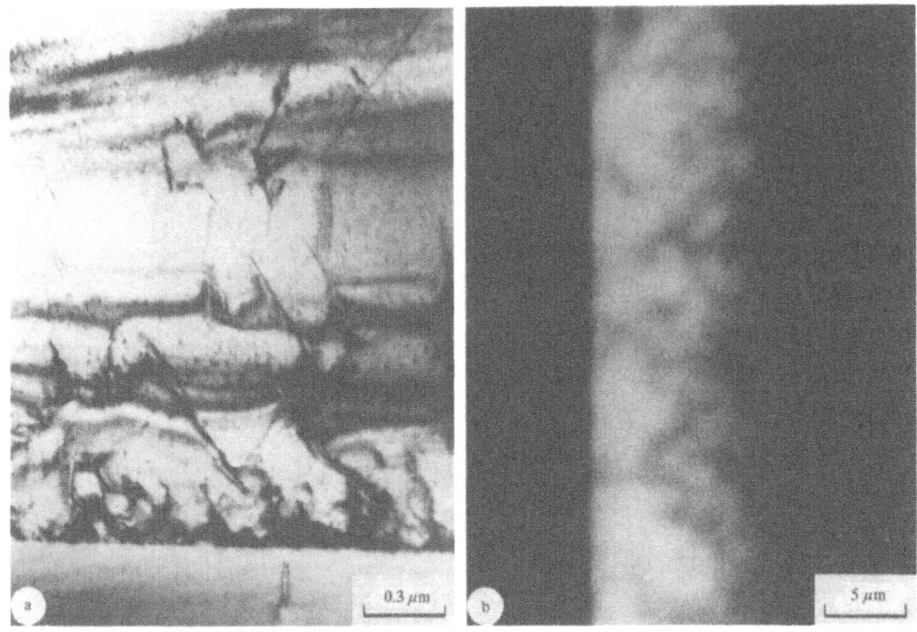

Fig. 10. Depiction of the transverse cleavage of the film–substrate (001) system at T_S = 600°C using reflected electrons (a) and CL (b).

order: substrate orientation, deposition temperature, and deposition rate [10]. Primarily, this relates to distribution of two-dimensional faults, found on the basis of electron diffraction data. The two-dimensional faults are microtwinned layers or packing faults that arise as a result of disruption of the ordering of atomic plane (111) alternation in the sphalerite lattice. As shown in [9], formation of twins or faults with neutral boundaries is common for ZnSe. It occurs due to shifting of the atoms of the upper sublayer of the (111) plane relative to the crystal matrix into the twinned position. The frequency of appearance of a twin is determined by the energy of its formation and by the conditions on the growth surface.

Definite fault stratification planes relative to the growth surface are characteristic for films of various orientations. For (001), they lie in two inclined {111} planes in which the upper sublayer is formed by Zn atoms. For {111}, the position of faults depends further on T_S. The faults lie parallel to the growth surface for (111) B at T_S < 320-350°C. At larger T_S, they are along the {111} planes. For (111) A, still another transition at $T_S \approx$ 450°C occurs above which faults primarily parallel to the growth surface are retained. The structural changes correlate well with the growth rate kinetics as functions of T_S for substrates of various orientations [4]. This seems natural since in both cases the basis of the observed features is the growth mechanism of the layers.

Sublimation of ZnSe, as for other compounds of the A^2B^6 group, is accompanied by its complete dissociation. The metallic component is found in the atomic state in the vapor, whereas the halogen is in the molecular state. Therefore, the fundamental growth process upon crystallization from the vapor should include adsorption of Se_x molecules, bond rupture in this molecule, and formation of the Zn—Se bond. In [10], formation of ZnSe was stated to occur significantly faster based on kinetic and morphological data if two uncompensated bonds per atom exist at the sites of incorporation of Se atoms into the crystal lattice. This condition is realized during growth of the (001) face, which has the largest rate and a unique growth mechanism over the whole T_S range. On the polar {111} faces, the surface atoms of which have one free bond, the kinetic and morphological changes in this same T_S range have a more complicated character. This can be explained by two competing growth mechanisms. The successive formation and growth of nuclei on the smooth portions of the surface, or primarily layered growth, can occur. Even at high saturations and T_S temperatures low in comparison with the melting point (~320°C), the normal growth mechanism is not realized. Instead a mechanism of successive nucleation is in effect. The undulating structure of the {111} surface growth with a nonuniformity dimension of ~100 nm on a microscopically smooth substrate attests to this.

Apparently, the nucleation mechanism reflects the principal feature of the growth of the {111} ZnSe face. Formation of ZnSe in the step-free surface portions in the adlayer, which consists of Zn atoms and Se_x molecules, will occur when activated redistribution of bonds in the aggregate of several adparticles having identical configuration is realized. As a result, from two to four bonds are formed in the two-dimensional nucleus for each Se atom regardless of the polarity of the face. The probability of nucleus formation in the matrix and the twinned orientation is approximately unity since the difference in energy of nucleation is determined only by the different mutual position of the second-order neighbors. The high concentration of twinned layers parallel to the growth surface in the layers with the {111} orientation over the whole T_S range can be explained by this.

With a decrease of saturation and an increase of T_S, the normal growth rate of {111} films falls and the development of twins along the inclined {111} planes and the growth features related to them is observed simultaneously. They have the shape of tripyramids (Fig. 4) on the (111) B face and a less clear shape on the (111) A face. The shapes are located at the apices of growth hillocks at high T_S. The question of the structure and the reasons for formation of complex faults and tripyramids has already been examined in the literature [11]. We can only add that both the appearance of inclined twins and their disappearance on the (111) A face is related to the definite T_S values at which the change of the nature of the kinetic function v (T_S) occurs.

In the present work, two methods of quantitative evaluation of the structural perfection of the films were applied. These were electron diffraction and x-ray diffractometry. These methods are sensitive to structural disruptions on different scales, on the order of nanometers for electron diffraction and micrometers for x rays. The data obtained by these methods complement each other as, for example, in the case of ZnSe films grown on GaAs(001) substrates at $T_S = 600°C$. For these, electron diffraction shows a relatively low concentration of microfaults whereas x-ray diffractometry confirms that an average lattice is practically absent in these films at distances on the order of 1 μm. As already noted, the growth mechanism for the (001) orientation remained constant over the whole T_S interval. The face structure is such that two free bonds arise at each atom that is incorporated into the lattice. Therefore, the probability for localization of an atom in the lattice in the correct position is higher than for growth from many centers with subsequent nucleation on the (111) A face.

At high T_S, the difference in the width of the oscillation curves for films of both orientations is contracted, suggesting that the main parameter that determines the lattice deformation is the deposition temperature. The thermodynamic equilibrium concentration of native point defects increases with an increase of T_S. These have a tendency to form dopant—defect complexes. We note that in the epitaxial ZnSe layers that are deposited on GaAs substrates from organometallic compounds, the optical quality of the layers is also determined by the deposition temperature at $T_S > 500°C$ [12].

A direct correlation between the structural ideality and the luminescence characteristics is found upon comparison of the CL spectra (Fig. 6) and the x-ray curves (Fig. 2) from films prepared at various T_S and substrate orientations. In layers with ideal structure, for which twinned reflections are absent in the electron diffraction patterns and the width of oscillation curves is 2-3', emission occurs only in the blue spectral region and is edge emission. Degradation of layer structure, which is accompanied by broadening of the x-ray oscillation curves up to 20' [for the (001) orientation at $T_S > 450°C$ and for (111) A at all T_S], leads to a decrease of edge emission intensity and an increase of deep center density. The recombination of these gives red emission with a maximum at 610-640 nm.

The broad luminescence band in the $\lambda = 610$ nm region is usually associated with self-activated emission. The emission model assumes that the transition occurs between a shallow donor, for example, a group III element dopant, and a deep acceptor, for example, a Zn vacancy [8]. The data obtained in our work on the heterojunction structure (by transmission electron microscopy) indicate, by the distribution of elements through the transverse cross section of the film and on the growth surface (by electron probe microanalysis), that the local increase of red CL intensity near the heterojunction is related to the increased Ga content. At $T_S =$ 600°C, diffusion of Ga into the growing layer and diffusion of Zn into the substrate occur. Thus, the position of the heterojunction is clouded. The rate and extent of diffusion depends on T_S and the duration of deposition. An increased Ga concentration and intensified red CL are observed not only near the heterojunction but also in regions that adjoin the structural faults. This signifies accelerated transfer of dopant from the substrate along the faults that pass through the entire film thickness.

In conclusion, it should be noted that the propensity of ZnSe to formation of self-activated centers creates substantial difficulties in the preparation of layers that emit only in the blue spectral region. The concentration of point defects that are responsible for the decline of luminescence characteristics is determined not only by the deposition temperature but also by the layer growth mechanism. The studies carried out allow the optimal conditions for growth of ZnSe layers in a quasisealed container to be determined. The most ideal layers are prepared on substrates of (001) orientation at $T_S =$ 350-400°C and growth rates above 100 Å/sec. Such layers are characterized by a low packing fault concentration, a width of x-ray diffraction of 2-3', and an intense edge emission without a red band in the CL spectra.

REFERENCES

1.   A. N. Georgobiani, "Wide band $A^{II}B^{VI}$ semiconductors and prospects for their use," *Usp. Fiz. Nauk*, 113, 129-155 (1974).

2.   I. P. Kalinkin, B. V. Aleskovskii, and A. V. Simashkevich, *Epitaxial Films of $A^{II}B^{VI}$ Compounds* [in Russian], Leningrad State Univ. (1978).

3.   T. Yao and S. Maekava, "Molecular beam epitaxy of zinc chalcogenides," *J. Cryst. Growth*, 53, 423-431 (1981).

4.   S. A. Semiletov, V. Ch. Stankevich, and A. A. Tikhonova, "Heteroepitaxy of ZnSe on GaAs," *Izv. Akad. Nauk SSSR, Ser. Fiz.*, 50, No. 3, 505-508 (1986).

5.   S. A. Semiletov, V. Ch. Stankevich, and A. A. Tikhonova, "Electron diffraction study of epitaxial ZnSe films," *Izv. Akad. Nauk SSSR, Ser. Fiz.*, 48, No. 9, 1744-1747 (1984).

6.   D. W. Pushley and M. J. Stowell, "Electron microscopy and diffraction of twinned structures in evaporated films of gold," *Philos. Mag.*, 8, 1605-1632 (1963).

7.   Y. Tarui, Y. Komiga, and Y. Harada, "Preferential etching and etched profile of GaAs," *J. Electrochem. Soc.*, 118, 48 (1971).

8.   T. Yao, M. Ogura, P. Natsuoka, et al., "Electrical and photoluminescence properties of ZnSe thin films grown by molecular beam epitaxy: substrate temperature effect," *Jpn. J. Appl. Phys.*, 22, No. 3, L144 (1983).

9.   J. O. Williams, T. L. Ng, A. C. Wright, et al., "High resolution transmission and analytical electron microscopy for characterization of epilayers and interfaces in MOVPE growth of $ZnSe_{1-y}S_y$," *J. Cryst. Growth*, 72, No. 1/2, 155-161 (1985).

10.   V. Ch. Stankevich, "Features of growth and structure of epitaxial ZnSe layers on GaAs prepared in a quasisealed container," Candidate's Dissertation in Physical–Mathematical Sciences, Moscow (1986).

11.   *Faults in Crystalline Semiconductors*: A Collection of Articles [Russian translation], Mir, Moscow (1969).

12.   B. Cockayne, P. J. Wright, M. S. Scolnick, et al., "The growth by MOCVD using new group VI sources and assessment by HRTEM and CL of Zn-based II–VI single crystal layers," *J. Cryst. Growth*, 12, No. 1/2, 17-22 (1985).

KINETICS OF GALLIUM ARSENIDE DOPING
IN GAS-PHASE EPITAXY

L. G. Lavrent'eva

INTRODUCTION

Multilayered structures prepared by gas-phase epitaxy (GPE) are used in the fabrication of devices and microcircuits based on gallium arsenide. Control of the doping level is one of the commercially important problems in materials science. Since the introduction of gas-phase epitaxy of GaAs, many experimental results on the problem of doping group II, IV, and VI elements, as well as transition metals, have been published. Brief reviews of these efforts can be found in [1-3]. Significantly less attention has been paid to the mechanism of the process. The following tendencies in modelling of the doping process can be identified: description of the equilibrium and trapping of impurity limited by external diffusion, surface diffusion, and adsorption (including the shift of the Fermi level on the surface). A brief review of the model is given in [3, 4]. At present, surface kinetics is considered the principal step limiting crystallization and doping during GPE of GaAs. In this case, the rate should depend on the surface structure. Analysis of this function allows information on the micromechanisms of the process to be obtained.

In the present article, the relationship of the doping kinetics to the surface structure is examined. The basic principles of the method used for analysis of results were stated earlier in [4-6].

In order to unify the description of the doping kinetics and the deposition kinetics, we introduce the impurity trapping rate j_i, a quantity analogous in physical meaning to the deposition rate. To this end, we write the flux densities of the principal j_0 and dopant j_i atoms from the gas phase into the solid using the experimentally measured values V, the layer deposition rate, and N_i, the impurity concentration in the layer:

$$j_0 = V\Omega, \quad j_i = N_i V. \tag{1}$$

Here Ω is the molecular volume for the crystalline substance. It is assumed that $j_0 \gg j_i$. From (1), j_0 can be seen to differ from V only by a constant factor. With respect to the dopant, use of N_i as a kinetic parameter is possible only at a constant deposition rate, which is not valid for GPE of GaAs.

The basic features will be examined for the example of GaAs layers doped with tellurium. Data for layers doped with zinc, tin, and sulfur will be deduced.

Epitaxial layers of GaAs were grown in two systems, Ga–AsCl$_3$–H$_2$ and GaAs–AsCl$_3$–H$_2$. In the first system, the dopant in a mixture with hydrogen was supplied as a gas stream through an independent channel (for example, SF$_6$) or was dissolved in the gallium source (for example, Sn). In the second system, single-crystalline GaAs doped (for example, with Te) to the required level was used as the source. The value of the

Table 1. Crystal Chemical Properties of Some Compounds

Compound	Enthalpy of formation, kJ/mole	Structure	Lattice constants, nm
GaAs	74	$F\bar{4}3m$	0.56534
Ga_2Te_3	273	$F\bar{4}3m$	0.5887
GaTe	120	$P6_3/mmc$	0.406; 1.696
Ga_2S_3	510	$F\bar{4}3m$	0.5181
GaS	196	$P6_3/mmc$	0.3585; 1.515
Zn_3As_2	136	$P4_2/nmc$	0.833; 1.178
As_2Te_3	38	$C2/m$	1.144; 0.405; 0.992

systems with the doped source was that the ratios of dopant/principal substance concentrations are maintained strictly constant in the gas phase. This simplifies analysis of the kinetic functions.

The standard conditions for deposition of GaAs layers were substrate temperature 750°C, source temperature 830°C, pressure of $AsCl_3$ at the reactor entrance 200 Pa, calculated rate of gas mixture entry into the reactor 0.5 cm/sec (300 K), and total pressure $1.01 \cdot 10^5$ Pa. Other experimental conditions are indicated in the text.

1. MACROSCOPIC LIMITING STEPS
OF THE DEPOSITION AND DOPING PROCESSES

The chemical transport of GaAs in a chloride system is known to proceed by the reaction

$$GaAs(s) + HCl(g) = GaCl(g) + \frac{1}{4} As_4(g) + \frac{1}{2} H_2(g). \qquad (2)$$

The formation of the molecular complex $AsGaCl^*$ in the adsorbed layer or in the deposition step is an intermediate stage of crystallization, i.e., GaAs deposition is a result of a reaction in the adlayer [7]:

$$AsGaCl^* + H^* = GaAs(s) + HCl(g). \qquad (3)$$

The principal dopants in GaAs are elements of groups II (Zn, Cd), VI (S, Se, Te), and IV (Sn). Some of these are transported as chlorides (Zn, Cd, Sn), whereas others travel as hydrides (S, Se, Te) or in the free state [8].

By analogy with the crystallization of gallium arsenide, the transport of impurities from the gas phase into the solid occurs through formation of the corresponding molecular complexes:

$$Te^* + GaCl^* = TeGaCl^*, \qquad (4)$$

$$TeGaCl^* + H^* = Ga_{Ga}Te_{As} + HCl(g), \qquad (5)$$

$$ZnCl^* + As^* = AsZnCl^*, \qquad (6)$$

$$AsZnCl^* + H^* = Zn_{Ga}As_{As} + HCl(g), \qquad (7)$$

i.e., the doping is by nature the dissolution in GaAs of a binary compound that contains the dopant and one of the basic components.

Elements of groups II and V, as well as III and VI, are known to form at least two compounds of different stoichiometric composition: GaTe and Ga_2Te_3, GaS and Ga_2S_3, and $ZnAs_2$ and Zn_3As_2 [9]. The more stable and structurally similar to GaAs are the high-valence compounds. Some physicochemical data on the listed compounds are given in Table 1. We note that the enthalpy of formation for all the compounds included is larger than for GaAs.

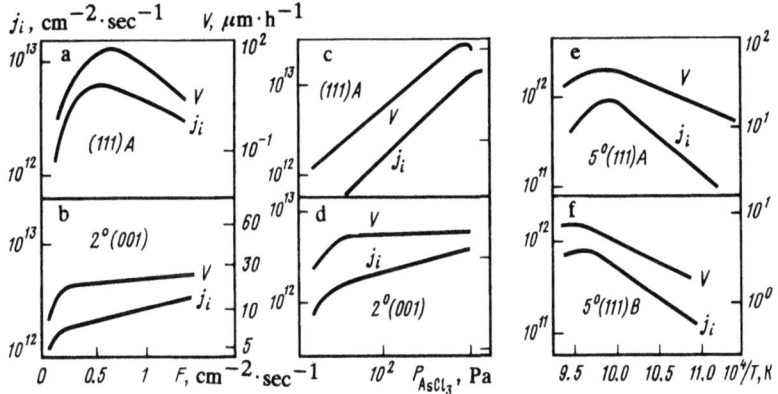

Fig. 1. GaAs layer deposition rate and the impurity trapping rate as functions of the system parameters for GaAs(Te)–AsCl$_3$–H$_2$. The flux of gaseous mixture (a, b), entering AsCl$_3$ pressure (c, d), and deposition temperature (e, f) for several crystallographic orientations are shown.

The GPE process includes three macroscopic stages: diffusion of the starting substances to the surface, a heterogeneous deposition reaction of the substance, and diffusion of the reaction products from the surface through the boundary layer [10, 11].

The general expression for the atomic flux through the gas–solid interface can be written in the form [11]

$$j = P \bigg/ \left(\frac{\delta}{D_{init}} + \frac{\delta}{K_p D_{pr}} + \frac{1}{\vec{k}} \right). \qquad (8)$$

Here P is the entering pressure of the component which ensures deposition of the substance, D_{init} and D_{pr} are diffusion coefficients of the starting component and reaction product, respectively, δ is the thickness of the boundary layer, $K_{eq} = \vec{k}/\overleftarrow{k}$ is the equilibrium constant of the given process, and $\vec{k}$ and $\overleftarrow{k}$ are the rate constants for the forward and reverse reactions, respectively. Deposition is considered the forward reaction.

Applied to actual processes, the expression can be complicated; for example, the stoichiometric condition of the deposited compound must be considered keeping in mind that the thickness of the boundary layer diminishes in proportion to $F^{-1/2}$ (F is the gas flux) and other peculiarities.

Without delving into the details of the analysis (see [10, 11]) we note that the conditions for realization of the surface–kinetic regime are: a) large fluxes when the thickness of the boundary layer is close to zero, the composition of the gas phase near the surface is close to that entering, or the crystallization rate reaches a maximum; and b) low crystallization temperatures when the atomic exchange processes in the adsorbed layer and the subsurface layer of the solid are very slow or their rate depends on the surface structure.

A similarity in the deposition rate and the impurity trapping rate as functions of the basic process parameters should be expected with common macromechanisms that limit the deposition of the basic and dopant components.

Figure 1 shows V and j_i as functions of the flux of the gaseous mixture, pressure of AsCl$_3$ at the system entrance, and the crystallization temperature for several crystallographic orientations of a growth surface. The data are taken from [12-15].

The listed functions in all cases turn out to be symbatic for V and j_i. Since the macromechanisms that limit the deposition rate of GaAs are well known [10, 11, 16], it can be concluded from the interrelationship that the stages that limit both the deposition and the doping are: external diffusion for low fluxes of gaseous mixtures, and surface kinetics for high fluxes; the deposition rate of the substance on the surface for low chlorine pressures, and the desorption rate of reaction products for high chlorine pressures; and surface kinetics for low deposition temperatures, and transport thermodynamics for high deposition temperatures.

An important feature of this analysis is that a change of the steps that limit the deposition rate and impurity trapping occurs usually at similar values of the (F, P, T) parameters, although other situations are possible if the interaction energy of the impurity with the matrix (GaAs) is substantially larger than the energy of the bond in the GaAs crystal. This comment refers in the first place to sulfur, as well as to dopants of transition metals [17, 18].

In the region where the surface kinetics are limiting, it was noticed during analysis of the macrokinetic functions that the derivative of the impurity trapping rate with respect to the corresponding independent variable is not equal to the derivative of the deposition rate:

$$\frac{\partial j_i}{\partial P_{Cl}} > \frac{\partial V}{\partial P_{Cl}}, \quad \frac{\partial j_i}{\partial F} > \frac{\partial V}{\partial F}, \quad \frac{\partial \ln j_i}{\partial (1/T)} \gtrless \frac{\partial \ln V}{\partial (1/T)}. \tag{9}$$

Since the experiment was set up such that the ratio of the impurity pressure to that of the main substance in the gas phase was constant, it follows from the inequalities that the impurity trapping process cannot be considered completely independent from the crystallization of the main substance. This is quite reasonable if the doping is viewed as dissolution of dopant-containing compounds in GaAs. In this case, the interaction of impurities and main atoms in the adsorbed layer and the effect of these interactions on the kinetics and mechanism of crystallization must be considered.

2. KINETICS OF IMPURITY TRAPPING BY STEPPED SURFACES

Even the first investigations showed that impurity trapping during GPE of GaAs depends on the crystal structure of the main faces [19-21]. The "facet effect" was detected, i.e., a sharp elevation of impurity concentration in layers during transport from the vicinal to the main faces [22, 23]. The results demonstrated the existence of a connection between the kinetics of doping and the surface structure. Attempts to explain the observed features due to the influence of only the deposition rate were not successful.

Progress in understanding the mechanisms that control surface kinetics of impurity trapping was made possible with the use of vicinal surfaces with a regulated angular deviation from a singular face in the modelling of the structure. Data on the anisotropy of the doping level of GaAs by Te, Sn, Si, and Zn are found in [4-6, 24, 25].

According to current concepts, the vicinal faces contain an echelon of microsteps, the ends of which have a structure similar (in the direction of the deviation) to a singular face and a density of steps which is related to the angular deviation φ by the relation $N_{st} = h^{-1}\sin \varphi$, where h is the step height.

The kinetics of crystallization on these surfaces can be described using the Burton—Cabrera—Frank model (BCF model) [26]. Two parallel crystallization processes exist: through formation of nuclei on the smooth face and through binding to the steps in the echelon.

In [5, 6], we proposed use of the BCF model for describing the kinetics of impurity trapping. Chernov and Stoyanov [27] calculated the kinetics of doping for the principal substance, using the BCF model in a system with a constant deposition rate. They accounted for the limiting effect on the doping kinetics of dopant atom surface diffusion and its layering onto the step.

With the condition that the rate of atomic transport from the adlayer into the echelon of steps is limited by surface diffusion, the following expression for the atomic flux density through the gas—solid interface can be written using the calculations of [6, 27]:

$$j_1 = A p N_{st} \lambda_s \tanh \lambda/2\lambda_s = A \bar{\beta} P = \beta P, \tag{10}$$

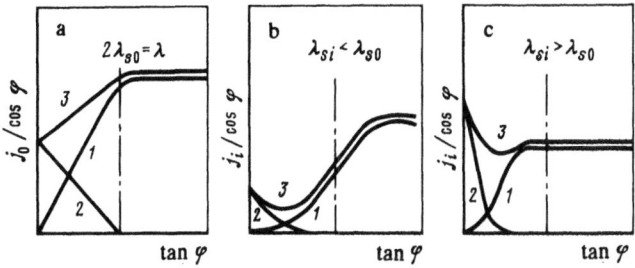

Fig. 2. Scheme which illustrates the contribution of the echelon of steps (1) and the nucleation process (2) to the total crystallization rate (3) of the principal (a) and impurity (b, c) atoms. The coordinates were chosen based on [28, 35].

where A is a constant, P is the vapor pressure, λ_s is the diffusion length of the adsorbed particle, β is the kinetic coefficient of crystallization, and $\bar{\beta}$ is its part that depends on λ_s and λ; $\lambda = N_{st}^{-1}$, where N_{st} is the density of steps and λ is the mean distance between them.

Equation (10) expresses the crystallization rate anisotropy, which is related to the echelon of steps and can be applied to both the main and the impurity atoms.

The second component of the atomic flux density through the interface is related to nucleation and can be written in the form

$$j_2 = j_{2m} (1 - 2\lambda_s N_{st}).$$

$$(11)$$

Here j_{2m} is the maximal value of j_2 which corresponds to the stepless surface. Each step on the surface excludes a band of width $2\lambda_s$ from the nucleation processes.

Summation of the two components of flux density is shown schematically in Fig. 2.

For the principal atoms, j_{01} reaches saturation with an angular deviation of φ_0, at which $2\lambda_{s0} = \lambda$ and j_{02} reverts to zero. This condition allows the components of the deposition rate to be separated and analyzed individually [28].

The task is more complicated for the flux of impurity atoms. First, the condition $\lambda = 2\lambda_{si}$ is fulfilled for a single angular deviation which is larger or smaller in comparison to φ_0 depending on the ratio of energies of the adsorbed impurity and main atoms. Second, the effectiveness of steps as an outlet for impurity atoms is lower than for the main atoms. This is related to the existence of an energetic and purely geometric barrier for trapping of a foreign atom into the lattice [29]. Apparently, this should lead to appearance of discontinuities in the path of the functions $j_i(\varphi) = j_{i1}(\varphi) + j_{i2}(\varphi)$.

The relative anisotropy of the doping level, which is related only to the step relief, is equal to

$$\eta \cong j_{i1}/j_{01} = C \frac{\lambda_{si} \tanh \lambda/2\lambda_{si}}{\lambda_{s0} \tanh \lambda/2\lambda_{s0}} = C\eta^*,$$

$$(12)$$

where C is a constant which does not depend on the surface structure, and the indices 0 and 1 refer, as before, to the main and impurity atoms, respectively.

It follows from an analysis of the anisotropy factor η^* that if the atomic adsorption energies are related as $E_{ads}^{(i)} > E_{ads}^{(0)}$, i.e., $\lambda_{si} > \lambda_{s0}$, then the doping level increases with an increase of step density. For the reverse relation ($E_{ads}^{(i)} < E_{ads}^{(0)}$, $\lambda_{si} < \lambda_{s0}$), the doping level is lowered with a deviation from a smooth face.

Estimates show [6] that deceleration of the exchange processes on the step acts in the examined dependence in the same manner as a decrease of diffusion length. If the trapping into the step of impurity and main atoms is the slowest stage, then doping anisotropy should not be observed.

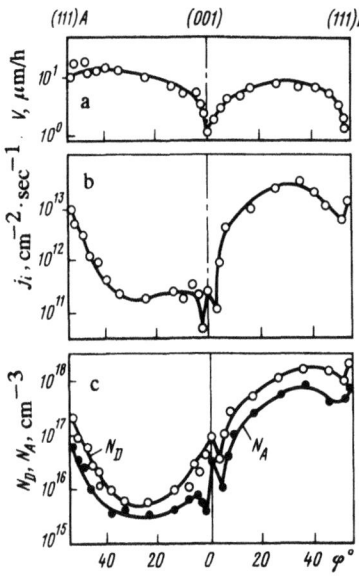

Fig. 3. Deposition rate (a), rate of impurity trapping (b), and concentrations of donors N_D and acceptors N_A in GaAs layers (c) as functions of crystallographic orientation for the GaAs(Te)–AsCl$_3$–H$_2$ system.

Figure 3 for the GaAs(Te)–HI–H$_2$ system presents data on the anisotropy of GaAs deposition rate, the anisotropy of doping level, and the tellurium trapping rate. Qualitative agreement with calculation is seen. A lowering of doping level with an increase of deviation angle occurs in the vicinity of the (111) A face where the Te–Ga bond energy is larger than As–Ga and, consequently, $\lambda_{si} > \lambda_{s0}$. In the vicinity of (111) B, where a relatively weak Te–As bond is formed upon adsorption on the terrace, the doping level and impurity trapping rate rise with an increase of step density. However, the relative position of the curves in the A- and B-ranges is not obvious overall. At first glance it seems that the A-surfaces, which provide a more stable bond with Te adatoms in comparison with the B-surfaces, should correspond with the higher doping level. This is not confirmed in reality. This fact requires a more detailed study accounting for the actual structure of adsorbed layers on the stepped faces.

Figure 4 shows data on the GaAs doping anisotropy for a number of dopants in the range of orientations (111) A and B-(001) and the anisotropy of impurity trapping rate in this range. The superposition of graphs, corresponding to the ranges (111) A-(001) and (111) B-(001), allow the features of the impurity trapping related to the polarity effect of the ⟨111⟩ direction in the sphalerite structure to be discerned. Evidently, the dopant concentration in the GaAs layers with orientations (111) A and (111) B differ by an order of magnitude. For the donor dopants (Te, Sn, Si), we have N_i(111) B/N_i(111) A > 1 whereas for the acceptors (Zn), we have N_i(111) B/N_i(111) A < 1. This result confirmed that the (111) B surface incorporates primarily donor and (111) A acceptor dopants [19].

If the direct influence of the deposition rate is excluded and the anisotropy of impurity trapping rate is treated, then the polar effect in the impurity trapping on the (111) A and (111) B faces is observed to be expressed well only for zinc and relatively poorly for the donor dopants. Within the range, trapping of donor impurity on the B-surface proceeds significantly more effectively than on the A-surface.

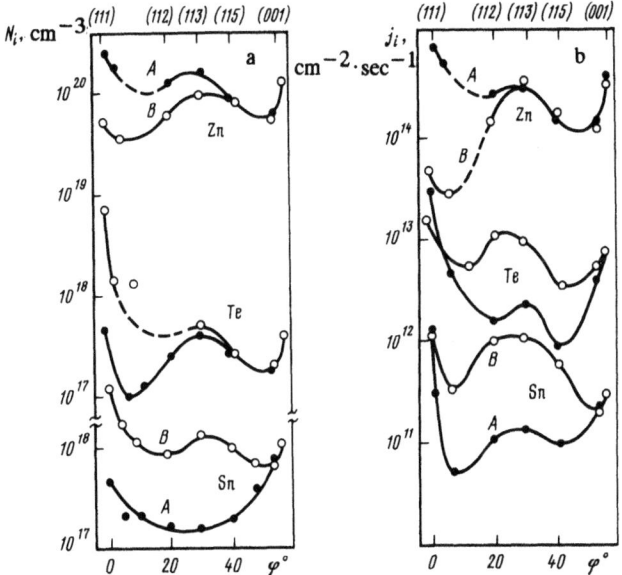

Fig. 4. Zn, Te, and Sn dopant concentrations in GaAs layers
(a) and the trapping rate of these impurities (b) as functions
of the crystallographic orientation.

Apparently, the form in which the dopant approaches the adsorbed layer plays the determining role. In particular, crystallization of atoms that form stable chlorides (Ga, Zn, Fe, Mn, etc.) proceeds more effectively on the (111) A and (001) faces where chloride forms an outer adsorbed layer.*

Besides this, the possible rearrangement of the adsorbed layer structure with an increase of step density should be noted. This is especially true in the A-range where the surface is sensitive to the effect of an impurity [30, 31]. The possibility of nonequilibrium conditions at the gas–adsorbed layer interface is also noteworthy.

We return to the activation energies of the surface processes (Table 2). The activation energy of the deposition E_0 and doping E_i processes for the vicinal faces are seen to form the series $E_0 2°(001) > E_0 5°(111)$ B $> E_0 5°(111)$ A and $E_i 2°(001) > E_i 5°(111)$ A $> E_i 5°(111)$ B, in which the surfaces (111) A and (111) B exchange places. In [32], an analogous hierarchy of activation energies of GaAs doping by Si and Ge dopants was obtained.

The analysis in [15] showed that exchange of (111) A and (111) B places in the energetic series is a result of the inequality $\lambda_{si} \gtrless \lambda_{s0}$ (> on the A-faces and < on the B-faces).

Equation (10) shows that the crystallization rate as a function of temperature for atoms of a given type at constant step density is determined by the functions $\lambda_s(T)$ and $P(T)$. The vapor tension in the crystallization zone increases with a rise in the source temperature. For the GaAs(Te)–AsCl$_3$–H$_2$ system, the enthalpies of vaporization of GaAs and Te were estimated at 111 and 158 kJ/mole, respectively [15]. The diffusional length decreases with increasing temperature $\lambda_s = (D_s \tau_s)^{1/2} \sim \exp(E_{des} - E_{sd})/2kT$ (E_{des} and E_{sd} are the activation energies of desorption and surface diffusion, respectively). The contribution of λ_s to the temperature dependence can be determined only for the condition $2\lambda_s < \lambda$, i.e., a part of the atoms revaporize without reaching the growth stage.

---

*In this respect, data on the transition metals that form stable chlorides (Fe, Mn) are especially indicative. Concentrations 1-3 orders of magnitude lower than for diffusion [18] can be attained upon doping of GaAs by these dopants under GPE conditions.

Table 2. Activation Energy of Deposition Rate E_0 and Rate of Impurity Trapping E_i [15]

Orientation	(111) A	5° (111) A	(111) B	5° (111) B	(001)	2° (001)
E_0, kJ/mole	92	67	146	109	205	172
E_i, kJ/mole	–	151	–	100	–	184

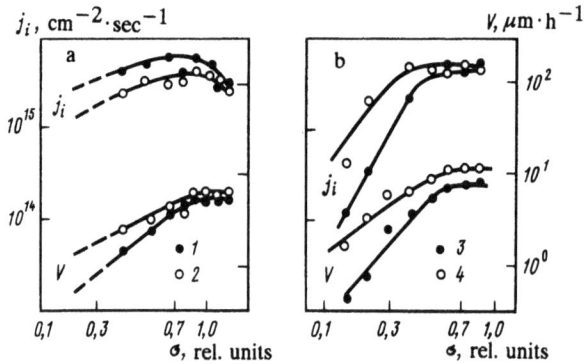

Fig. 5. Deposition rate and impurity trapping rate as functions of supersaturation for the GaAs(Te)–AsCl$_3$–H$_2$ system. The crystallization temperatures are 675 (a) and 800°C (b) for orientations: 1) (111) A; 2) 5°(111) A; 3) (111) B; and 4) 5°(111) B.

On the surface 5°(111) A, we have $\lambda/2 < \lambda_{s0} < \lambda_{si}$, i.e., surface diffusion affects crystallization of the principal atoms but does not limit trapping of impurity. As a result, $E_i \approx \Delta H$ and $E_0 < \Delta H_0$.

On the surface 5°(111) B, we have $\lambda_{si} < \lambda_{s0}$. The condition $\lambda = 2\lambda_{s0}$ is fulfilled at $\varphi > 5°$ (the deposition rate reaches a plateau), i.e., the surface 5°(111) B is coarse relative to the main component ($E_0 \approx \Delta H_0$), but is not sufficiently coarse to the impurity ($E_i < \Delta H_i$). A part of the impurity atoms can revaporize without reaching the growth stage. Lowering of the activation energy by 58 kJ/mole for the impurity trapping rate approximately corresponds to the activation energy for surface diffusion of Te dopant on (111) B of GaAs [15].

For the (001) and its vicinal surface, an elevation of activation energy is observed in comparison with the enthalpies of the corresponding processes. This indicates the existence of additional energy-intensive processes.

Figure 5 shows the deposition rate and the impurity trapping rate by these same faces as functions of the supersaturation in the system. The supersaturation is regulated by the source temperature change and is calculated by the usual method [33]:

$$\sigma = \ln K_{eq}^{(s)} - \ln K_{eq}^{(d)},\qquad(13)$$

where K_{eq} is the equilibrium constant for reaction (2) and the indices "s" and "d" refer to the source and deposition zones, respectively.

As expected, the small-angle anisotropy of the deposition rate and impurity trapping rate is exhibited weakly at large supersaturations (in the dense adsorbed layers). The role of the growth stage in crystallization increases with lowered supersaturation. With respect to the impurity, the trapping rate increases with a deviation from the face, in agreement with the scheme of Fig. 2, if $\lambda_{si} < \lambda_{s0}$. This is realized in the vicinity of (111) B. If $\lambda_{si} > \lambda_{s0}$, then the smooth face guarantees better conditions for impurity trapping.

However, it must be considered that the condition $E_{des}^{(i)} > E_{des}^{(0)}$ leads not only to the relation $\lambda_{si} > \lambda_{s0}$, but also to an increase of the characteristic time for establishment of adsorption–desorption equilibrium on the surface. The time for establishment of adsorption equilibrium, which is estimated as $\tau = \nu^{-1} \exp E_{des}/kT$, in-

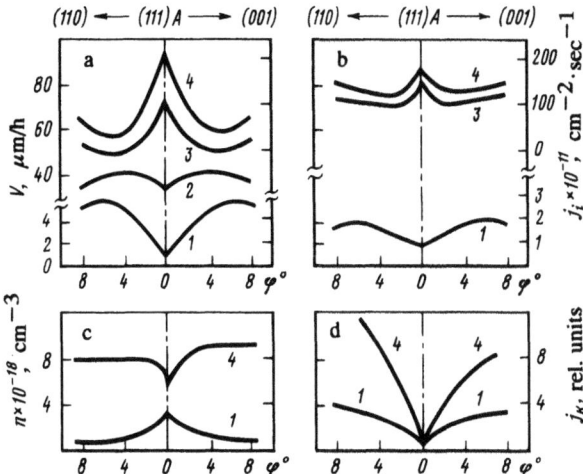

Fig. 6. Deposition rate (a), Te trapping rate (b), electron concentration (c), and rate of $Te_{As}V_{Ga}$ complex trapping (d) at elevated P_{AsCl_3} in the GaAs(Te)–AsCl$_3$ system as functions of a change of orientation in the vicinity of the (111) A face. Concentration of Te = $4 \cdot 10^{18}$ cm^{-3}. P_{AsCl_3} = 1) $1 \cdot 10^1$; 2) $2.2 \cdot 10^2$; 3) $4 \cdot 10^2$; and 4) $8 \cdot 10^2$ Pa.

creases from 10^{-8} to 10^{-1} sec for an increase of E_{des} from 1 to 2.5 eV (from 96 to 240 kJ/mole) for the same values of the other parameters ($v = 10^{13}$ sec^{-1} and $T = 1000$ K). The average time between successive occurrence of growth stages on the surface at a layer deposition rate of 10 nm/sec and a step thickness of 0.5 nm will be about 1/20 sec, i.e., completely comparable to the time for formation of a complicated molecular complex on the surface.

Under these conditions, the dopant concentration in the layers should be expected to be a function of the deposition rate. Such functions are in fact observed for both tellurium [6] and sulfur [17].

3. FEATURES OF THE KINETICS OF DEPOSITION AND DOPING AT LARGE CONCENTRATIONS OF CHLORINE AND IMPURITY

We now examine in more detail the anisotropy of deposition and doping at small deviations from the (111) A and (001) faces. These faces are unique in that their surface is covered by chlorine in a chloride–hydrogen system. The gallium atom layer is located beneath this coating. Recalling that group VI elements form stronger bonds to gallium than arsenic, it should be expected that the presence of tightly bound atoms in the adlayer can affect the kinetics of crystallization and impurity trapping.

In [34], the change of GaAs deposition rate anisotropy near the principal faces under the influence of tellurium was studied. At small Te concentration in the GaAs source, the deposition rate of the (111) A and its vicinal face increased with increasing AsCl$_3$ pressure in the system. The depth of the deposition rate minimum at the (111) A point gradually decreased but the nature of the deposition rate anisotropy was retained at all AsCl$_3$ pressures.

A different pattern is observed at large Te concentration in GaAs (Fig. 6a). As the AsCl$_3$ pressure is increased, a qualitative change occurs. The deposition rate minimum at the (111) A point is replaced by a sharp maximum. The (111) A face loses stability. Comparing the deposition rate of the lightly and heavily doped GaAs layers, we see [34] that such a change arises as a result of a different rate of increase with AsCl$_3$ pressure of the deposition rate for (111) A and its vicinal face: $\partial V/\partial P_{AsCl_3}$ (111) A > $\partial V/\partial P_{AsCl_3} \, n^\circ$(111) A. This means that the relative increase of j_{02} is larger than j_{01}, according to the scheme of Fig. 2.

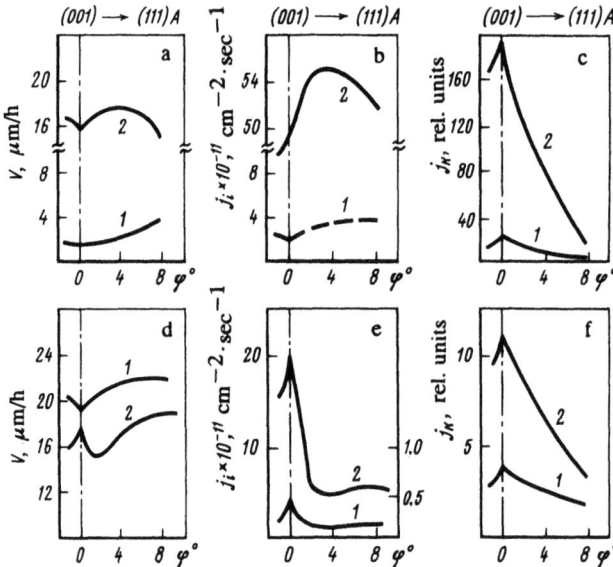

Fig. 7. Effect of P_{AsCl_3} and P_{SF_6} on small-angle anisotropy of deposition processes in the vicinity of the (001) face in the GaAs(Te)–AsCl₃–H₂ (a-c) and GaAs–AsCl₃–H₂–SF₆ (d-f) systems. AsCl₃ pressure, Pa (upper row): 1) $1.1 \cdot 10^1$ and 2) $4 \cdot 10^2$. $N_{Te} = 4 \cdot 10^{18}$ cm^{-3}. SF₆ pressure, Pa (lower row): 1) $3 \cdot 10^{-3}$ and 2) $9 \cdot 10^{-2}$. $P_{AsCl_3} = 4 \cdot 10^2$ Pa.

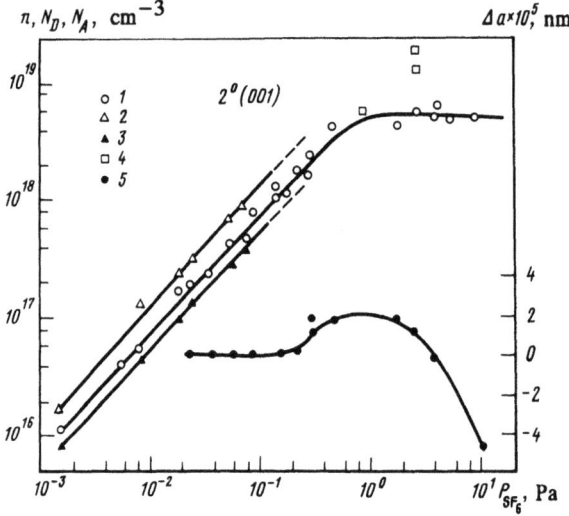

Fig. 8. Concentrations of electrons (1), donors (2), acceptors (3), dopants (4), and lattice constants (5) for GaAs layers doped with sulfur as functions of SF₆ pressure.

The predominant sites of impurity adsorption on the growing faces are known to be fractures on the steps. The deactivation mechanism of steps was examined in [29, 35]. If this mechanism were unique, then a decrease of the depth of the minimum at the (111) A point up to its complete disappearance would be expected. The deposition rate maximum on the (111) A face can appear only if the impurity atoms initiate the nucleation processes. In this case, the tellurium trapping rate by the (111) A face also has a sharp maximum (Fig. 6b). Convergence of the minima of deposition rate and impurity trapping rate on the (001) face upon tellurium doping is not observed (Fig. 7a and b). However, when tellurium is replaced by sulfur, the corresponding effects are seen not only on (111) A but also on (001) B (Fig. 7d and e), i.e., a large difference in dopant—gallium bond energies is required for appearance of this effect on the (001) face.

The entry of group VI elements into GaAs was examined in Section 1 as the dissolution of binary compounds of the $A^{III}B^{VI}$ and $A_2^{III}B_3^{VI}$ types. The initial data (Table 1) show that the $A_2^{III}B_3^{VI}$ compounds are thermodynamically more stable and that their crystal lattices (defect sphalerite) are similar to the GaAs lattice not only in type but also in constants.

Comparing the stoichiometric formulas, we notice that dissolution of $A_2^{III}B_3^{VI}$ type compounds into GaAs should be accompanied by formation of vacancies in the Ga sublattice. Since the gallium vacancies are electrically active (acceptors), and their complexes with Te and S atoms are optically active, then study of the correlations between $[Te_{As}]$ (donors) and N_A (acceptors) or $[Te_{AS}]$ and $[V_{Ga}Te_{As}]$ can provide information as to whether the dissolution of dopant should be examined as dissolution of Ga_2Te_3 (or Ga_2S_3) in GaAs (similar to the way it is examined in single-crystalline GaAs [36]). Experiments show the existence of such correlations between N_D and N_A for Te and S (Fig. 3c and Fig. 8) and between $[V_{Ga}Te_{AS}]$ and N_D for tellurium [37].

The change of GaAs lattice constant under the influence of sulfur was analyzed in [38]. A decrease of constant $\Delta a = -4 \cdot 10^{-5}$ nm was observed at $[S] = 2 \cdot 10^{19}$ cm^{-3}. Estimates showed that such a decrease correlates with a calculated Ga_2S_3 content in GaAs of about 7 mole % (the ratio in the gas phase [Ga]:[S] = 40). However, a variable function with a hump in the range of S concentrations from 10^{18} to 10^{19} cm^{-3} (Fig. 8) is observed instead of the expected smooth decline of lattice constant with increased P_{SF_6}. The appearance of the hump cannot be explained by the formation of a substitution solid solution. Interstitial sites must be considered. Both dopant and principal atoms can occupy the interstices.

The size of the tetrahedral interstices in diamond lattices reaches 0.85 of the size of the principal atom, i.e., they have a radius of 0.104 nm in this case. The covalent radii of the atoms under consideration are (nm): Ga, 0.126; As, 0.118; S, 0.104; and Te, 0.137. It is rather obvious that exceedingly favorable conditions for distribution of the dopant in the interstices are created upon doping with sulfur. There, the atoms will exhibit acceptor properties. Formation of gallium vacancies is more probable upon doping with tellurium. Also, shift of arsenic from its sublattice into the gallium sublattice with formation of As_{Ga} antistructural defects (i.e., additional donors) or trapping of As into interstitial positions can occur.

Molecular dimers such as S_2 and As_2 can play a definite role in formation of atom pairs at a lattice site—atom in the interstice. The site—interstice distance on the GaAs (001) surface is ~0.200 nm whereas the bond length in the S_2 molecule is 0.190 nm. The geometric correspondence engenders additional favorable conditions for distribution of the S_{As}–S_M pairs even upon impingement of the S_2 molecule into the adlayer on the GaAs (001) surface. However, it is not clear why the sulfur dissolution mechanism in GaAs depends on the sulfur concentration in a nonmonotonic manner.

Comparison of the impurity trapping rate anisotropy in the elemental form (Te_{As} and S_{As}) and in the complex form ($V_{Ga}Te_{As}$) and ($V_{Ga}S_{As}$) is interesting since the kinetics of deposition and doping of GaAs in the GPE systems is determined primarily by the kinetics of processes in the adsorbed layer. The corresponding data for the vicinity of the (111) A and (001) faces are shown in Figs. 6d and 7c and f [17, 39, 40].

Evidently, a deviation from (111) A increases the trapping of Te and $V_{Ga}Te_{As}$. The dependence is exhibited more strongly for the latter, i.e., the appearance of the echelon of steps fosters trapping of the complex. In the vicinity of the (001) face the situation is different. The trapping of $V_{Ga}Te$ and $V_{Ga}S$ complexes declines sharply with deviation from the (001) face whereas the tellurium trapping increases under these same conditions.

Together, these data reveal the existence of a connection between the surface structural elements and the mechanism of impurity trapping in various forms.

4. CONCLUSIONS

Comparison of the GaAs crystallization rate and the impurity trapping rate as functions of the basic parameters of the GPE system shows that these processes should be viewed as cocrystallization of components. The processes characteristically have common macroscopic limiting steps: transport by convective diffusion and surface kinetics.

Use of the Burton—Cabrera—Frank layered growth model enables the surface kinetics of impurity trapping during GPE of GaAs to be described to a first approximation. Kinetic features of doping related to the reaction of dopant and principal components in the adsorbed layer can also be identified. Theoretical models which more adequately reflect the essence of the effects occurring in the dense adsorbed layer are necessary for description of crystallization and doping with participation of chemical reactions. At present, this represents one of the pressing problems in the theory of crystal growth.

REFERENCES

1. *Current Problems in Materials Science* [Russian translation], No. 2, Mir, Moscow (1980).
2. M. Heyen and P. Balk, "Epitaxial deposition of GaAs in chloride transport systems," *Prog. Cryst. Growth Charact.*, **6**, 265-303 (1983).
3. J. B. Mullin, "Element trapping in vapor grown III-V compounds," *J. Cryst. Growth*, **42**, 77-89 (1977).
4. L. G. Lavrent'eva, "Impurity trapping during vapor-phase epitaxy of gallium arsenide," *Izv. Vyssh. Uchebn. Zaved., Fiz.*, No. 10, 31-44 (1983).
5. L. G. Lavrent'eva and Yu. G. Kataev, "Epitaxial gallium arsenide. Anisotropy of impurity distribution," in: *Gallium Arsenide* [in Russian], No. 2, Tomsk State Univ. (1969), pp. 46-52.
6. L. G. Lavrentyeva, Ju. G. Kataev, V. A. Moskovkin, and M. P. Jakubenya, "Effect of substrate orientation on deposition rate and doping level of vapor grown gallium arsenide," *Krist. Tech.*, **6**, 607-622 (1971).
7. R. Cadoret, "Application of the theory of rate process in CVD GaAs," *Curr. Top. Mater. Sci.*, **5**, No. 1, 219-279 (1980).
8. S. E. Toropov and M. P. Ruzaikin, "Calculation of equilibrium coefficients of tellurium and zinc distributions during gas-phase deposition of gallium arsenide," *Izv. Akad. Nauk SSSR, Neorg. Mater.*, **17**, No. 8, 2122-2125.
9. *Physicochemical Properties of Semiconducting Substances* (Handbook) [in Russian], Nauka, Moscow (1979).
10. F. A. Kuznetsov, "Some physicochemical problems of deposition of single-crystalline layers from the gas phase," in: *Deposition Processes and Structure of Single-crystalline Semiconductor Layers*, Nauka, Novosibirsk (1968), pp. 50-62.
11. D. W. Shaw, "Mechanism of epitaxial growth of semiconductors from the vapor phase," in: *Crystal Growth* [Russian translation], Vol. 1, Mir, Moscow (1977), pp. 11-74.
12. L. G. Lavrent'eva, S. E. Toropov, and L. P. Porokhovnichenko, "Effect of gas mixture flux and chlorine concentration on tellurium trapping during GPE of gallium arsenide," *Izv. Vyssh. Uchebn. Zaved., Fiz.*, No. 11, 18-21 (1982).
13. L. G. Lavrent'eva, V. G. Ivanov, I. V. Ivonin, and V. A. Moskovkin, "Effect of crystallization temperature on the structure of growing surface epitaxial layers of gallium arsenide in the GaAs–AsCl$_3$–H$_2$ system," *Izv. Vyssh. Uchebn. Zaved., Fiz.*, No. 9, 105-108 (1982).
14. L. G. Lavrent'eva, V. G. Ivanov, and I. V. Ivonin, "Effect of crystallization temperature on deposition rate of epitaxial gallium arsenide layers in the GaAs–AsCl$_3$–H$_2$ system," *Izv. Vyssh. Uchebn. Zaved., Fiz.*, No. 9, 101-104 (1982).
15. L. G. Lavrent'eva, M. D. Vilisova, V. A. Moskovkin, and S. E. Toropov, "Effect of deposition temperature and substrate orientation on tellurium deposition in epitaxial gallium arsenide layers," *Izv. Vyssh. Uchebn. Zaved., Fiz.*, No. 11, 12-17 (1982).
16. L. G. Lavrent'eva, "Kinetics and deposition mechanism of gallium arsenide crystals during gas-phase epitaxy," *Izv. Vyssh. Uchebn. Zaved.*, No. 1, 23-37 (1980).
17. I. A. Bobrovnikova, L. G. Lavrent'eva, and S. E. Toropov, "Trapping of impurity complexes during the vapor-phase epitaxy of gallium arsenide," *Fiz. Tekh. Poluprovodn.*, **20**, 1701-1703 (1986).
18. M. D. Vilisova and L. G. Lavrent'eva, "Gas-phase epitaxy of semi-isolated gallium arsenide," *Obz. Élektron. Tekh.*, Ser. 6, No. 3, 1-32 (1985).
19. F. V. Williams, "The effect of orientation on the electrical properties of epitaxial gallium arsenide," *J. Electrochem. Soc.*, **3**, 886-888 (1964).
20. A. V. Rzhanov, B. S. Lisenier, et al., "Epitaxial gallium arsenide layers with high electron mobility," *Fiz. Tekh. Poluprovodn.*, **2**, 714-719 (1968).
21. L. G. Lavrent'eva, Yu. G. Kataev, L. G. Nestoryuk, and V. V. Silyarenko, "Anisotropy of impurity distribution coefficient in epitaxial gallium arsenide," in: *Deposition Processes and Structure of Single-Crystalline Semiconductor Layers* [in Russian], Nauka, Novosibirsk (1968), pp. 238-248.
22. B. D. Joyce and J. B. Mullin, "The facet effect in epitaxial GaAs," *Solid State Commun.*, **5**, 727-729 (1967).
23. A. E. Blakeslee, "Effect of substrate misorientation on epitaxial GaAs," *Trans. Metall. Soc. AIME*, **245**, 577-581 (1969).
24. L. G. Lavrent'eva, L. P. Porokhovnichenko, and P. N. Tymchishin, "Effect of zinc on gallium arsenide deposition rate anisotropy in the GaAs–AsCl$_3$–H$_2$ system," *Izv. Vyssh. Uchebn. Zaved., Fiz.*, No. 2, 143-145 (1974).
25. L. G. Lavrent'eva, L. P. Porokhovnichenko, I. V. Ivonin, et al., "Effect of dopant type on the anisotropy of deposition rate and doping level of epitaxial gallium arsenide," in: *Gallium Arsenide* [in Russian], No. 5, Tomsk State Univ. (1974), pp. 88-91.

26.   W. Burton, N. Cabrera, and F. Frank, "Growth of crystals and equilibrium structure of their surfaces," in: *Fundamental Processes of Crystal Growth* [Russian translation], Izd. Inostr. Lit., Moscow (1959), pp. 11-109.

27.   A. A. Chernov and S. S. Stoyanov, "On impurity transport during condensation of molecular beams," *Kristallografiya*, **22**, 248-255 (1977).

28.   L. G. Lavrent'eva, "Anisotropy of deposition rate and mechanism of gallium arsenide in gas-transport systems," *Kristallografiya*, **25**, 1273-1279 (1980); ibid., **27**, 818-821 (1982).

29.   V. V. Voronkov, "Crystal surface structure in the Kossel model," *Growth of Crystals*, Vol. 10, Consultants Bureau, New York (1978).

30.   L. G. Lavrent'eva, I. V. Ivonin, L. M. Krasil'nikova, et al., "Growth rate and surface structure of autoepitaxial layers of gallium arsenide," *Deposition and Doping of Semiconducting Crystals and Films* [in Russian], Nauka, Novosibirsk (1977), Part 2, pp. 84-98.

31.   L. G. Lavrent'eva and I. V. Ivonin, "Effect of dopant type on formation of growth relief in gallium arsenide," *Izv. Vyssh. Uchebn. Zaved., Fiz.*, No. 1, 44-48 (1976).

32.   T. Nakanisi and M. Kasiwaga, "Effect of growth temperature and substrate orientation on the trapping of Si, Ge, and Sn into vapor epitaxial GaAs," *Jpn. J. Appl. Phys.*, **13**, 484-494 (1974).

33.   L. G. Lavrent'eva, M. D. Vilisova, and V. A. Moskovkin, "Kinetics of tellurium trapping by (111) faces of gallium arsenide in the GaAs–AsCl$_3$–H$_2$ system at various supersaturations," *Izv. Vyssh. Uchebn. Zaved., Fiz.*, No. 6, 42-46 (1983).

34.   S. E. Toropov, L. P. Porokhovnichenko, and L. G. Lavrent'eva, "Growth of gallium arsenide epitaxial layers near singular faces at various pressures of chlorine and impurity (tellurium)," *Izv. Vyssh. Uchebn. Zaved., Fiz.*, No. 6, 46-50 (1983).

35.   A. A. Chernov and B. Ya. Lyubov, "Questions of crystal growth theory," *Growth of Crystals*, Vol. 5, Consultants Bureau, New York (1968).

36.   M. G. Mil'vidskii, O. V. Pelevin, and B. A. Sakharov, *Physicochemical Principles of Obtaining Dissociating Semiconductor Compounds* [in Russian], Metallurgiya, Moscow (1974).

37.   S. E. Toropov and L. G. Lavrent'eva, "Anisotropy of impurity trapping by epitaxial gallium arsenide layers near singular faces at various pressures of chlorine and impurity (tellurium)," *Izv. Vyssh. Uchebn. Zaved., Fiz.*, No. 2, 92-96 (1985).

38.   M. D. Vilisova and O. M. Ivleva, "Study of structure and properties of epitaxial gallium arsenide layers doped with sulfur," *Izv. Vyssh. Uchebn. Zaved., Fiz.*, No. 7, 24-27 (1985).

39.   I. A. Bobrovnikova, L. G. Lavrent'eva, and S. E. Toropov, "Complexation during gas epitaxy of gallium arsenide doped with tellurium," *Izv. Vyssh. Uchebn. Zaved., Fiz.*, No. 2, 96-100 (1985).

PHOTOSTIMULATED GROWTH OF DOPED LEAD TELLURIDE

Yu. I. Gorina and G. A. Kalyuzhnaya

INTRODUCTION

The great interest in A^4B^6 semiconductors is stimulated not only by their properties but also by their important applications. The compound PbTe and solid solutions based on it belong to this class.

The A^4B^6 compounds, in contrast to elemental semiconductors and A^3B^5 compounds, are phases of variable composition with homogeneous region dimensions (in the solid) from 10^{18} to 10^{20} cm^{-3} (0.01-1 at. %). Deviations from stoichiometry in these take the form of vacancies at the metal and chalcogen positions. The width of the homogeneous region of PbTe and solid solutions based on it narrows with decreasing temperature (retrograde solubility). This feature leads to disruption of the structural uniformity of the material upon cooling and separation of microprecipitates. Strictly speaking, the composition of PbTe according to the phase diagram is conveniently represented as $Pb_{1-\delta}Te_{1-\gamma}$, and that of the PbTe–Sn solid solution as $(Pb_{1-x}Sn_x)_{1-\delta}Te_{1-\gamma}$, where x is the mole fraction of SnTe and the values of δ and γ characterize the deviation from stoichiometry. The molarity x determines such properties of the solid solution as the forbidden band gap whereas the deviation from stoichiometry (γ and δ) determines the type of conductivity, concentration, mobility, and lifetime of carriers, photosensitivity, and luminescence. A characteristic feature of these semiconductors, for example PbTe, which are grown by various methods, is the high concentration of free carriers ($\sim 10^{18}$ cm^{-3}) which is governed by the large quantity of equilibrium point defects — vacancies at the Pb and Te positions. Thus, control of PbTe properties, primarily a lowered vacancy concentration, is important for practical applications and is possible only by control of the deviation of its composition from stoichiometry. Until recently, reduction of the carrier concentration in such materials was accomplished by prolonged homogenization annealing in a metal-enriched medium for several weeks. The annealing is a complicated process, as a result of which structural changes as well as a reduction of carrier concentration occurs [1]. Prolonged annealings cannot be carried out technologically whereas for epitaxial structures this method is in principle impossible due to reciprocal diffusion of layer and substrate components.

The present article is a review of efforts made at the Institute of Physics of the Academy of Sciences over the last 10 years to control the composition of A^4B^6 compounds using PbTe as an example. This substance is widely used for fabrication of optoelectronic devices operating in the IR region.

At present, the following possibilities for control of the composition and properties of A^4B^6 compounds and PbTe are contemplated:

1. Epitaxial growth at low temperatures. One of these methods is molecular epitaxy with hot walls, in which growth conditions are near equilibrium [2, 3]. The PbSe and PbTe layers are grown at 350°C at a rate

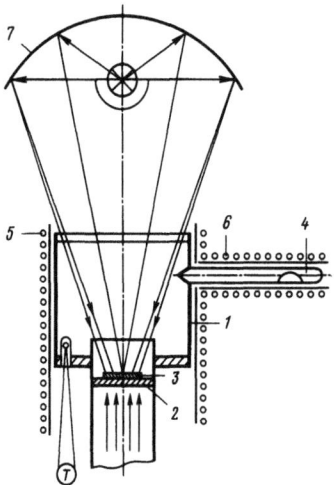

Fig. 1. Light path in the photostimulated epitaxy apparatus: 1) reaction chamber; 2) cooled plate; 3) substrate; 4) side arm with doping component; 5) resistance furnace I; 6) resistance furnace II; 7) emission source with mirror.

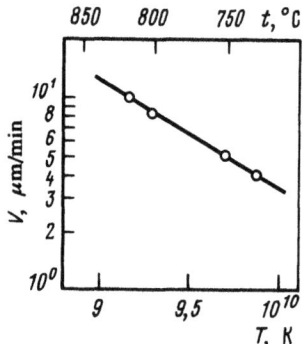

Fig. 2. Growth rate as a function of substrate temperature during growth of $Pb_{0.8}Sn_{0.2}Te$ by photostimulated epitaxy.

of 10-15 μm/h on a freshly cleaved KCl surface. High mobilities of $(2-2.5) \cdot 10^4$ cm^2/(V$\cdot$sec) (77 K) at electron concentrations of $5 \cdot 10^{17}$ cm^{-3} attest to the perfection of the layers obtained.

2. Doping of bulk single crystals with suitable dopants in a quantity which is in accord with the homogeneous region dimensions. The present work contains a description of the methods and experimental results on preparation and study of doped lead—tin tellurides.

3. Use of electromagnetic irradiation of the growth zone in gas-phase epitaxy, the photostimulated epitaxy method, to which special attention is paid in the present work.

THE PHOTOSTIMULATED EPITAXY METHOD

Epitaxial layers of compounds of variable composition over a wide composition range can be prepared by photostimulated epitaxy. Doping during growth can also be carried out.

The essence of the method consists of a light effect during gas-phase epitaxial growth, in which the components of the deposited material are in the gas phase [4]. Regulation of the composition of compounds,

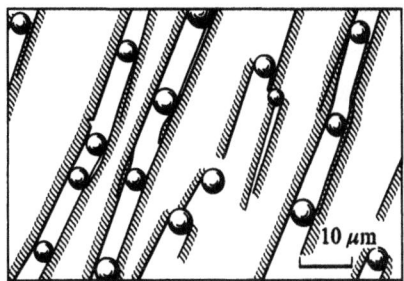

Fig. 3. Morphological features of the surface layer during growth by a vapor– liquid–solid mechanism.

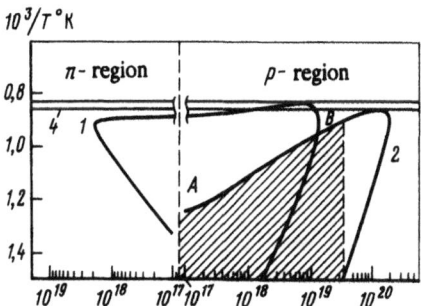

Fig. 4. Diagram of $Pb_{1-x}Sn_xTe$ existence region with the portion where cooling of the material should not be accompanied by separation of precipitates marked by dashes. 1) $x = 0$; 2) $x = 0.2$; 3, 4) liquidus lines of PbTe and $Pb_{0.8}Sn_{0.2}Te$, respectively.

the region of existence of which is inferred from the intrinsic defects or vacancies, was found to be achieved by light excitation at suitable wavelengths and intensities of the electronic shells of atoms of a chosen component during epitaxy. The valence properties of atoms that are incorporated into the lattice and, therefore, the potentials of their interaction can be readily changed by such excitation. Control of the quantity of intrinsic defects is connected with this fact [5].

The main components of the apparatus used in photostimulated epitaxy are the emission source, the optical window in the reaction chamber, and the optical system for admitting radiation into the reaction volume (Fig. 1). A xenon lamp serves as the emission source. An example of photostimulated growth is the preparation of $Pb_{0.8}Sn_{0.2}Te$ epitaxial layers using a metal-enriched source. Excitation of lead atoms in the growth zone is effected by radiation of 10 W/cm^2 intensity and 1.27 eV energy quanta. At working temperatures of 700-850°C, the growth rate of single-crystalline layers of $Pb_{1-x}Sn_xTe$ ($x = 0$-0.25) doped with Pb depends weakly on temperature (Fig. 2). This implies a diffusion mechanism of photostimulated deposition which is determined by mass transport of the substance in the gas phase. This conclusion is supported by the small values of the change of calculated chemical potential during deposition ($\Delta\mu \approx$ 3-4 kcal/mole).

In this case, epitaxy involving a liquid phase based on Pb–Sn occurs, i.e., through a vapor–liquid–solid mechanism. The metal, which is in the form of a thin film on the surface of the growing layer, acts as a solvent, constantly receding as a drop from the growing layers, i.e., the growth occurs under conditions which maximally approach equilibrium (Fig. 3). The metal enters the lattice of the growing crystal only in a quantity which is determined by the width of the homogeneous region of solid $Pb_{1-x}Sn_xTe$ at a given growth temperature such that the composition of the layer corresponds to the AB portion (Fig. 4) of the phase diagram solidus

Table 1. X-Ray Structural and Luminescence Characteristics of $Pb_{1-x}Sn_xTe$

Epitaxial layer/substrate	Line half-width,[*] deg	
	B_θ, (600) reflection	B_W, (200) reflection
$Pb_{0.8}Sn_{0.2}Te$ (Pb + Sn)/PbTe	0.07	0.07
$Pb_{0.8}Sn_{0.2}Te$ (Pb + Sn)/BaF_2	0.085	0.25
$Pb_{0.8}Sn_{0.2}Te$, doped with In/BaF_2	0.07	0.17
PbTe, doped with Ag/BaF_2	0.06	–
$Pb_{0.8}Sn_{0.2}Te$ prepared by liquid-phase epitaxy	0.05	0.1
Unannealed crystal of $Pb_{0.8}Sn_{0.2}Te$ [†]	0.09	0.2
Annealed crystal of $Pb_{0.8}Sn_{0.2}Te$	0.15	1.0

[*]B_θ and B_W are experimental half-widths of diffraction $(\theta/2\theta)$ and oscillation (W) lines, respectively.
[†]Luminescence is absent in this case only.

line. Figure 4 shows that the $Pb_{1-x}Sn_xTe$ layers that correspond to the AB portion do not undergo retrograde separation of precipitates upon cooling from the growth temperature at any cooling rate.

The concentration of carriers in the $Pb_{0.8}Sn_{0.2}Te$ layers was determined by the amount of excess metal (Pb + Sn) added to the PbSnTe source mixture without a change of the ratio of these components in the layer. A change of excess Pb + Sn from 1 to 11% in the source mixture decreases the concentration of acceptors in the layer from $2 \cdot 10^{19}$ to $3 \cdot 10^{17}$ cm^{-3} (77 K) up to the conductivity type inversion and allows preparation of $Pb_{0.8}Sn_{0.2}Te$ layers of n-type conductivity with a carrier concentration of $1.5 \cdot 10^{18}$ cm^{-3} (77 K) during growth under illumination conditions.

X-ray structural studies showed high crystalline perfection of the grown layers [6]. Broadening of diffraction and oscillation lines in them was not observed. This does occur in single crystals annealed in a metal-enriched medium (Table 1).

Epitaxial layers of $Pb_{0.8}Sn_{0.2}Te$ grown on PbTe substrate have effective luminescence which corresponds to that of the initial composition $Pb_{0.8}Sn_{0.2}Te$. This enables using them as an active lasing region without additional homogenizing annealing [6]. Thus, the presence of Pb and Sn atoms excited by light on the irradiated $Pb_{0.8}Sn_{0.2}Te$ surface can be used to widen the range of prepared compositions up to n-type material that cannot be prepared by any other methods, i.e., to increase the homogeneous region of the material.

2. DOPING OF LEAD-TIN TELLURIDE FROM THE GAS PHASE

The search for directed changes of the PbSnTe homogeneous region led us to study the doping processes of this material by extraneous impurities. Photostimulated epitaxy permits doping of layers to be effected during their growth either from a source independent of temperature or from a source mixture with the corresponding dopant additions. Elements of groups I (Ag and Au), III (In, Tl, and V), and V (Bi) were chosen as dopants. Plates of BaF_2 with orientations (111) and (100) and fresh chips of BaF_2 were used as substrates.

Bulk-doped lead–tin telluride crystals were grown by directed gas-phase crystallization for comparison under the same temperature conditions but without illumination. The starting material used as the source for growth of crystals and films was prepared by direct ampul synthesis with vibrational mixing of the melt of starting components Pb and Te taken in the ratio 1:1 with addition of the dopant in an amount from 0.1 to 0.5%. The metal dopant was added with an equiatomic amount of tellurium. The total concentration of uncontrolled dopants was $\sim 10^{16}$ cm^{-3}. The monophasic nature and homogeneity of the synthesized material was established by x-ray diffraction. Single crystals were grown in evacuated ampuls placed in a vertical furnace

Table 2. Electrophysical Properties and Luminescence as Functions of Dopant Content in $Pb_{1-x}Sn_xTe$ Single Crystals at $T_g = 840°C$

Dopant/material	Dopant content		Conductivity type	Carrier concentration, 77 K	Carrier mobility, 77 K	Corresponding luminescence intensity, rel. units
	Mixture, mole %	Crystal, at. %				
PbTe	0	0	p	$5 \cdot 10^{18}$	$1 \cdot 10^3$	12
Bi /PbTe	0.53	0.53	n	$2.1 \cdot 10^{19}$	$2.8 \cdot 10^3$	–
	0.67	0.67	n	$6.7 \cdot 10^{19}$	–	–
	No data	0.87	n	$7.2 \cdot 10^{19}$	356	–
	The same	1.8	n	$1.6 \cdot 10^{20}$	–	–
	–"–	2.6	n	$2.12 \cdot 10^{20}$	1.3	–
$Pb_{0.8}Sn_{0.2}Te$	0	0	p	$6.2 \cdot 10^{19}$	$0.4 \cdot 10^3$	–
$Bi/Pb_{0.9}Sn_{0.1}Te$	0.005	No data	p	$7 \cdot 10^{18}$	$6.2 \cdot 10^3$	–
$Bi/Pb_{0.78}Sn_{0.22}Te$	0.05	The same	p	$3.5 \cdot 10^{19}$	$1 \cdot 10^3$	–
	0.5	–"–	p	$2 \cdot 10^{19}$	$2.3 \cdot 10^3$	–
	0.1	–"–	n,p	$1 \cdot 10^{18}$	$2.2 \cdot 10^4$	–
	1	–"–	n	$2.3 \cdot 10^{19}$	$2.3 \cdot 10^3$	–
	5	–"–	n	$9 \cdot 10^{19}$	$0.1 \cdot 10^3$	–
Tl/PbTe	0.1	No data	p	$1.1 \cdot 10^{19}$	$5.8 \cdot 10^3$	–
	0.5	0.3	p	$2.5 \cdot 10^{19}$	$1.67 \cdot 10^3$	–
	0.7	No data	p	$5 \cdot 10^{20}$	$3.6 \cdot 10^1$	–
Au/PbTe	Not determined	$1.28 \cdot 10^{-3}$	p	$2.6 \cdot 10^{18}$	$4 \cdot 10^3$	18
	The same	$1.2 \cdot 10^{-3}$	p	$5.9 \cdot 10^{18}$	$5.7 \cdot 10^3$	25
	–"–	$1.7 \cdot 10^{-3}$	p	$8.9 \cdot 10^{16}$	$2.4 \cdot 10^2$	14
	–"–	$2.7 \cdot 10^{-3}$	p	$2.6 \cdot 10^{18}$	$1.3 \cdot 10^4$	18
	–"–	$3.4 \cdot 10^{-3}$	p	$5.6 \cdot 10^{17}$	$5.8 \cdot 10^4$	–
	–"–	$1.2 \cdot 10^{-2}$	p	$1.5 \cdot 10^{19}$	$2.3 \cdot 10^3$	–
	–"–	$1.9 \cdot 10^{-2}$	p	$2.1 \cdot 10^{19}$	$1.6 \cdot 10^3$	–

with a temperature gradient of 0.5-1°C/cm. The prepared single crystals of 12 mm diameter and length up to 15 mm consisted of one or several blocks with the (100) face on the growth surface. The usual methods for studying the dopant behavior (determination of phase composition, microhardness, and crystal lattice constants of the doped material) were insensitive as a result of the small concentrations of added dopants (10^{16}-10^{18} at./cm³). Therefore, we used the following methods for studying the effect of dopants on the composition and properties of PbSnTe: determination of the type and concentration of carriers over a wide temperature range (300-8 K) by the Hall method, determination of the dislocation density by layered etching, and evaluation of luminescence properties.

The concentration of dopants in the grown material was determined by localized electron probe analysis on a MAR-2 instrument (sensitivity of the apparatus to the dopants was 0.1-0.5 at. %), by quantitative adsorption spectral analysis, and by laser mass spectral analysis. The sensitivity of the second and third methods were $5 \cdot 10^{15}$ and 10^{16} at./cm², respectively.

Specimens were cut from the grown crystals and films for electrophysical measurements. Then, the dislocation density, photoluminescence, and finally, the dopant content were determined on these same samples. The electrophysical parameters of single crystal $Pb_{1-x}Sn_xTe$ samples with Bi, Tl, and Au dopants are given in Table 2. The calculated concentrations of carriers and their mobility as functions of Bi, Tl, and Au content show that Bi is a donor dopant in PbTe. Good agreement between the amount of added Bi and the electron concentration in the material up to 1.8 at. % Bi (the apparent solubility limit of Bi in PbTe) is observed.

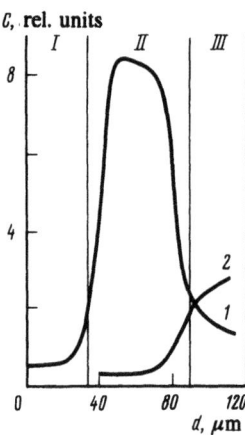

Fig. 5. Distribution curves of: 1) tin and 2) bismuth for laser structures of: I) PbTe(Tl); II) $Pb_{0.8}Sn_{0.2}Te$; and III) PbTe(Bi).

Traces of a second phase enriched in Bi are detected by x-ray phase analysis in crystals with higher Bi content. The observed correlation between the apparent Bi solubility limit, the width of the determined homogeneous region ($1.4 \cdot 10^{20}$ at./cm^3) [1], and the maximally attained carrier concentration ($n = 2 \cdot 10^{20}$) confirms that Bi at small concentrations ($N_{Bi} < 1.8$ at. %) in p-type PbTe occupies metal vacancy sites, imparting electronic conductivity.

Thallium behaves as a simple acceptor impurity. The PbTe crystals have stable p-type conductivity over the whole studied Tl concentration range from 0.07 to 0.7 at. %. The crystal lattice constant in this range of Tl concentrations increases from 6.460 to 6.471 Å (± 0.0006 Å). The PbTe single crystals with 0.1 at. % Tl content were used as substrate in epitaxial structures of n-PbTe(Bi)–p-$Pb_{0.8}Sn_{0.2}Te$–p-PbTe(Tl) [7, 8].

Microprobe studies of the component and dopant distributions were carried out on grown epitaxial structures of n-PbTe(Bi)–p-$Pb_{0.8}Sn_{0.2}Te$–p-PbTe(Tl). Diffusion of bismuth from the upper n-layer into the p-$Pb_{0.8}Sn_{0.2}Te$ layer was found to occur during growth (Fig. 5). Addition of Bi to the PbTe layer creates a p–n-transition in the structure and changes the mismatch of crystal lattice constants in the transition layer between PbTe and $Pb_{0.8}Sn_{0.2}Te$ by 0.02%.

The presence of Tl in the PbTe substrates prevents diffusion of Sn from the active $Pb_{0.8}Sn_{0.2}Te$ layer into the substrate and a shift of emission to shorter wavelengths.

Gold in all the PbTe samples studied, at a dopant content $\leq 10^{-3}$ at. %, which weakly affects the electrical properties of PbTe, increases the quantum yield of photoluminescence (Table 2). Addition of gold in an amount $\geq 10^{-2}$ at. % ($\sim 10^{18}$ at./cm^3) leads to an increased concentration of holes to a level of 10^{19} cm^{-3}. The hole conductivity in PbTe is determined by vacancies at the Me positions. Therefore, each added Au atom leads to the creation of at least ten electrically active metal–hole vacancies. Consequently, the Au dopant cannot be viewed as a simple acceptor dopant. The existence of a more complicated mechanism of atomic interaction of gold with the vacancy subsystem of the crystalline PbTe lattice may explain the results obtained.

3. THE BEHAVIOR OF INDIUM AND SILVER DOPANTS

Comparative analysis of material prepared by various methods was carried out based on the behavior of the two dopants In and Ag which strongly affect the electrical properties of PbTe. The Hall concentration and $Pb_{1-x}Sn_xTe$ mobility as functions of Ag and In dopant content are given in Table 3 and Figs. 6 and 7 [9-11]. The crystals and layers of PbTe doped with various amounts of Ag correspond to the two branches of the curves for n- and p-type conductivity with an inversion from n- to p-type at a Ag content of 4-$7 \cdot 10^{17}$ and a carrier

Table 3. Electrophysical Properties of Bulk Single Crystals and Epitaxial Layers of $Pb_{1-x}Sn_xTe$ as Functions of Ag and In Dopant Concentrations

Sample form	Material	Dopant concentration		Growth temperature.°C	Dislocation density, cm²	Conductivity type, 77 K	Carrier concentration, cm³, 77 K	Mobility, cm²/(V·sec), 77 K
		mixture, mole %	crystal, at./cm³, at. %					
bulk crystals	PbTe(Ag)	0	0	810	$2.3 \cdot 10^5$	p	$1.3 \cdot 10^{18}$	10^2
		0.15	$2.5 \cdot 10^{17}$	800	..	n	$2.7 \cdot 10^{17}$	$8 \cdot 10^3$
		0.15	$3.6 \cdot 10^{17}$	800	$2 \cdot 10^4$	n	$4.8 \cdot 10^{17}$	$3.8 \cdot 10^4$
		0.15	$4.3 \cdot 10^{17}$	800	—	n	$8.7 \cdot 10^{17}$	$3.2 \cdot 10^4$
		0.4	$6 \cdot 10^{17}$	800	—·	p	$5.6 \cdot 10^{17}$	$4.3 \cdot 10^2$
		5	$1.2 \cdot 10^{18}$	800	$7.6 \cdot 10^4$	p	$1.2 \cdot 10^{18}$	$1 \cdot 10^2$
		0.1	$4.65 \cdot 10^{16}$	840	—	n	$3.9 \cdot 10^{17}$	$3.8 \cdot 10^4$
		0.2	$9.3 \cdot 10^{16}$	840	—	p	$2.3 \cdot 10^{17}$	$2 \cdot 10^3$
		0.4	$4 \cdot 10^{17}$	840	—	p	$6 \cdot 10^{18}$	$1.2 \cdot 10^3$
		0.6	$1.3 \cdot 10^{18}$	840	—	p	$3.6 \cdot 10^{18}$	$1.4 \cdot 10^4$
epitaxial layers	PbTe(Ag)	0	0	800	$5 \cdot 10^5$	n	$4 \cdot 10^{18}$	$2 \cdot 10^4$
		0.1	$3.6 \cdot 10^{16}$	705	$3 \cdot 10^4$	n	$4.8 \cdot 10^{14}$	$4 \cdot 10^3$
		0.1	$3.5 \cdot 10^{17}$	810	..	n	$8.3 \cdot 10^{15}$	$3 \cdot 10^{2}$*
		0.4	$7.0 \cdot 10^{17}$	810	—	p	$2.1 \cdot 10^{18}$	$2.5 \cdot 10^{2}$**
		1	$8.4 \cdot 10^{17}$	780	$8 \cdot 10^4$	p	$1.4 \cdot 10^{18}$	$9 \cdot 10^3$
crystals	$Pb_{0.78}Sn_{0.22}Te(In)$	3	0.4	830	..	p	$4 \cdot 10^{19}$	$2 \cdot 10^2$
		5	0.8	830	..	n	$8 \cdot 10^{14}$	$1 \cdot 10^4$
		5	0.9	840	..	n	$3 \cdot 10^{14}$	$1 \cdot 10^4$
		7	1.4	830	—	n	$4 \cdot 10^{14}$	$8 \cdot 10^3$
		0	0	830	$1.5 \cdot 10^5$	p	$7 \cdot 10^{19}$	$5 \cdot 10^2$
layers	$Pb_{0.78}Sn_{0.22}Te(In)$	0.41		780	$1 \cdot 10^4$	p	$1.7 \cdot 10^{18}$	$6.5 \cdot 10^4$
		0.4		780	$1 \cdot 10^4$	p	$2.8 \cdot 10^{17}$	$4.1 \cdot 10^4$
		0.48		765	..	n	$1.0 \cdot 10^{17}$	$3.8 \cdot 10^4$
		0.7		770	$5 \cdot 10^4$	n	$1.3 \cdot 10^{14}$	$2.4 \cdot 10^3$
		1.35		780	$8 \cdot 10^4$	n	$6.7 \cdot 10^{14}$	$4.5 \cdot 10^3$
		1.6		775	—	n	$1.3 \cdot 10^{15}$	$3 \cdot 10^3$

For samples marked * and **, the mobility μ (6 K) reaches record levels for PbTe: $8.8 \cdot 10^5$ and $9.5 \cdot 10^4$ cm²/(V·sec), respectively.

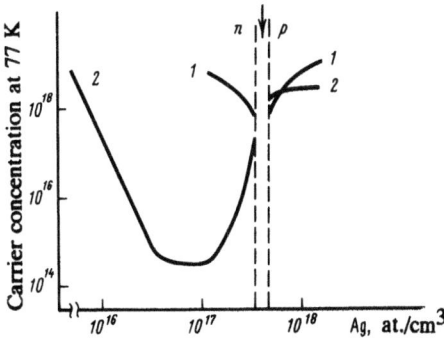

Fig. 6. Hall current carrier concentration at 77 K as a function of Ag concentration in: 1) crystals and 2) epitaxial layers of PbTe(Ag).

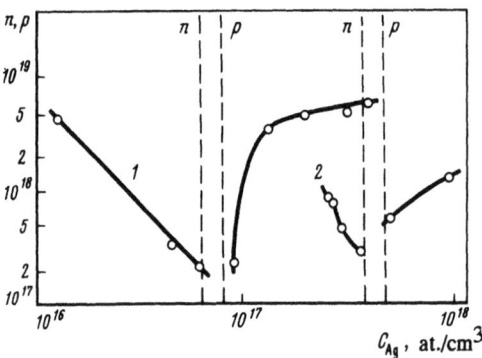

Fig. 7. Hall current carrier concentration at 77 K as a function of Ag concentration in crystals grown at: 1) 840 and 2) 800°C, respectively.

concentration of $(5-8) \cdot 10^{17}$ at./cm^3. The undoped layers had n-type conductivity and crystals had n- or p-type conductivity depending on the growth conditions (Fig. 6).

The effect of silver is very strong (an increase of Ag concentration by one order of magnitude from $\sim 10^{16}$ to $\sim 10^{17}$ at./cm^3 decreases the concentration of donors that are uncompensated by acceptors by three orders of magnitude) in the electronic conductivity region. A virtual multiplication of holes is observed under the influence of Ag [12].

Two series of bulk PbTe(Ag) single crystals were grown at 800 and 840°C. The inversion of conductivity type in samples of the first and second series was found to occur at different silver concentrations (Fig. 7): at $4-5 \cdot 10^{17}$ and $7 \cdot 10^{16}$ at./cm^3 for growth temperatures of 800 and 840°C, respectively. Since the growth temperature determines the degree of deviation of composition from stoichiometry, i.e., the amount of vacancies in thermodynamic equilibrium, the presence of two inversion points in this case should be related to the different degree of deviation from stoichiometry. The inversion of conductivity type in the material with the larger amount of vacancies in thermodynamic equilibrium occurs at a larger Ag content in the material.

The mobility function $\mu_H(T)$ in bulk PbTe(Ag) single crystals with both types of conductivity with a hole concentration of $\sim 10^{18}$ and an electron concentration of $\sim 10^{17}$ cm^{-3} was studied over the whole temperature range (7-300 K).

A significant increase of mobility with decreased temperature from 100 to 6 K is observed in bulk PbTe(Ag) single crystals of both n- and p-type conductivity (Table 3). The observed effect should apparently be assigned to a decreased defect level in the doped material.

Layered sputtering of samples in steps of 10-15 μm was used to study the dopant distribution with depth. The analysis showed that silver is distributed evenly on the surface and with depth. Measurement of the dislocation density in PbTe(Ag) layers was performed using a microscope. The dislocation density in PbTe(Ag) was found to remain constant with depth. It does not depend on silver concentration in the samples. The value of dislocation density in the doped samples is $(2-8) \cdot 10^4$ cm^{-2}, which is an order of magnitude smaller than the corresponding value in the undoped PbTe samples.

Photoluminescence of PbTe as a function of Ag content and the photoluminescence intensity as a function of temperature were studied by the method described in [13]. Figure 8 shows the relative photoluminescence intensity at 90 and 300 K as a function of Ag content in the crystals and layers. Doping with Ag clearly leads to an increase of luminescence intensity in the crystals by two times and in the epitaxial layers by five times in comparison to the undoped samples. In addition, the epitaxial layers have more uniform luminescence properties. The luminescence intensity within uncertainty limits does not change along the surface at a distance

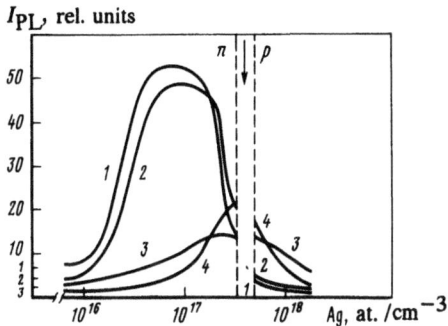

Fig. 8. Photoluminescence intensity of: 1, 2) epitaxial layers and 3, 4) crystals as functions of Ag content in PbTe. 1, 3) 300 K and 2, 4) 77 K.

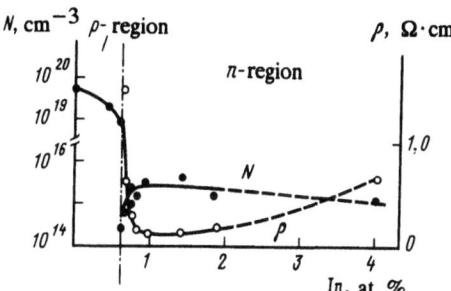

Fig. 9. Free carrier concentration and specific resistivity in $Pb_{0.78}Sn_{0.22}Te(In)$ single crystals as a function of In content (77 K).

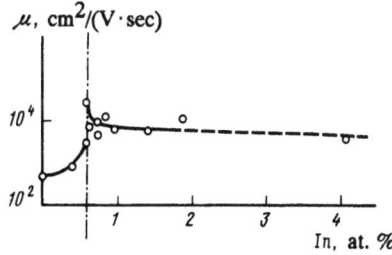

Fig. 10. Change of current carrier mobility in $Pb_{0.78}Sn_{0.22}Te(In)$ single crystals as a function of In content (77 K).

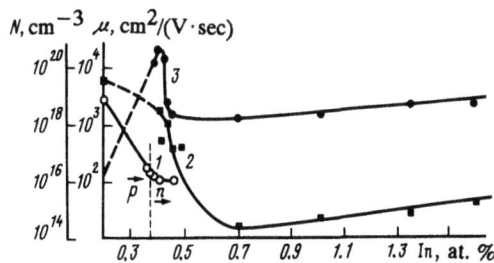

Fig. 11. Hall concentration (1, 2) and mobility (3) of carriers as functions of In content in $Pb_{1-x}Sn_xTe(In)$ layers. 1) $x = 0.2$; 2, 3) $x = 0.22$.

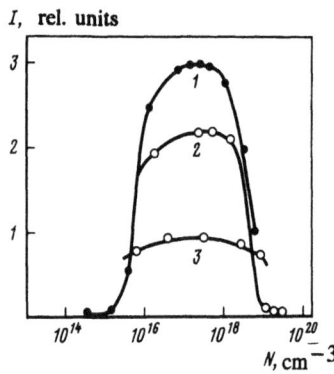

Fig. 12. Luminescence intensity of layers (1, 2) and undoped
(3) $Pb_{1-x}Sn_xTe(In)$ as a function of carrier concentration.

of several millimeters. The maximal luminescence intensity was observed in n-type epitaxial samples with Ag content of $(3-6) \cdot 10^{16}$ and a corresponding electron concentration of $n_{77} = 10^{14} - 10^{15}$ cm^{-3} [9]. This experimental fact may also be related to a possible decrease of radiationless recombination in this material. The decreased defect level of the crystal lattice is most probably due to formation of neutral complexes $2Ag^+ + Te^{--} + V_{Pb}^{--} + V_{Te}^{++} \rightarrow [2Ag^+V_{Pb}^{--}] + [Te^{--} + V_{Te}^{++}]$.

The behavior of Ag in the solid solution $Pb_{0.78}Sn_{0.22}Te$ showed that this dopant acts as an acceptor and has a multiplicative effect on the hole concentration, which also occurs in PbTe [14].

In contrast to silver, indium always acts as a donor in PbTe. All features of its doping are most clearly exhibited in $Pb_{0.78}Sn_{0.22}Te$. Addition of In dopants into $Pb_{0.78}Sn_{0.22}Te$ single crystals during growth from 0 to 0.6 at. % leads to a drop of hole concentration and an increase of Hall mobility (Figs. 9 and 10). This can be explained by a displacement of metal vacancies by dopant and a decreased concentration of charged defects related to this. Inversion of conductivity type from p to n occurs in the single crystals at an In content of 0.6-0.7 at. %. The concentration of free electrons in this composition region of $Pb_{0.78}Sn_{0.22}Te(In)$ reaches a record low value of $2 \cdot 10^{14}$ cm^{-3}. Further elevation of In content to 2% almost does not affect the carrier concentration [11]. The observed stabilization of the Fermi level in the bulk single crystals is explained by the generation of neutral dopant—vacancy complexes [15].

Doping $Pb_{0.78}Sn_{0.22}Te$ layers with indium during photostimulated growth allows the range of added dopant to be expanded to 2.7 at. % without destruction of the monophasic nature of the prepared material (Fig. 11). Moreover, the range of actual concentrations of free electron carriers is broadened to $n = 5 \cdot 10^{18}$ cm^{-3}. The luminescence quantum yield increases by several times in the $Pb_{1-x}Sn_xTe$ epitaxial layers ($x = 0.22, 0.20$) (Fig. 12). This is explained by the decreased concentration of vacancies in thermodynamic equilibrium. As a result of this, the number of precipitates which act as radiationless recombination centers is decreased [10, 16].

It is also important to note that the inversion of conductivity type from p to n caused by indium depends on the Sn concentration in $Pb_{1-x}Sn_xTe$. The p—n-transition in $Pb_{0.8}Sn_{0.2}Te$ occurs at 0.37 at. % In and in $Pb_{0.78}Sn_{0.22}Te$ at 0.45 at. % since the width of the homogeneous region in these solid solutions increases with increasing x.

Measurements showed that the structural perfection of layers doped with silver and indium lies at the level of luminescing epitaxial layers that are grown at much lower temperatures (Table 1).

Figure 13 presents measurements of the dislocation density distribution as a function of In content in $Pb_{0.78}Sn_{0.22}Te(In)$ structures (epitaxial layer)/PbTe (substrate) and $Pb_{0.78}Sn_{0.22}Te(In)$ (epitaxial layer)/ $Pb_{0.78}Sn_{0.22}Te$ (substrate). The functions obtained indicate that:

1. The dislocation density D in the epitaxial layers depends on the amount of added In. Thus, the dislocation density increases from $1 \cdot 10^4$ cm^{-2} for 0.2 at. % In to $8 \cdot 10^4$ cm^{-2} for 1.2 at. % In at identical D in the substrates.

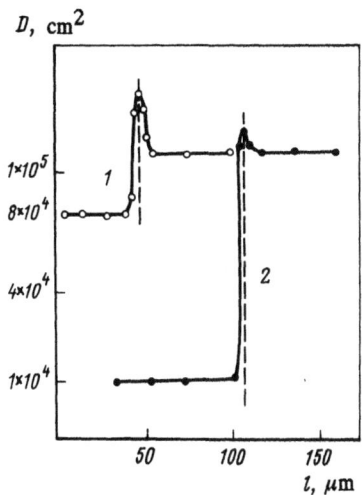

Fig. 13. Dislocation density distribution in heterostructures:
1) n-$Pb_{0.78}Sn_{0.22}Te + 1.2$ at. % In (epilayer)/p-PbTe (substrate); 2) p-$Pb_{0.78}Sn_{0.22}Te + 0.2$ at. % In (epilayer)/p-PbTe (substrate).

2. The epitaxial layers have different dislocation densities depending on the In concentration, but for all the structures studied, D in the layer is significantly lower than in the substrate as a result of the growth features under electromagnetic irradiation conditions.

The increased dislocation density observed at the heterojunctions occurs as a result of the mismatch of lattice constants. The width of the transition layer depends on the jump of the lattice constant at the heterojunction.

The dependence of the free carrier and dislocation density concentrations on the amount of added indium was estimated during preparation and practical use of heterostructures based on $Pb_{1-x}Sn_xTe(In)$. Dielectric properties at low temperatures, extremely high photoconductivity with large relaxation times [17], and ferroelectric polarization related to a phase transition [18] were detected in $Pb_{0.78}Sn_{0.22}Te$ doped with indium.

4. CONCLUSIONS

Analysis of the experimental results demonstrates that directed doping of $Pb_{1-x}Sn_xTe$ by dopants in an amount in accord with the homogeneous region allows the type and concentration of current carriers to be controlled. A record low electron concentration of $n_{77} \approx 10^{14}$ cm^{-3} can be attained. More importantly, the luminescence quantum yield can be raised severalfold when the material is used in optoelectronic devices. The position of the conductivity type inversion point as a function of growth temperature and mole fraction x can be observed in the cases of doping with silver and indium, respectively. In both cases this fact is related to a change of the homogeneous region dimensions and confirms that the principal type of interaction of dopant atoms with the $Pb_{1-x}Sn_xTe$ crystal lattice is interaction with vacancies. The discontinuous nature of the carrier concentration as a function of dopant content in $Pb_{1-x}Sn_xTe$ indicates the existence of different mechanisms of interaction with Pb and Te vacancies at different dopant amounts. An important result of doping is a decrease of crystal lattice defects. The increased luminescence quantum yield upon doping $Pb_{1-x}Sn_xTe$ with Ag and In, obtained for the first time, attests to this.

Comparison of electrophysical and luminescence properties of the prepared doped crystals and epitaxial layers clearly shows that the $Pb_{1-x}Sn_xTe(Ag)$ layers themselves have a minimal carrier concentration of $n = 5 \cdot 10^{14}$ cm^{-3} (77 K) and a maximal luminescence quantum yield value. Consequently, incorporation of dopant into the crystal lattice during doping is most effective for photostimulated growth. Since radiation encompassing the region of ≤ 5 eV energy quanta impinges on the growth zone, it should be assumed that the lamp radiation can excite the electronic subsystem of dopant atoms and thereby increase their activity for interaction with the PbTe lattice.

Thus, studies on the doping of bulk single crystals and the epitaxial growth process under photostimulation conditions based on the example of $Pb_{1-x}Sn_xTe$ demonstrate the advantage of combining the two methods for controlling the composition of the of A^4B^6 compound into the photostimulated doping method. In this process, the dopant interacts more effectively with the crystal lattice.

REFERENCES

1. Yu. I. Gorina, G. A. Kalyuzhnaya, K. V. Kiseleva, et al., "On the nature of defects in undoped PbTe," *Kratk. Soobshch. Fiz.*, No. 11, 24-27 (1975).
2. A. P. Shotov, K. V. Vyatkin, and A. A. Sinyatynskii, "Injection lasers with a double heterostructure based on $Pb_{1-x}Sn_xSe$ prepared by the method of molecular epitaxy," *Pis'ma Zh. Éksp. Teor. Fiz.*, 6, No. 16, 983-986 (1980).
3. K. V. Vyatkin and A. P. Shotov, "Optical properties of epitaxial films of PbSe," *Fiz. Tekh. Poluprovodn.*, 14, No. 7, 1331-1334 (1980).
4. Yu. I. Gorina, G. A. Kalyuzhnaya, A. V. Kuznetsov, et al., "Method of film growth," Inventor's Certificate 466816 (USSR), *Byull. Izobret.*, No. 29 (1978).
5. G. M. Guro, G. A. Kalyuzhnaya, T. S. Mamedov, and L. A. Shelepin, "Effect of irradiation on crystal growth kinetics," in: *Statistical and Coherent Methods for Studying Physical Systems*, Proc. P. N. Lebedev Phys. Inst., Acad. Sci. USSR, Vol. 24 (1980), pp. 127-140.
6. Yu. I. Gorina, G. A. Kalyuzhnaya, K. V. Kiseleva, et al., "Study of laser epitaxial structures of the PbTe(Bi)–PbSnTe–PbTe type," *Fiz. Tekh. Poluprovodn.*, 13, No. 2, 305-310 (1979).
7. G. A. Kalyuzhnaya, K. V. Kiseleva, Yu. I. Gorina, et al., "Study of crystal chemical characteristics of heterostructures based on $Pb_{1-x}Sn_xTe$ prepared by gas epitaxy," *Kratk. Soobshch. Fiz.*, No. 8, 15-19 (1976).
8. Yu. I. Gorina, S. Zainudinov, G. A. Kalyuzhnaya, et al., "Growth and doping from the gas phase of lead–tin telluride," *Izv. Akad. Nauk SSSR, Neorg. Mater.*, 20, No. 7, 1112-1116 (1984).
9. Yu. I. Gorina, S. Zainudinov, S. I. Zolotov, et al., "Effect of silver on the electrical properties and photoluminescence of lead telluride," *Izv. Akad. Nauk SSSR, Neorg. Mater.*, 22, No. 7, 1105-1108 (1986).
10. O. V. Aleksandrov, G. A. Kalyuzhnaya, K. V. Kiseleva, and N. I. Strogankova, "Study of the solid solution $Pb_{1-x}Sn_xTe$ doped with indium with low current carrier concentration," *Izv. Akad. Nauk SSSR, Neorg. Mater.*, 14, No. 7, 1277-1279 (1978).
11. G. A. Kalyuzhnaya, T. S. Mamedov, K. V. Kiseleva, and A. D. Britov, "Properties of epitaxial layers of solid solutions doped with indium," *Izv. Akad. Nauk SSSR, Neorg. Mater.*, 15, No. 2, 231-234 (1979).
12. O. V. Aleksandrov, Yu. I. Gorina, G. A. Kalyuzhnaya, et al., *Kratk. Soobshch. Fiz.*, No. 7, 35-41 (1981).
13. S. I. Zolotov and A. É. Yunovich, "Method for studying the photoluminescent power and quantum yield in narrow-band semiconductors," *Prib. Tekh. Éksp.*, No. 6, 185-187 (1985).
14. Yu. I. Gorina, G. A. Kalyuzhnaya, K. V. Kiseleva, et al., "Effect of silver dopant on the electrical properties of a lead–tin telluride solution," *Kratk. Soobshch. Fiz.*, No. 7, 42-47 (1981).
15. O. V. Aleksandrov, K. V. Kiseleva, and S. V. Kuchaev, "Chemical bonding of In–Te in the $Pb_{0.78}Sn_{0.22}Te(In)$ lattice," *Kratk. Soobshch. Fiz.*, No. 1, 60-63 (1985).
16. A. D. Britov, S. M. Karavaev, G. A. Kalyuzhnaya, et al., "Laser diodes based on PbSnTe in the 5-15 μm region," *Kvantovaya Élektron.* (Moscow), 3, No. 10, 2238-2241 (1976).
17. B. M. Vul, I. D. Voronova, G. A. Kalyuzhnaya, et al., "Features of transition effects in $Pb_{1-x}Sn_xTe$ with large indium concentrations," *Pis'ma Zh. Éksp. Teor. Fiz.*, 29, No. 1, 21-25 (1979).
18. K. H. Herrman, G. A. Kaljuzhnaja, K. P. Mollman, and M. Wendt, "Photoferroelectric behavior in $Pb_{0.78}Sn_{0.22}Te:In$," *Phys. Status Solidi A*, 71, K21-K23 (1982).

Part IV

CRYSTALLIZATION OF BIOLOGICAL SUBSTANCES

CRYSTALLIZATION OF RIBOSOMES, RIBOSOMAL SUBUNITS, AND INDIVIDUAL RIBOSOMAL PROTEINS

B. K. Vainshtein and S. D. Trakhanov

1. CRYSTALLIZATION AND STRUCTURAL STUDIES
OF RIBOSOMAL PARTICLES

There is interest in the three-dimensional structure of the ribosome since it is a unique multicomponent biological system responsible for the biosynthesis of protein in the cell. The exceedingly complex structure of ribosomes (for example, the *Escherichia coli* ribosome, the molecular weight of which is $2.15 \cdot 10^6$ Da, consists of three types of RNA and 52 individual ribosomal proteins) makes it difficult to study them using classical methods of x-ray structural analysis.

Significant progress in determination of the integral form of the ribosome and ribosomal subunits has been achieved due mainly to application of constantly improving electron microscopy methods including immuno electron microscopy. These studies have demonstrated the architectural role of ribosomal RNA and the peripheral disposition of individual ribosomal proteins. Chemical modification and neutron scattering methods were employed to reveal the approximate topography of individual ribosomal proteins completely for the 30S ribosomal subunit of *E. coli* and partially for the 50S subunit. These studies are collected in the monograph of A. S. Spirin [1] and in a series of subsequent reviews [2-5].

In this respect, a quantum leap in our understanding of the three-dimensional structure of the ribosome, based on physicochemical methods, cannot be expected without application of x-ray structural analysis and, in turn, without growth of single crystals of ribosomes as the first very important stage of the crystallographic study.

This problem until recently seemed difficult to surmount in spite of the prospects of relating work on ribosomes to similar lines of work. First, other nucleoprotein particles, the nucleosomes [6, 7], which have exceedingly complex structure although not to the same degree as ribosomes, have been successfully crystallized. Virus particles have also been crystallized [8]. Viruses, similar to ribosomes, are nucleoproteins although they differ from the latter by a significantly smaller collection of proteins, a multiplicity of individual protein copies, and a symmetric distribution of them about the nucleoprotein periphery.

Second, ribosomes of some eukaryotic cells *in vivo* under certain conditions form ordered structures that can be viewed as two-dimensional crystals. Periodic packing of the ribosome was observed in *Entamoeba histolytica* [9], *En. invadens* [10], in chick embryos [11, 12], in amphibians [13], and in the human brain during pathology [14]. This periodic packing appears as tubes for *En. invadens* [10] or as double layers of ribosomes associated with the membrane for ovocytes of the amphibian *Lacerta sicula* [13]. A similar association is ob-

169

Table 1. Crystallographic Constants of Prokaryotic Ribosomes and Subunits from Electron Microscopy

Specimen	Source			Unit cell constants (Å)						Reference
				a	b	c	α	β	γ	
70S	Escherichia coli			340±30	340±30	590±30	90°	90°	120°	[38]
70S	Thermus thermophilus			490±20	490±20	–	–	–	90°	[39]
50S	Escherichia coli			330±20	330±30	–	–	–	123±5°	[15]
50S	Bacillus stearothermophilus	crystalline form	I	134	256	–	–	–	96°	[18]
			II	158	288	–	–	–	97°	[18]
			III	260	288	–	–	–	104°	[18]
			IV	405	405	256	–	–	120°	[18]
			V	240±20	230±20	310±30	111°	90°	120°	[19]
30S	Thermus thermophilus			310	360	160	90°	90°	90°	[57]

served as a result of prolonged cooling of ovocytes. Ribosomes in the last case are organized into tetramers. Their periodic packing in a plane is characterized by unit cell constants $a = b = 595$ Å.

Ribosomal tetramers are also observed in ordered structures formed in chick embryos upon slow cooling [12]. The two-dimensional distribution of electron-dense material in electron photomicrographs coincides with the distribution for the ribosomes of *L. sicula*, and the unit cell constants are the same ($a = b = 593$ Å).

Layers of membrane-bound ribosomes which are stacked on each other and form ordered structures in this manner were observed in specimens of human brain [14]. Optical diffraction of electron microscopic images of cross sections of these sandwich structures gave the packing constants $a = b = 130$ Å and $\gamma = 56°$.

The ability to form ordered two-dimensional layers was also detected for ribosomes isolated from bacteria. Conditions for ultraenrichment of ribosomes are apparently not realized *in vivo* in prokaryotic cells, in contrast to several eukaryotic cells. Ribosomes under conditions unfavorable for biosynthesis (prolonged cooling) are stored as ordered structures, which ensure their maximal packing density. Crystals of large *E. coli* ribosomal subunits prepared in experiments *in vitro* [15] are closed tubes or crystalline sheets of one or more ribosomal layers. The unit cell constants of the crystals are given in Table 1.

Thus, the ability of ribosomes and their subunits (at least the 50S subunits) to form two-dimensional crystals is a persuasive argument in favor of the feasibility of preparing crystals that are periodic in all three directions.

The first successful step was preparation of three-dimensional crystals of the large subunit of *Bacillus stearothermophilus* ribosomes [16]. The positive result of the experiments begun by Yonath was due primarily to the choice of source for ribosome isolation. The bacterium *B. stearothermophilus* is a moderately thermophilic microorganism with an optimal growth temperature of 60°C. Obviously, the proteins, nucleic acids, and nucleoproteins of this microorganism in the isolated form will have increased stability in comparison to macromolecules isolated from traditional sources (*E. coli*), both with respect to temperature and to the effect of denaturing agents.

Conditions for growth of crystals of the large ribosomal subunit of *B. stearothermophilus* were investigated in detail further in [17, 20]. Crystals were observed to be formed using organic solvents as precipitating agents.

Table 2. Crystallographic Constants of the Large Prokaryotic Ribosomal Subunit from X-Ray Structural Analysis

Specimen	Source	Space group	Unit cell constants, Å						Reference
			a	b	c	α	β	γ	
50S	Bacillus stearothermophilus	P222$_1$ (P2$_1$2$_1$2; P2$_1$2$_1$2$_1$)	350±10	670±10	950±10	90°	90°	90°	[22]
50S	Halobacterium marismortui	P2$_1$	182±5	584±10	186±5	90°	109°	90°	[26]

Monatomic and diatomic low-molecular-weight alcohols were used (methanol, ethanol, ethylene glycol, 2-methyl-2,4-pentanediol, etc.). Exchange of one alcohol for another as well as the use of two-component mixtures was noticed in a number of cases to affect only the packing of 50S subunits in the crystal and not the appearance of the crystal itself. The morphology of crystals and the internal crystalline packing did not change with a change of pH in the stable range of the ribosomal subunits (pH 6-9) with exchange of one buffer by another (MES, HEPES, Tris, Gly-NaOH). Crystals were grown by both gaseous diffusion (hanging and sitting drops) and free diffusion. The authors developed their own modification of the methods described in [17]. The presence of crystals in the samples was visualized under a microscope. Their macromolecular nature was demonstrated by polymerization of crystals crosslinked with glutaric aldehyde in epoxide resin with subsequent investigation of thin sections under an electron microscope. In subsequent publications [18-20], the results of such investigations were reported and the constants of the prepared crystalline forms of the *B. stearothermophilus* 50S subunit grown in various organic solvents were given. These data are summarized in Table 1.

It should be noted that a strong correlation exists between the ability of the 50S subunit specimens to crystallize and their biological activity [19, 20]. Specimens having low activity could not be crystallized. In all cases, the crystalline material maintained its biological activity for several months whereas the initial specimen of isolated subunits was quickly inactivated in solution.

The internal crystalline packing of the 50S subunits varies depending on the growth conditions of the crystals. This is illustrated by the data of Table 1 (crystalline forms I-V) although the possibility that at least some of the constants are obtained from different sections of the same crystalline modification is not excluded.

Optimization of the growth conditions enabled crystals of the 50S subunit of dimensions up to 1.0 mm to be prepared (the two other dimensions are substantially smaller and do not exceed 0.1-0.2 mm) [22]. Although Laue patterns could be taken using synchrotron radiation [21] and later x-ray precession photos were obtained [22] from the crystals, they are unsuitable for systematic structural studies due to their extreme brittleness and the very large unit cell constants (Table 2). The propensity of the crystals to crack with any mechanical handling is so large that they should be grown directly in x-ray capillaries and photos should be taken without removing the mother liquor [22].

The large ribosomal subunit from *Halobacterium marismortui*, an avid halophile which grows only at high salt concentrations in the culture medium, turned out to be a more promising specimen for crystallographic studies. Polyethyleneglycol with a molecular weight of 6000 Da was used for crystallization of the 50S subunit from this source. This is a precipitating agent which has recently been successfully and widely used for crystallization of proteins [23]. Three crystalline modifications of the 50S subunit were prepared depending on the

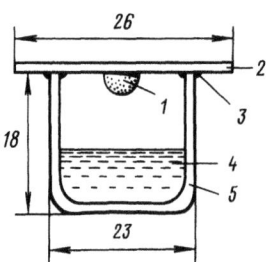

Fig. 1. Crystallization cell for macromolecules by the hanging drop method. 1) ribosome solution (5-15 μl); 2) cover glass; 3) vacuum grease; 4) countersolution; and 5) lower part of a penicillin vial. Dimensions are shown in millimeters.

Mg^{2+} concentration in the crystallized sample and the salt concentration in the countersolution [24-26]. Crystals of dimensions 0.6 × 0.6 × 0.2 mm (crystal form III) suitable for x-ray structural analysis were grown by introduction of seed crystals of form II (thin plates) into a drop of the 50S subunit crystallizing solution.*

The unit cell constants of the form III crystals determined using synchrotron radiation are given in Table 2. In spite of the fact that the unit cell volume in this case is significantly smaller than that for the crystals of the *B. stearothermophilus* 50S subunit, the asymmetric unit of the cell contains two 50S subunit molecules of molecular weight $3.2 \cdot 10^6$ Da. Thus, determination of the three-dimensional structure of the 50S subunit with the resolution which the crystals provide (about 10 Å) is a uniquely complex problem and is fraught with difficulties both during data collection from the native crystal and in the search for heavy-atom derivatives. Possible means for solution of the phase problem were analyzed in [27]. One of these consists of inclusion of neutron scattering data for the crystals. This was done for nucleosomes [28].

Thus, investigation of the three-dimensional structure of a small ribosomal subunit was much more attractive. The mass of the small 30S ribosomal subunit of prokaryotic cells is only 850,000 Da. This is approximately two times smaller than the mass of the large subunit. Besides this, the 30S subunit is simpler in protein composition (20-21 proteins) and contains only one molecule of ribosomal RNA. From the viewpoint of protein—protein and RNA—protein interactions, the 30S subunit is also much more studied than the 50S subunit. This can be significant for interpretation of the electron density map.

Recently, efforts to crystallize the small ribosomal subunit of *T. thermophilus* have been made at the Institute of Proteins and the Institute of Crystallography of the Academy of Sciences of the USSR. This extremely thermophilic microorganism has an optimal growth temperature of 75°C. This makes it a source of nucleic acid proteins and their complexes which have high stability even relative to the macromolecules isolated from *B. stearothermophilus*.

Actually, a number of proteins from this source [31-34], including such complicated ones as RNA-polymerase [31] and catalase [32], have been crystallized and structural studies of several of these (polypeptide chain elongation factor G [35] and catalase [36]) have been completed.

Crystals of the *T. thermophilus* 30S ribosomal subunit were prepared by two methods: the first method uses dialysis in microdialysis cells and gaseous diffusion. The second is a mechanical method using a variation of the hanging drop method shown in Fig. 1. The crystallization cell consisted of a container of countersolution, the lower part of a penicillin vial with a polished edge, and a cover glass prepared from a standard micro-

---

*The method of introduction of microseeds [29], as well as macroseeds [30], is commonly used for growth of large protein crystals. It turned out to be very useful in our work on crystallization of small ribosomal subunits of *T. thermophilus* (see below).

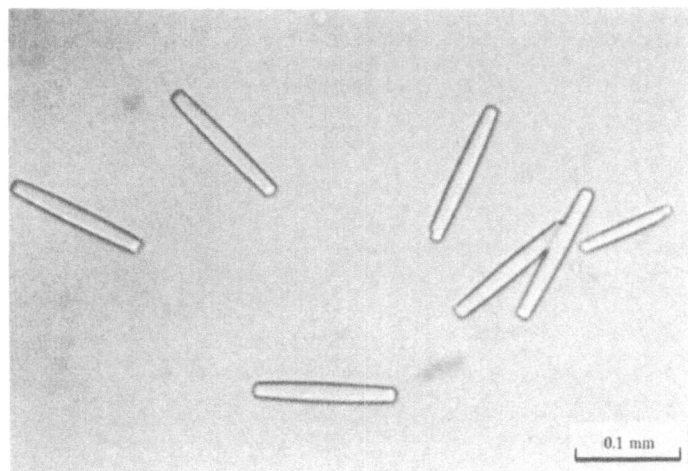

Fig. 2. Photomicrograph of crystals of the small *T. thermophilus* ribosomal subunit grown as described in [57].

scope slide cut transversely into three parts. Equilibrium will be established in the enclosed volume, which is hermetically sealed with vacuum grease and contains the countersolution and hanging drop of the crystallizing solution, due to diffusion of organic solvent molecules into the drop. The solubility of macromolecules in the drop will be reduced and crystallization can begin upon attainment of supersaturation.

Figure 2 shows photomicrographs of 30S subunit crystals grown by the hanging drop method from a 15% solution of 2-methyl-2,4-pentanediol in a buffer of 25 mM $MgCl_2$ (pH 8.0). The crystals have a rodlike form with slightly curved long side faces. The length of the crystals reaches 0.2 mm with a width of 30-40 μm. Since crystals of this size are too small for *x*-ray structural analysis, their internal packing was studied on thin sections by electron microscopy. For this, crystals fixed overnight with 0.5% glutaric aldehyde were polymerized in Epon. Ultrathin sections were prepared on a microtome and stained with uranyl acetate. Figure 3 shows electron microscopic images of several sections of the 30S subunit crystals. The optical diffraction is given at the right of each picture. The computer-enhanced electron microscopic image of the corresponding crystal cross section is given in Fig. 3c.

The unit cell constants estimated from the optical diffraction data are 310 × 360 × 160 Å. The resolution attained in the sections is 35-40 Å.

The morphology of the crystals depends strongly on the concentration of 2-methyl-2,4-pentanediol in the countersolution. The crystals have a larger heterogeneity in size at concentrations of 18-20%. Their length is much smaller as a rule than for the optimal crystals at the same thickness (Fig. 2).

Introduction of a seed greatly facilitates crystallization, reducing the formation time of 30S crystals and improving their quality. In several identical samples that were left for crystallization, crystals formed in 1-2 days in only those samples in which seed crystals were previously introduced. In the remaining samples, the 30S specimen forms an amorphous precipitate. This is in complete agreement with the behavior exhibited during crystallization of the *B. stearothermophilus* 50S subunit according to which nonspecific aggregation of subunits occurs in the initial stage of 50S crystal growth. Formation of nuclei follows due to ordering of the packing inside the aggregates [27]. Actually, the amorphous precipitate of 30S subunits converts almost completely into a crystalline state after about a week of standing.

Study of the three-dimensional structure of the complete ribosomal particle is an exceedingly complex task. However, it remains on the agenda as a natural extension of crystallographic work on determination of the three-dimensional structures of the 30S and 50S subunits. An answer as to whether and how the three-

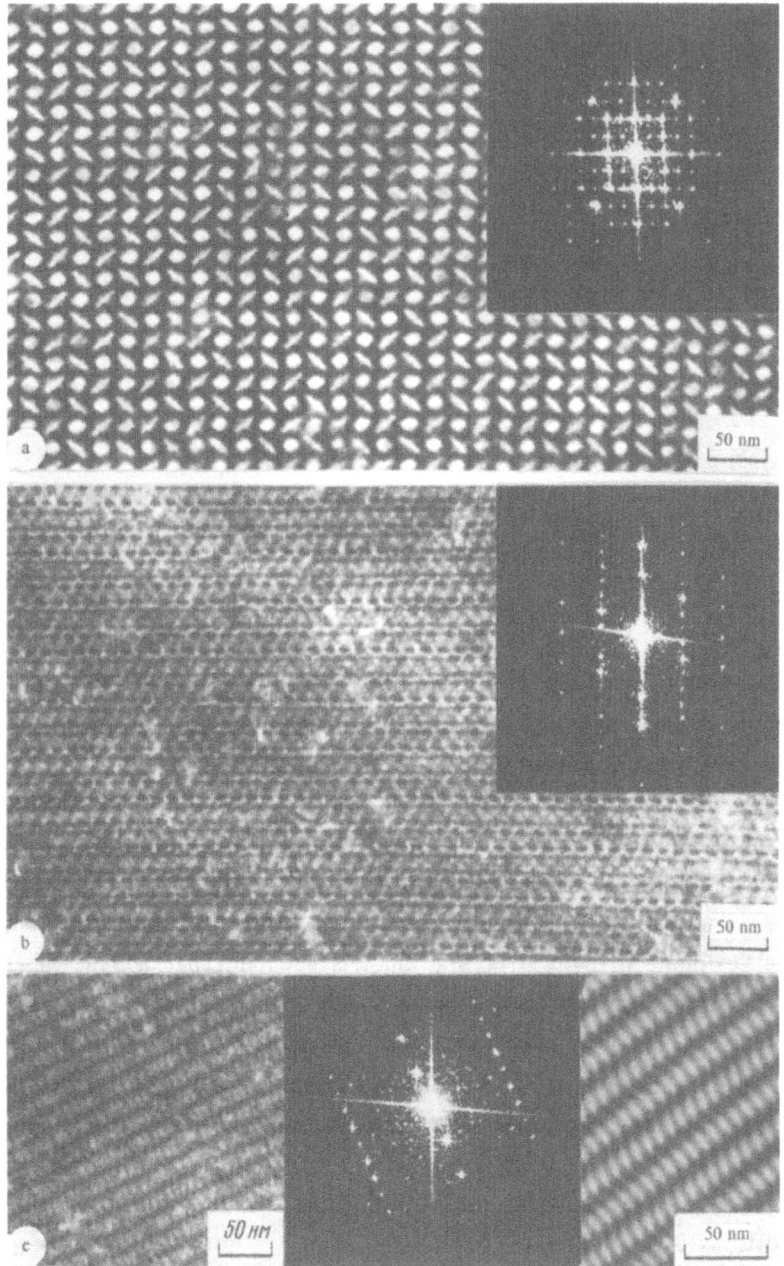

Fig. 3. Electron microscopic images of ultrathin sections of crystals of *T. thermophilus* 30S subunits. Optical diffraction patterns obtained from images a-c are on the right of a and b and in the center of c. The enhanced image of the corresponding crystal cross section is given on the right of c.

dimensional structures of the subunits in the free state differ from that in the complex, i.e., the ribosome, can already be obtained at low resolution.

The first three-dimensional crystals were prepared for the ribosome isolated from *E. coli* [38]. They are long (up to 0.2 mm) thin needles which have hexagonal packing. The unit cell constants (Table 1) were determined by electron microscopy during analysis of thin sections. Optical diffraction of electron microscopic image negatives gives 65 Å resolution. The dissolved crystals exhibit activity for polyphenylalanine synthesis on a poly(U) matrix in comparison to the activity of the starting specimen.

In parallel with our experiments on the search for crystallization conditions for the small *T. thermophilus* ribosomal subunit [39], work was carried out on growth of crystals of whole ribosomal particles from the same

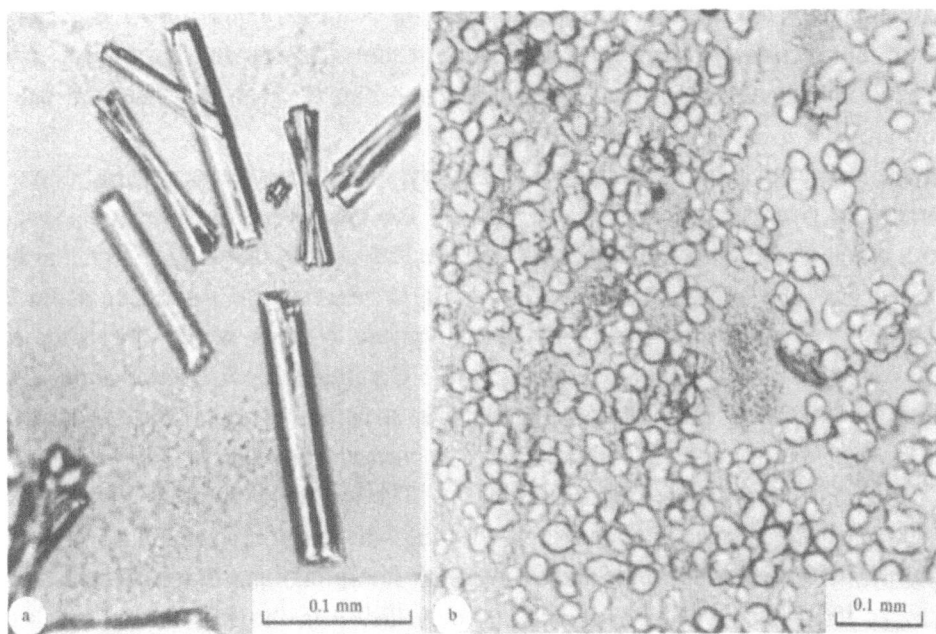

Fig. 4. Photomicrographs of crystals of *T. thermophilus* ribosomes (photographs supplied by the Institute of Proteins of the Academy of Sciences of the USSR). Crystals prepared by the hanging drop method [39] (a) and crystals grown by microdialysis (b). Ribosomes after the final purification step according to [58] were deposited on the bottom of a tube by ultracentrifugation and dissolved at a concentration of 8 mg/ml in buffer A (20 mM Tris-HCl, pH 8.0, 75 mM NH_4Cl, 25 mM $MgCl_2$, 5 mM 2-mercaptoethanol, and 0.2 mM EDTA). A 25-μl portion of ribosomes was mixed with an equal volume of countersolution (14% solution of 2-methyl-2,4-pentanediol in buffer A). Crystallization was carried out in microdialysis cells for 3-4 weeks at 4°C.

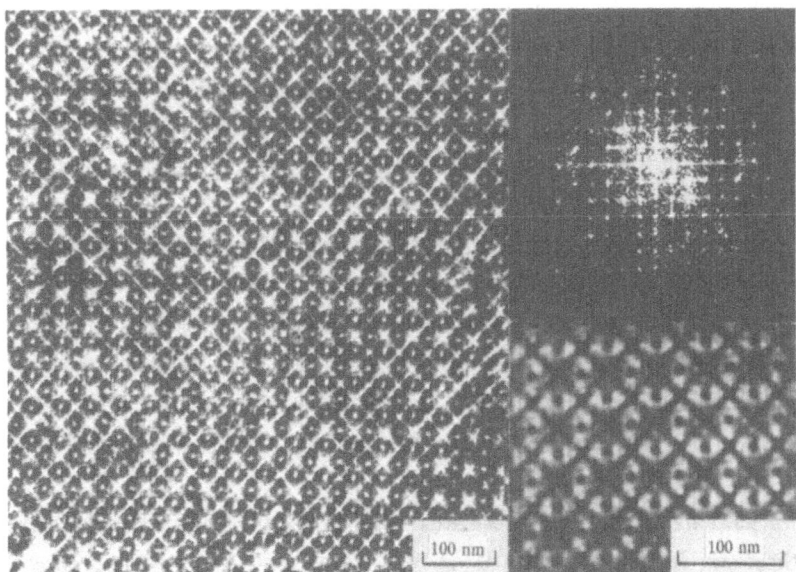

Fig. 5. Electron photomicrograph of a section of a *T. thermophilus* 70S ribosome crystal. Optical diffraction of the image (upper right) was carried out at the section. The enhanced image of the crystal cross section is given at the lower right.

source. Crystals in the shape of parallelepipeds (Fig. 4a) were prepared by the hanging drop method using a 10-15% methylpentanediol countersolution as precipitating agent. They appeared after 1-5 days at 5°C in a buffer of composition given in [39]. Ultrathin sections of the crystals after staining with uranyl acetate and lead citrate were analyzed under an electron microscope. Figure 5 shows a portion of the electron microscopic negative of the ultrathin section of one of the crystal cross-sectional planes and the optical diffraction from this section. The resolution estimated from the optical diffraction data is 70-80 Å. The unit cell constants in the section projection are $a = b = 490$ Å, $\gamma = 90°$.

Microdialysis produced *T. thermophilus* ribosome crystals of another crystalline form (Fig. 4b). The experimental details are given in the caption for Fig. 4. The volume of the prepared crystals with maximal dimension up to 0.2 mm is substantially larger than for the 70S crystals shown in Fig. 4a, although the resolution limits estimated from the optical diffraction data are approximately the same, i.e., about 70 Å.

In summary, we have prepared three-dimensional crystals of both whole ribosomes of *E. coli* and *T. thermophilus* and the 30S subunit of *T. thermophilus* and 50S subunits of *B. stearothermophilus* and *H. marismortui*. The crystals are still not suitable for systematic x-ray structural study in spite of the fact that precession x-ray photos can be obtained for the 50S subunits from *B. stearothermophilus* and *H. marismortui* and unit cell constants can be determined from them [22, 26] (Table 2). The quality of the x-ray photos is still not adequate for a detailed three-dimensional structural analysis of the ribosome and (or) its subunit, which at present require a voluminous quantity of experimental biochemical data for interpretation. Improvement of crystal quality by increasing the size of the crystal, which will apparently be achieved in the near future, will provide an impetus for further x-ray structural studies. Crystallization of the specimen under microgravity [37] may be a necessary step of the study if the required results are not obtained under normal gravitational conditions.

2. CRYSTALLIZATION AND STRUCTURAL STUDIES
OF INDIVIDUAL RIBOSOMAL PROTEINS

An alternative and complementary route of studying the structure of whole ribosomes and their large and small subunits *in situ* consists of fragmentation of the ribosomes and analysis of their constituent proteins, RNA, and intermediate complexes. Figuratively speaking, this is an approximation to the ribosomal structure "from below." The validity of such an approach will remain an open question until the spatial architecture of the ribosome is known with a high degree of accuracy.

If analysis of the ribosomal structure is carried out at the level of the isolated ribosomal proteins, then this question is equivalent to the following: Is the three-dimensional structure of the ribosomal proteins in the isolated state the same as in the intact ribosomal particles? At present, a positive answer to this question is tacitly suggested.

The rational basis for this assumption is that the ribosomal proteins in the isolated state have stable and well-determined secondary and tertiary structures. The initial opinion of a strongly distended and loose conformation for ribosomal proteins reflected the absence of methods for gentle isolation of individual proteins in the native state and apparently was also a result of the low conformational stability of several proteins outside the ribosome. Later, this viewpoint was reexamined. Evidence for the existence of a compact highly structured conformation of ribosomal proteins is given in the monograph of Spirin and the review of Liljas ([1, 4] and references therein).

An additional argument for the compactness of ribosomal proteins follows from x-ray structural analysis data [40-42]. The three-dimensional structures of three proteins of the *B. stearothermophilus* ribosome were determined at low resolution, proteins L6, L30, and S5. The three-dimensional structure of each protein can be approximated by a rotational ellipsoid with a ratio of axes not exceeding 2:1 in each case. This contradicts earlier work on the estimation of the inertial radii of the proteins by hydrodynamic methods according to which the protein molecules are highly asymmetric and can be represented as rotational ellipsoids with a ratio of axes from 5:1 to 7:1 [43, 44].

Table 3 contains crystallographic constants of these ribosomal proteins which at present are being studied by x-ray structural analysis. With the exception of the C-terminus of the L7/L12 protein, proteins which are

Table 3. Crystallographic Constants of Individual Ribosomal Proteins

Protein	Molecular weight, Daltons	Source of isolated protein	Space group	Unit cell constants, Å			Maximal resolution of crystals	Reference
				a	b	c		
L17	16068	Bacillus	$P2_1 2_1 2_1$	37.5	48.9	135.3	2.5	[50,51]
L30	7053	stearother- mophilus	$P4_3 2_1 2$	46.3	46.3	61.4	2.0(2.5)	[52, 56]
L6	19168		$P6_1 22$	72.7	72.7	124.9	2.5(6)	[51, 53]
S5	17628		$P3_1 21(P3_2 21)$	59.3	59.3	109.8	3.0(5)	[51, 52]
$(L13)_4 L$	66000		$P4_1 2_1 2(P4_3 2_1 2)$	108	108	242	3.5	[54]
L29	7262	Escherichia coli	$P2_1^*$	52	119	90	6	[51]
L7/L12	N-terminus, 1-36 amino acid residues		$P6_1(P6_5)$	72.7	72.7	137.4	4	[49]
	C-terminus, 47-120 amino acid residues		$P4_3 2_1 2$	54.7	54.7	42.5	1.5 (1.7)	[49]
	Complex of protein L25 with a 5S RNA fragment		$P6_1 22(P6_5 22)$	119	119	250	10	[55]

Note. The asterisk corresponds to $\gamma = 71°$. The resolution given in parentheses corresponds to that published with the structure.

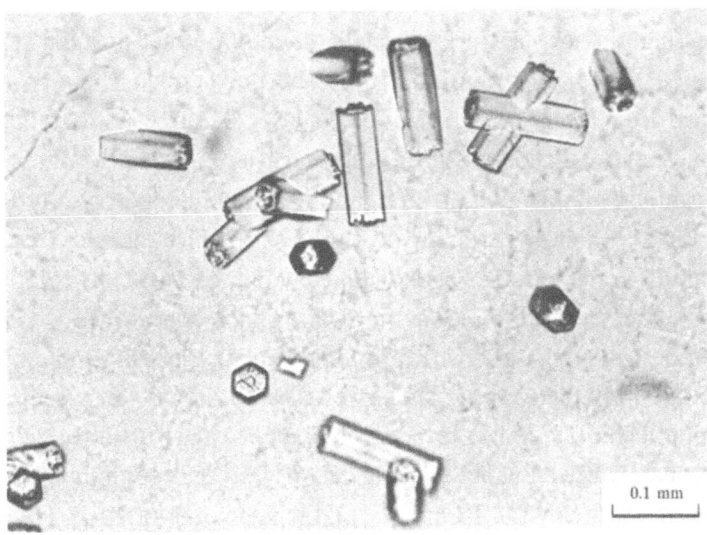

Fig. 6. Microphotographs of TL7 protein crystals from a large subunit of the *T. thermophilus* ribosome.

suitable for crystallographic study at high and medium resolution (i.e., at a resolution at which the path of the polypeptide chain can be traced from the electron density maps obtained) were isolated from the thermophilic microorganism *B. stearothermophilus.*

This microorganism, as well as other thermophilic bacteria, is obviously a source of ribosomal proteins and nucleic acids which are much more promising for structural investigations. This is supported by the report on the successful crystallization of 5S RNA from the ribosome of extremely thermophilic *T. thermophilus* [45].

The methods used for crystallization of ribosomal proteins given in Table 3 are standard and well described in many reviews, for example the monograph of McPherson [46]. Both organic compounds and inorganic salts were used as precipitating agents. Addition of 5% dioxane to a crystallization medium containing 1.8 M phosphate improves the crystal morphology for protein L6 [40].

Eight proteins of the 50S subunit were isolated using a complex approach to the study of the three-dimensional structure of *T. thermophilus* ribosomal proteins. One of these proteins TL7 (in all probability an analog to protein L6 from *E. coli* ribosomes) was crystallized. Crystals of hexagonal shape (Fig. 6) were grown by the hanging drop method under conditions published in [47].

A very important achievement in the crystallographic study of ribosomal proteins was the determination of the spatial organization of the L7/L12 protein of *E. coli*, which was resolved to 2.6 Å [48] and then to 1.7 Å [4]. The L7/L12 protein, which consists of 120 amino acid residues, undergoes spontaneous proteolysis during crystallization [49]. The two types of crystals obtained are the N-terminus (amino acid residues 1-36) and the C-terminus (amino acid residues 47-120) fragments of the molecule. Crystals of the latter are suitable for high-resolution x-ray structural study.

The three-dimensional structure of the C-terminus of the L7/L12* protein consists of three α-helices (residues 64-76, 79-89, and 100-114) which form one layer of a compact protein globule whereas three antiparallel β-sheets form the other layer. Analysis of the spatial location of the amino acid residues reveals that a large part of them are tied up in formation of hydrogen bonds. This makes it similar to structures that are extremely stable to denaturation. The exception is the N-terminus of the peptide domain (residues 47-52) which is poorly defined in the electron density maps and indicates mobility [4].

Recently, the three-dimensional structure of yet another ribosomal protein, protein L30 from *B. stearothermophilus,* was determined to 2.5 Å resolution [56]. The polypeptide chain, consisting of 61 amino acid residues, is formed into two α-helices to which a β-folded layer of three antiparallel β-sheets abuts from one side. Thus, in spite of the very weak homology of amino acid sequences, the three-dimensional structures of protein L30 and the C-terminus of the L7/L12 protein fragment are topologically very similar, with the sole exception that one of the three α-helices is absent in the L30 protein molecule.

In conclusion, it should be emphasized that significant improvement of isolation methods for individual ribosomal proteins, which is characterized by their preparation in a homogeneous and structured (native) state and a shift of emphasis into structural work with proteins from thermophilic microorganisms in the next ten years, will guarantee intensified crystallographic studies of these proteins. Besides the proteins listed in Table 3, for which detailed structural information have been obtained, x-ray structural studies for a number of others should be expected in the near future. Possible candidates may be thermophilic analogs of proteins S2, S3, S7, S13, S15... (small subunits) and proteins L9, L17, L27... (large subunits) of the *E. coli* ribosome. Preparation of suitable single crystals of them will be a necessary stage of the study, which will be successful as follows from an analysis of the experimental data of the present review.

---

*L7 and L12 are designations of two electrophoretically identified forms of the same protein. The difference consists in acetylation of the N-terminus amino acids. This modification leads to a change of the electrophoretic mobility of the protein.

REFERENCES

1. A. S. Spirin, *Molecular Biology. Structure of Ribosomes and Biosynthesis of Protein* [in Russian], Vysshaya Shkola, Moscow (1986), pp. 62-133.
2. G. Chamberlis, G. R. Craven, J. Davies, K. Davies, L. Kahan, and M. Nomura (eds.), *Ribosomes: Structure, Function and Genetics*, Univ. Park Press, Baltimore (1980).
3. V. D. Vasil'ev, "Electron microscopy of ribosomes of *Escherichia coli*," Author's Abstract of Doctoral Dissertation, Biological Sciences, Moscow (1984).
4. A. Liljas, "Structural studies of ribosomes," *Prog. Biophys. Mol. Biol.*, **40**, 101-228 (1982).
5. H. G. Wittmann, "Architecture of procaryotic ribosomes," *Annu. Rev. Biochem.*, **52**, 35-65 (1983).
6. J. T. Finch, L. C. Lutter, D. Rhodes, et al., "Structure of nucleosome core particles of chromatin," *Nature* (London), **269**, 29-36 (1977).
7. J. T. Finch, R. S. Brown, D. Rhodes, et al., "X-ray diffraction study of a new crystal form of the nucleosome core showing higher resolution," *J. Mol. Biol.*, **145**, 757-769 (1981).
8. F. A. Jurnak and A. McPherson (eds.), *Biological Macromolecules and Assemblies*, Vol. 1, Virus Structure, Wiley, New York (1984).
9. Y. Kress, M. Wittner, and R. M. Rosenbaum, "Sites of cytoplasmic ribonucleoprotein-filament assembly in relation to helical body formation in axenic trophozoites of *Entamoeba histolytica*," *J. Cell Biol.*, **49**, 773-784 (1971).
10. J. A. Lake and H. S. Slayter, "Three-dimensional structure of the chromatoid body helix of *Entamoeba invadens*," *J. Mol. Biol.*, **66**, 271-282 (1972).
11. R. A. Milligan and P. N. T. Unvin, "*In vitro* crystallization of ribosomes from chick embryos," *J. Cell Biol.*, **95**, 648-653 (1982).
12. P. N. T. Unvin, "Three-dimensional model of membrane-bound ribosome obtained by electron microscopy," *Nature* (London), **269**, 118-122 (1977).
13. W. Kuhlbrandt and P. N. T. Unvin, "Distribution of RNA and protein in crystalline eucaryotic ribosomes," *J. Mol. Biol.*, **156**, 431-448 (1982).
14. M. W. Clark, K. Leonard, and J. A. Lake, "Ribosomal crystalline arrays of large subunits from *Escherichia coli*," *Science*, **216**, 999-1001 (1982).
15. L. O'Brien, K. Shelley, J. Towfighi, and A. McPherson, "Crystalline ribosomes are present in brains from senile humans," *Proc. Natl. Acad. Sci. USA*, **77**, 2260-2264 (1980).
16. A. Yonath, J. Mussig, B. Tesche, et al., "Crystallization of the large ribosomal subunits from *Bacillus stearothermophilus*," *Biochem. Int.*, **1**, 428-435 (1980).
17. A. Yonath, J. Mussig, and H. G. Wittmann, "Parameters for crystal growth of ribosomal subunits," *J. Cell Biochem.*, **19**, 629-639 (1982).
18. A. Yonath, B. Tesche, S. Lorenz, et al., "Several crystal forms of the *Bacillus stearothermophilus* 50S ribosomal particles," *FEBS Lett.*, **154**, 15-20 (1983).
19. A. Yonath, J. Piefke, J. Mussig, et al., "A compact three-dimensional form of the large ribosomal subunits from *Bacillus stearothermophilus*," *FEBS Lett.*, **163**, 69-72 (1983).
20. A. Yonath, "Three-dimensional crystals of ribosomal particles," *Trends Biochem. Sci.*, **9**, 227-230 (1984).
21. A. Yonath, H. D. Bartunik, K. S. Bartels, and H. G. Wittmann, "Some x-ray diffraction patterns from single crystals of the large ribosomal subunits from *Bacillus stearothermophilus*," *J. Mol. Biol.*, **177**, 201-206 (1984).
22. A. Yonath, M. A. Saper, I. Makowsky, et al., "Characterization of single crystals of the large ribosomal particles from *Bacillus stearothermophilus*," *J. Mol. Biol.*, **187**, 633-636 (1986).
23. A. McPherson, "Crystallization of proteins from polyethylene glycol," *J. Biol. Chem.*, **251**, 6300-6303 (1976).
24. A. Shevack, H. S. Gevitz, B. Hennemann, et al., "Characterization and crystallization of ribosomal particles from *Halobacterium marismortui*," *FEBS Lett.*, **184**, 68-71 (1985).
25. M. Shoham, J. Mussig, A. Shevack, et al., "A new crystal form of large ribosomal subunits from *Halobacterium marismortui*," *FEBS Lett.*, **208**, 321-324 (1986).
26. I. Makowski, F. Frolow, M. A. Saper, et al., "Single crystals of large ribosomal particles from *Halobacterium marismortui* diffract to 7 Å," *J. Mol. Biol.*, **193**, 819-822 (1987).
27. A. Yonath and H. G. Wittmann, "Crystallographic and image reconstitution studies on ribosomal particles from bacterial sources," in: *Structure, Function, and Genetics of Ribosomes*, B. Hardesty and G. Kramer (eds.), Springer, Heidelberg–New York (1986), pp. 112-127.
28. G. A. Bentley, J. T. Finch, and A. Lewit-Bentley, "Neutron diffraction studies on crystals of nucleosome cores using contrast variation," *J. Mol. Biol.*, **145**, 771-784 (1981).
29. D. R. Davies and D. M. Segal, "Protein crystallization: microtechniques involving vapor diffusion," *Methods Enzymol.*, **22**, 266-269 (1971).
30. C. Thaller, L. H. Weaver, G. Eichele, et al., "Repeated seeding technique for growing large single crystals of proteins," *J. Mol. Biol.*, **147**, 465-469 (1981).
31. S. Tsuji, K. Suzuki, and K. Imahori, "Crystallization of DNA-dependent RNA polymerase from *Thermus thermophilus* HB8," *Nature* (London), **261**, 725-726 (1976).
32. V. V. Barynin and A. I. Grebenko, "T-catalase, a nonheme catalase of extremely thermophilic bacteria," *Dokl. Akad. Nauk SSSR*, **286**, 461-464 (1986).

33. L. S. Reshetnikova and M. B. Garber, "Crystallization of elongation factor G from an extreme thermophile, *Thermus thermophilus* HB8," *FEBS Lett.*, **154**, 149-150 (1983).

34. L. S. Reshetnikova, M. M. Chernaya, and V. N. Ampilova, "Isolation and crystallization of phenylalanyl-*t*RNA-synthetase from *Thermus thermophilus* HB8," *Bioorg. Chem.*, **13**, No. 4, 546-549 (1987).

35. V. V. Barynin, A. A. Bagin, V. R. Melk-Adamyan, et al., "Three-dimensional structure of T-catalase with 3 Å resolution," *Dokl. Akad. Nauk SSSR*, **288**, 877-880 (1986).

36. Yu. N. Chirgadze, S. V. Nikonov, E. V. Braznikov, et al., "Crystallographic study of elongation factor G from *Thermus thermophilus* HB8," *J. Mol. Biol.*, **168**, 449-450 (1983).

37. W. Littke and C. John, "Protein single crystal growth under microgravity," *J. Cryst. Growth*, **76**, 663-672 (1986).

38. H. G. Wittmann, J. Mussig, J. Piefke, et al., "Crystallization of *Escherichia coli* ribosomes," *FEBS Lett.*, **146**, 217-220 (1982).

39. E. A. Karpova, I. N. Serdyuk, Yu. S. Tarkhovskii, et al., "Crystallization of ribosomes from *Thermus thermophilus*," *Dokl. Akad. Nauk SSSR*, **289**, 1263-1265 (1986).

40. K. Appelt, I. Tanaka, S. W. White, and K. S. Wilson, "Proteins of the *Bacillus stearothermophilus* ribosome. The structure of L6 at 6 Å resolution," *FEBS Lett.*, **165**, 43-45 (1984).

41. K. Appelt, J. Dijk, S. White, K. Wilson, and K. Bartels, "Proteins of the *Bacillus stearothermophilus* ribosome. A low resolution crystal analysis of protein L30," *FEBS Lett.*, **160**, 72-74 (1983).

42. S. White, K. Appelt, J. Dijk, and K. S. Wilson, "Proteins of the *Bacillus stearothermophilus* ribosome. A 5 Å structure analysis of protein S5," *FEBS Lett.*, **163**, 73-75 (1983).

43. L. Giri, J. Littlechild, and J. Dijk, "Hydrodynamic studies on the *Escherichia coli* ribosomal proteins S8 and L6 prepared by two different methods," *FEBS Lett.*, **79**, 238-244 (1977).

44. Y. Georgalis and L. Giri, "Shape of protein S5 from the S30 subunit of *Escherichia coli* ribosome determined in two different environments," *FEBS Lett.*, **95**, 99-102 (1978).

45. K. Morikawa, M. Kawakami, and S. Takemura, "Crystallization and preliminary x-ray diffraction study of S5 from *Thermus thermophilus* HB8," *FEBS Lett.*, **145**, 194-196 (1982).

46. A. McPherson, *Preparation and Analysis of Protein Crystals*, Wiley, New York (1982).

47. S. É. Sedel'nikova, "Isolation and crystallization of protein TL7 of the large subunit ribosomes of *Thermus thermophilus*," *Biopolim. Kletka*, **3**, No. 4, 163-166 (1987).

48. M. Leijonmarck, S. Eriksson, and A. Liljas, "Crystal structure of a ribosomal component at 2.6 Å resolution," *Nature (London)*, **286**, 824-826 (1980).

49. A. Liljas, S. Eriksson, D. Donner, and C. G. Kurland, "Isolation and crystallization of stable domains of the protein L7/L12 from *Escherichia coli*," *FEBS Lett.*, **88**, 300-304 (1978).

50. K. Appelt, J. Dijk, and O. Epp, "The crystallization of protein BL17 from 50S ribosomal subunits of *Bacillus stearothermophilus*," *FEBS Lett.*, **103**, 66-70 (1979).

51. K. Appelt, J. Dijk, R. Reinhardt, et al., "The crystallization of ribosomal proteins from 50S subunit of *Escherichia coli* and *Bacillus stearothermophilus* ribosome," *J. Biol. Chem.*, **256**, 11787-11790 (1981).

52. K. Appelt, S. W. White, and K. S. Wilson, "Proteins of the *Bacillus stearothermophilus* ribosome. Crystallization of proteins L30 and S5," *J. Biol. Chem.*, **258**, 13328-13330 (1983).

53. K. Appelt, J. Dijk, S. W. White, and K. S. Wilson, "Proteins of the *Bacillus stearothermophilus* ribosome. Crystallization of protein L6," *FEBS Lett.*, **160**, 75-77 (1983).

54. A. Liljas and M. E. Newcomer, "Purification and crystallization of a protein complex from *Bacillus stearothermophilus* ribosomes," *J. Mol. Biol.*, **153**, 393-398 (1981).

55. S. S. Abdel-Meguid, P. B. Moore, and T. A. Steitz, "Crystallization of a ribonuclease-resistant fragment of *Escherichia coli* 5S ribosomal RNA and its complex with protein L25," *J. Mol. Biol.*, **171**, 207-215 (1983).

56. K. S. Wilson, K. Appelt, J. Badger, et al., "Crystal structure of a prokaryotic ribosomal protein," *Proc. Natl. Acad. Sci. USA*, **83**, 7251-7255 (1986).

57. M. M. Yusupov, S. D. Trakhanov, V. V. Barynin, et al., "Crystallization of the 30S subunit of the *Thermus thermophilus* ribosome," *Dokl. Akad. Nauk SSSR*, **292**, 1271-1274 (1987).

58. Z. V. Gogiya, M. M. Yusupov, and T. N. Spirina, "Structure of the *Thermus thermophilus* ribosome. 1. Method of isolation and purification of ribosomes," *Mol. Biol.* (Moscow), **20**, 519-526 (1986).

CHROMATOGRAPHIC METHODS OF PURIFICATION
AND CRYSTALLIZATION OF SPHERICAL VIRUSES

B. V. Mchedlishvili, E. Yu. Morgunova, and A. M. Mikhailov

Viruses are highly organized submolecular systems. Their basic components are nucleic acid and protein. More than half of the infectious diseases known at present are caused by them. The nucleic acid, which carries the whole complement of information necessary for generation of the virion, is enclosed for the majority of spherical viruses (Fig. 1) in an icosahedral capsid composed of $60T$ quasiequivalent protein molecules. Here T is the number of triangulations. Besides the protective function, the capsid is responsible for the sphericity of the virion. Thus, x-ray structural analysis of the biocrystals is a fundamental technique and one of the most informative methods for study of the structure—activity relation of viruses.

Preparation of a highly pure homogeneous virus specimen is the first step in studying the properties and structure of the virion. Its availability is necessary for preparation of large highly regular virus single crystals.

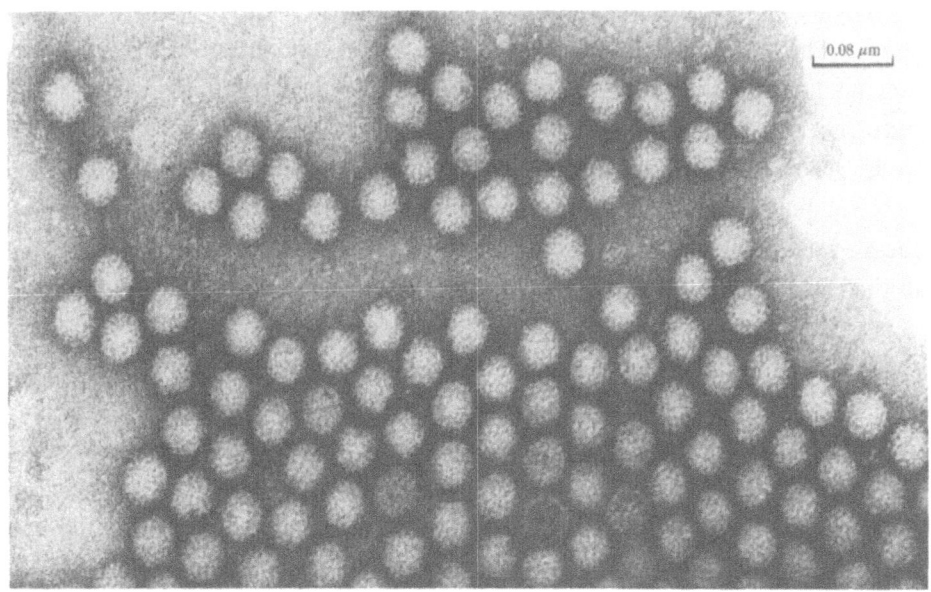

Fig. 1. Electron microscopic image of negatively stained carnation mottle virus. The diameter of the virion is 300 Å.

181

Simple viruses, which include the small icosahedral viruses that are examined here, are composed of a nucleoprotein with a molecular weight of about 1 million daltons. Packing of the protein molecules in the capsid follows icosahedral symmetry. It predetermines their spatial equivalence at $T = 1$ and quasiequivalence at $T > 3$. Here $T = Pk^2$, where $P = a^2 + ac + c^2$, a and c are whole numbers which do not have a common factor, and k is any whole number.

1. CHROMATOGRAPHIC METHODS OF PURIFICATION ON SILICA

Chromatography is one of the current methods for purification of crystallization solutions of biopolymers, especially viruses. This method is practically the only alternative to sedimentary purification. Procedures which use synthetic macroporous silicas — macroporous glasses, silichromes, and silica gels [1] — occupy a special position among the various forms of chromatography.

In the present report, we examine in detail the chromatographic purification of crystallization solutions of viruses on macroporous glasses. These are one of the most promising supports in current liquid chromatography of colloidal solutions [2]. In this respect, only one of the types of chromatography will be discussed. That type is gel-filtration, which allows precise separation of mixture components according to differences in their molecular weights [3].

The rigidity of the silicate structure of porous glass, its chemical inertness, the wide variation of pore sizes from 0.01 to 1 μm with a narrow size distribution ($\pm 10\%$), and the marked values of total porosity (up to 2.2 cm^3/g) all favorably distinguish porous glasses from such traditional gel-chromatographic materials as sephadexes, sepharoses, etc.

The advantages of porous glasses permit them to be used not only in laboratory purification of virus solutions but also in industrial purification of virus specimens. A number of problems arise during development of methods and technical procedures of purification. These are determined by the peculiarities of the porous glass as support (gel-chromatographic material) and the physicochemical properties of the virus suspensions. The virus suspensions, which are obtained from plant leaves or in cell culture, contain an enormous quantity of impurities: extraneous proteins and other cell components. The virus component itself is usually 10^{-2}-$10^{-3}\%$ of the total mixture (not counting solvent).

Macroporous glass under certain conditions exhibits significant adsorption of virions and proteins. Adsorption of these components of the separated mixture on the glass surface not only degrades but also affects the net effectiveness of the gel-filtration process. Therefore, development of a gel-filtration purification method entails first of all the choice of conditions under which the adsorptivity of the porous glass is absent or minimal. These conditions can vary widely for virions and proteins of the mixtures. Thus, an unusual modification of the adsorbent occurs if part of the proteins remains on the glass surface after complete elution of viruses from the column. Modifiers can endow virus particles with surface properties other than those of the principal strain if the population is heterogeneous. Multiple use of the porous support necessitates a regeneration procedure for such "modified" glass, i.e., renewal of the initial surface properties. In many cases, the surface of the glass particles is intentionally modified by various physical or chemical methods for elmination of virion adsorption.

The pore size distribution in macroporous glass does not exceed 10%. The pores under an electron microscope appear as a system of channels of constant diameter. Penetration into them of mixture components to different depths depending on the ratio of sizes of the narrowing pores and diffusing particles, which occurs for sephadexes, is practically impossible. Therefore, separation of components of the virus suspension into two groups occurs in the majority of cases on porous glasses. The first of these consists of particles which are not able to penetrate the glass pores. The second includes molecules of smaller sizes which diffuse freely into the pores. The purification process proceeds most effectively if the pores are slightly smaller than the size of the virus particles and suitable for diffusion of the majority of mixture component impurities, which vary in size. Here, effective sizes rather than geometric sizes are considered.

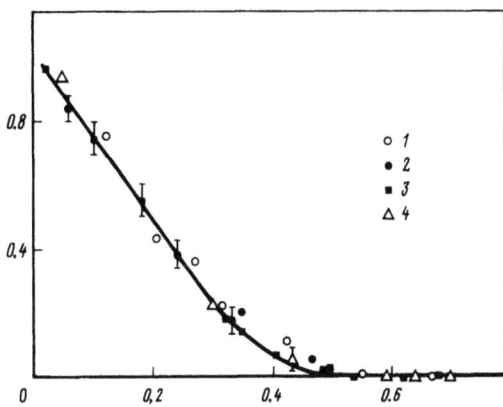

Fig. 2. Universal calibration curve for spherical virus particles. The curve was developed for gel-chromatography on porous glasses (without accounting for pore size distribution). Eluent is 0.05 M phosphate or Tris buffer with 0.15 M NaCl at pH 7.4-7.8. Eluent flow rate is 0.1-0.01 cm/min. Viruses: 1) ReO; 2) tick-borne encephalitis; 3) bacteriophage MS2; and 4) polystyrene latex (test sample).

Choice of the ratio between geometric virion sizes α and the pore diameter D at which maximal purification is possible is an obligatory step in development of the gel-filtration method. Then, the macroporous glass particles are prepared by grinding and sieving, i.e., as pieces of various irregular shapes. Therefore, methods for filling columns with this granulated material must be found. They should guarantee not only uniform column filling by adsorbent but also reproducible and effective packing, i.e., a certain and constant value of the quantity V_0/V_t, where V_0 is the void volume of the column and V_t is the volume of adsorbent in the column (total column volume). The presence in the column of adsorbent particles of various shapes and sizes must not be reflected in the shape of the chromatographic zones. The bands in such a column will be further broadened due to a lower quality of adsorbent packing, as well as due to different diffusion time of mixture components inside the glass parts of various sizes and shapes. These peculiarities necessitate the determination of the optimal eluent flow rate, maximal volume of starting mixture placed on the column, and optimal adsorbent particle size α_p.

1.1 Choice of Optimal Pore Size

Even without adsorption of virus and impurity on the macroporous glass surface (or other types of silica), a high degree of virus suspension purification can only be attained with correct choice of the ratio between the sizes of virion and average support pore diameter. The criterion for correctness of such a choice in our experiments was the equality of the virus elution volume in gel-chromatography to the column void volume (i.e., the ratio $V_e/V_0 = 1$). In selecting the optimal pore size, the virus retention volumes for glasses with different pore sizes were determined. The function $V_e/V_0 = f(D)$ or $K_\alpha = f(\lambda)$, where λ is the ratio of virus particle and glass pore sizes, is found from the experimental results (Fig. 2). The values of virus and protein impurity elution volumes $V_e = V_0 + K_\alpha V_i$ are calculated from the position of their peak maxima. Here V_0 is that quantity of eluent required for elution from the column of molecules that do not enter the pores, K_α is the distribution coefficient, and V_i is the glass pore volume.

The position of the virus peak maximum can shift at high eluent flow rates and with adsorptive interaction between the virus and glass. Therefore, the functions (for example, those in Fig. 2) are obtained under conditions where adsorptive interaction of the virions with the glass is absent and the flow rate is such (0.01-0.1 cm/min) that the virus and impurity peaks are symmetric.

If the virions do not penetrate the pores, then $V_e/V_0 = 1$ ($K_\alpha = 0$). As soon as the support pores are available to the virions ($K_\alpha \neq 0$), their elution volume begins to increase. The maximal pore diameter at which a deviation of the quantity V_e/V_0 from unity and the quantity K_α from 0 is not observed for the viruses is considered optimal for gel-chromatographic purification of the crystallization solution of the given virus.

Functions of the form $K_\alpha = f(\lambda)$ are called calibration functions. The calibration functions for the viruses show that free diffusion of viruses into the glass pores ($K_\alpha = 1$) occurs only when the average pore diameter exceeds the average virus particle size (for spherical viruses this is the average diameter) by 10 or more times, i.e., for $\lambda \leq 0.1$. However, penetration of virions into the glass pores begins at $\lambda = 0.55 \pm 0.05$. This means that the glass pore diameters that are optimal for gel-chromatographic purification of viruses should be determined from the relation $D^0 = 2\bar{\alpha}$, where $\bar{\alpha}$ is the average size of the virus population from a histogram. In concluding this section, we note that penetration into the glass pores of such rigid colloidal particles as virus particles differs sharply from the diffusion into these pores of macromolecules of globular proteins and especially linear macromolecules. In the last case, penetration of the macromolecules into pores begins already for $\lambda = 1$-1.5 (as can be seen from the calibration curve in Fig. 3). This condition is explained by the many conformations of linear molecules that are allowed inside the support pores.

Within the limits $0 \leq \lambda \leq 0.5$, all experimental points (Fig. 2) lie within the region defined by the family of curves $K_\alpha = (1-2\lambda)^a$, where $a = 1$-2. The best agreement with experiment occurs at $a = 1.5$. A similar semiempirical function is in full agreement with one of the principal tenets of gel-chromatography theory according to which K_α is a function only of the ratio of macromolecule and support pore geometric sizes [7].

1.2. Porosity (Pore Volume) and Column Separation Factor

The effectiveness of gel-filtration purification of virus solutions is determined not only by the degree of purity of the bands obtained as a result of the chromatography but also by the reproducibility of the procedure. The specific pore volume of the glass used as support (porosity) has a substantial effect on both these parameters. The determining factor in the degree of virus suspension and separation coefficient depends mainly on the distance ΔV_e between the maxima of the separated substances. Thus, ΔV_e (and also the separation coefficient) is higher the greater the internal column volume V_i (total pore volume of support in the column), i.e., in the final analysis, the higher the glass porosity. Moreover, the porosity is directly related to still another column "volume" parameter, the amount of starting virus sample V. The volume of the separated sample together with the eluent flow rate determine the reproducibility of gel-filtration. If the weight of glass g in the column is constant, then an increase of V_i should be directly proportional to the increase of porosity (Π), i.e., $V_i = g\Pi$. Figure 4 gives calculated functions and experimental points for values of V_i and V_t (with $g = $ const). The col-

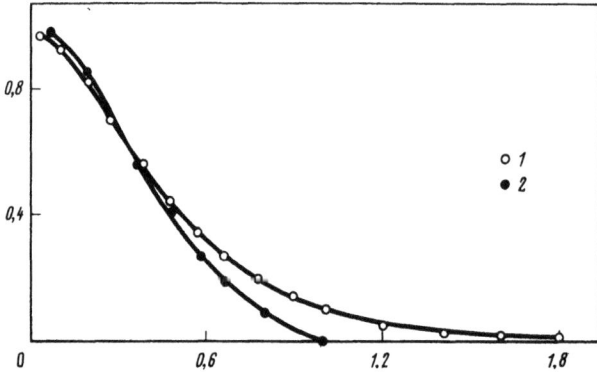

Fig. 3. Calibration curve for: 1) a homologous series of proteins and 2) flexible linear polymers.

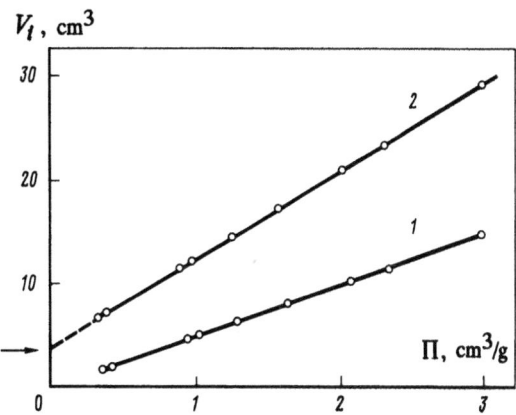

Fig. 4. 1) Internal and 2) total column volumes as functions of macroporous glass pore volume (D = 0.18-0.22 μm). V_m is the volume of glass matrix in the column. The mass of glass in the column is constant (5 g).

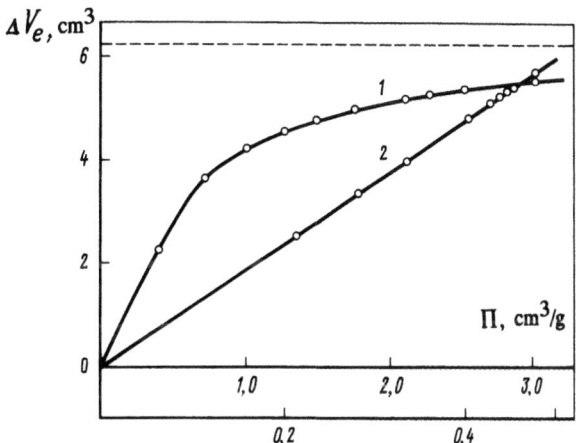

Fig. 5. Column separation factor as a function of: 1) macroporous glass pore volume and 2) fraction of glass pore volume per 1 cm^3 of column. The total constant column volume is 11.1 cm^3 and D = 0.18-0.22 μm. Solid lines are calculated, points are experimental, and the dashed line is the $(1 - \kappa)$ level.

umns that are filled with high-porosity glasses visibly have a large volume V_i. The increase of separation factor of the column (i.e., the value of $\Delta V_e = V_i$) under these conditions is directly proportional to the increased glass porosity.

In practice, investigators are interested in another question: how much does the column separation factor increase if the glass porosity is increased under otherwise equal conditions (column volume, grain size, elution rate, starting sample size). Figure 5 shows the experimental points that define the column separation factor ΔV_e as a function of glass porosity. The calculated curve $V_i = f(\Pi)$ is also shown. The semiempirical formula by which this function was defined was derived from the following relations:

$$V_t = V_0 + V_i + V_m = V_0 + g(\Pi + 1/\rho).$$

Here ρ and V_m are the density and volume of silica porous glass matrix and V_i is the total pore volume of support in the column.

Since the void volume V_0 comprises a certain part κ of the total column volume, we have $V_i = V_t \kappa \Pi/(\Pi + 1/\rho)$. This equation has a simple physical meaning: increasing (at V_t = const) the glass porosity, we decrease the fraction of column volume occupied by the glass matrix, i.e., at the limit $\lim_{\Pi \to \infty} V_i/V_t = 1 - \kappa$. This value is also the limit for the function $\Delta V_e = f(V_i')$, which shows that up to $(1 - \kappa)$ the separation factor of the column increases linearly with increasing total pore volume per 1 cm³ column, V_i'. In this case it is essential that the principal increase of column separation factor occurs only in the range of porosity values up to 2 cm³/g. At larger porosity values, the separation factor increases insignificantly (by 3% for a porosity change from 2.5 to 3.0 cm³/g), approaching the limiting value $(1 - \kappa)V_t$. In this part of the curve, the increase of protein impurity retention volumes becomes comparable to the uncertainty in peak position determination.

The criterion for effective purification is the usual degree of purification (in %): $n = 1 - (c/c_0)$, where c is the concentration of protein impurities in the chromatographic fraction which contains 90% of the virus particles and c_0 is the concentration of these impurities in the virus suspension before purification. A 100-fold purification (i.e., removal from solution of not less than 99% of the protein impurities) is usually considered satisfactory. Therefore, the optimal V_{opt} here means the maximal volume of separated sample for which $n \geq$ 99%, i.e., the limiting permissible starting solution volume which is purified by gel-filtration to this degree.

Experience shows that use of glass of porosity 2-2.5 cm³/g for the majority of virus suspensions gives a V_{opt} value up to 65-75% of V_t, i.e., up to values which are characteristic for traditional organic polymeric supports [9]. Thus, gel-filtration procedures on macroporous silica can compete successfully in reproducibility with procedures which use traditional supports.

1.3. Porous Glass Particle Size

Mass transfer between stationary and mobile phases (i.e., between solution outside and inside the support particles) in chromatographic columns is improved with decreasing granule size. Therefore, silica with a particle size from 5 to 100 μm is used in current analytical chromatography of biopolymers. Use of these particles improves the chromatographic characteristics of the process. For example, the time for gel-filtration analysis of the molecular-mass distribution is reduced by 300-400 times with a column change from 100 μm particles to 5 μm particles. However, particles of micron size are unsuitable for preparative chromatography, which is used in purification of virus crystallization solutions, due to difficulties in apparatus fabrication (high pressures) and the possibility of clogging the upper layer with colloidal particles that occur in working solutions and, as a result, the increased hydrodynamic resistance of the layer to eluent flow. Therefore, optimal results in preparative chromatography of virus suspensions are obtained through use of glass with particle sizes between 50 to 300 μm. The column effectiveness falls sharply for particles of larger size (greater than 300 μm).

1.4. Eluent Flow Rate

The nonequilibrium nature of the process increases with increased flow rate v of the chromatographic bands along the column. As a result of this, the peaks shift, broaden, and become unsymmetrical. Attempts are made to avoid such procedures in the chromatography of biopolymers and viruses. The optimal eluent flow rates in equilibrium chromatographic procedures on cross-linked gels are about 10^{-2} cm/min whereas the maximal, which are determined by the stability characteristics of these materials, do not exceed 1 cm/min [9]. The rates are much higher for chromatography of proteins and viruses on volume-porous silicas (including porous glasses), up to 10 cm/min [4], and reaches 100 cm/min on surface-porous silicas [10]. Moreover, equilibrium or quasiequilibrium separation conditions exist at $v \leq 0.1$ cm/min for macromolecules and particles which do not enter the pores, or, conversely, which diffuse freely into them. However, in this case, the mass transfer process between the mobile and stationary phases is not disrupted, as seen in Fig. 6. At higher rates, a certain part of

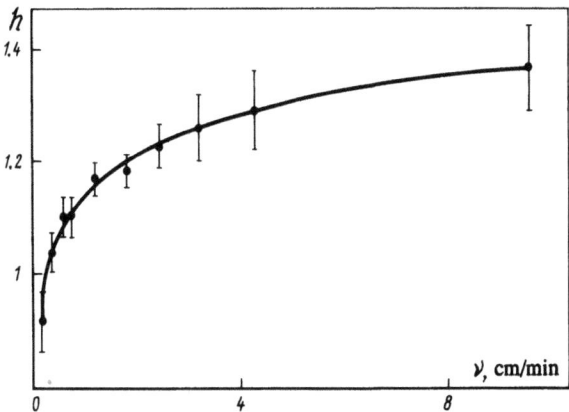

Fig. 6. Effect of linear rate of movement of chromatographic bands on the column with porous glass on the value of the ratio of peak heights h of virus particles and protein impurities. In quasiequilibrium regimes ($v = 10^{-2}$ cm/min), the conditions of the gel-chromatographic experiment were chosen so that $h = 0.9$.

the impurity macromolecules does not succeed in entering the glass pores and contaminates the virus particle band, eluting together with them. However, it should be stated that an increase of eluent flow rate up to 10 cm/min is apparently too small to render the procedure distinctly nonequilibrium. This occurs at higher rates when the peak of impurity macromolecules is practically completely absent and both substances elute as one band.

The optimal range of rates for gel-filtration purification of virus solutions from impurity macromolecules and their aggregates was chosen based on the data obtained. For this, the following were considered: 1) the separation process should not be distinctly nonequilibrium (symmetric peaks, absence of long tails, presence in the virus band of not less than 50% of the initial virus activity); 2) the degree of solution purification should be not less than 99%; and 3) the eluent flow rate should be sufficient for preparative purification of virus suspensions [eluent flows up to 100 ml/(h · cm^2)]. Eluent flow rates of 2-4 cm/min, i.e., values which are characteristic for nonequilibrium regimes of gel-chromatography, are consistent with these conditions.

1.5. Modified Porous Glass

Glasses are coated with polymer layers which are physically or covalently bonded to the support matrix for prevention of virus adsorption on the surface. Several of the more simple and reliable methods of modification of macroporous glasses are given here [1].

Polyvinylpyrrolidone (PVP) Adsorption. A portion of porous glass is treated with a 4% aqueous solution of PVP. The glass:PVP ratio is 1:3 and contact time for adsorption is 4 h. The polymer solution is poured off and the moist adsorbent is dried for 20-24 h at 130°C. The operation is repeated 2-4 times. The modified support is washed with distilled water until traces of polymer in the eluate are absent.

Grafting of Carbohydrates. A portion of porous glass is treated with γ-aminopropyltriethoxysilane (APTES) for 20 h at 120°C. The reaction proceeds according to the scheme

$$-Si-OH + (CH_3CH_2O)_3-Si-(CH_2)_3-NH_2 \rightarrow -Si-O-Si-(CH_2)_3-NH_2 .$$

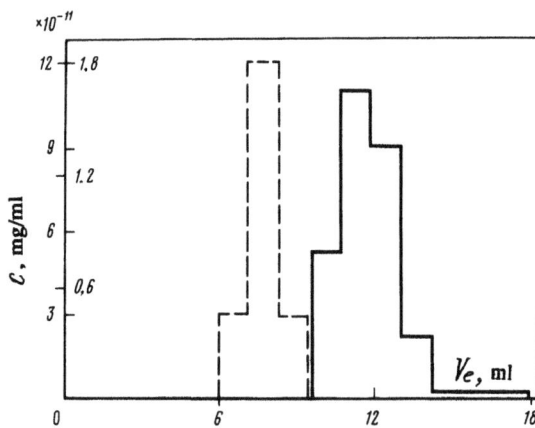

Fig. 7. Gel-chromatography of a bacteriophage MS_2 suspension on porous glass (separation conditions given in the text). The virus was preliminarily purified and concentrated by microfiltration on particle filters. The initial virus titer was $1.5 \cdot 10^{12}$ PFU/ml. Protein content was 8.7 mg/ml. Here and later the dashed line is the virus peak; the solid line is the impurity peak.

A 10% solution of glycerine aldehyde or glucose in 0.1 M phosphate buffer at pH 6-7 is added to the aminated silica. The suspension is mixed at 40°C for 3 h, washed with distilled water, and dried. The formulas for the synthesized silicas are

glycerosilica

$$-Si-O-Si-(CH_2)_3-N=CH-CH-CH_2 ;$$
$$\qquad\qquad\qquad\qquad\qquad | \quad |$$
$$\qquad\qquad\qquad\qquad\qquad OH \ OH$$

glucososilica

$$-Si-O-Si-(CH_2)_3-N=CH-(CHOH)_4-CH_2-OH$$

The Schiff bases which are hydrolyzed in acidic media can be reduced with sodium borohydride.

Grafting of Proteins. The aminated silica is treated with aqueous solutions of glutaric aldehyde and then protein (usually serum albumin): 4-5 ml protein solution with a concentration of 8-10 mg/ml per 1 g glass for 24-48 h at room temperature. After rinsing unbound protein from the support with distilled water, the modified absorbent is treated for 20-24 h with a 5-10% solution of formaldehyde at room temperature for fixation of modifier bound to the surface.

1.6. Regeneration of Porous Glasses

A single portion of macroporous glass can be used for a very long time without changing its properties. However, regeneration of the support must be carried out after purification of a certain quantity of virus solutions ($V \approx 10V_i$). This operation is accomplished by unloading the glass from the column. The regeneration consists of successive treatment of the modified silica by distilled water and a 1 M NaCl solution at pH 8-9. As a result, the chromatographic properties of the support are practically completely restored.

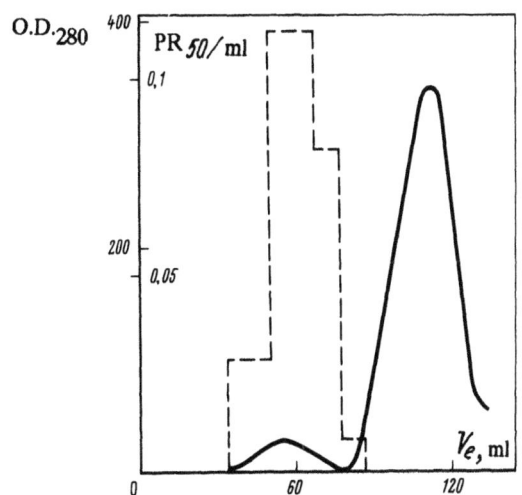

Fig. 8. Gel-chromatography of inactivated tick encephalitis virus on porous glass with average pore diameter of 0.1 μm modified by covalently bound human albumin. Column 2.2 × 35 cm, v = 1.5–2.0 cm/min, V = 10 ml, and eluent 0.1 M Tris buffer, 0.15 M NaCl, pH 7.8. PR is biological activity (protective) of the virus.

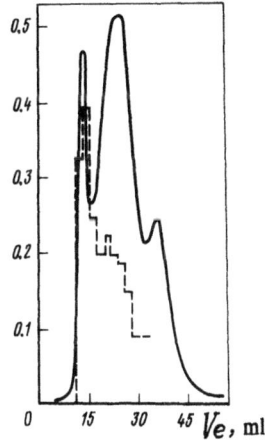

Fig. 9. Gel chromatography of carnation mottle virus on porous glass.

1.7. Purification of Virus Suspensions

Examples of gel-chromatographic purifications of several spherical bacterial viruses, plant viruses, and animal and human viruses will be given below.

Purification of Bacterial Viruses. The spherical bacteriophage MS_2 has dimensions near 0.026 μm. Therefore, MPS-700GKh porous glass with average pore diameter 0.06 μm modified with PVP was used for its purification. A 3-ml portion of phagolysate filtered through a 0.1-μm particle filter was placed on a 0.9 × 25 cm column. The eluent was 0.05 M phosphate buffer at pH 7.5. The bacteriophage MS_2 was isolated practically free from impurity proteins as a result of gel-chromatographic purification (Fig. 7). The ratio of optical densities at UV wavelengths of 260 and 280 nm in the purified virus specimens is comparable to those obtained from virus purification by ultracentrifugation, i.e., the gel-chromatographic method is comparable in effectiveness to the traditional method of differential centrifugation. The yield of phage as measured by biological activity is quantitative. Here, it is appropriate to compare the purification of this virus on porous glasses with another traditional method, gel-chromatography on polymeric gel. In [11], which is concerned with crystallization of the bacteriophage MS_2, purification on a sepharose CL-4B column is described. The loading of virus on the column was about $0.01V_t$, whereas for chromatography on porous glasses it is $V = 0.2V_t$. The rate of chromatographic separation on porous silicas ($v = 2$ cm/min) is approximately 30 times higher than for separation on sepharose ($v = 0.07$ cm/min). The yield of virus according to activity was not more than 50% in the sepharose case.

Thus, the high effectiveness of the purification of virus crystallization solutions on porous silicas is indubitable.

Purification of Animal and Human Viruses [12]. Figure 8 shows the gel-chromatography of suspensions of tick encephalitis virus (0.046 μm average size of spherical particles). Purification proceeds effectively on porous glasses whose surfaces are modified with sugars, PVP, proteins, and polymers with terminal carboxyl groups. The optimal pore size of modified supports for purification of tick encephalitis virus was 0.1 μm. The virions do not enter pores of this size whereas the impurity proteins diffuse freely into them. The content of impurity proteins was reduced from 0.5 mg/ml (in the initial specimen) to 2-10 (in the purified) μg/ml. The yield of virus was 70-90%.

Purification of Plant Viruses. Here we examine in detail the gel-chromatography of carnation mottle virus, a small spherical virus of plants with average diameter 0.028 nm. The virus was isolated from cells of young leaves of *Dianthus caryophyllus* L. The frozen leaves were ground, homogenized with an equal volume of 0.1 M Tris-HCl buffer at pH 7.4 and, after 30-60 min, squeezed through cloth. Triton X-100 up to 1% (m/v), 0.2% metcaptoethanol, 8% butanol, and chloroform were added to the extract obtained. The suspension was shaken. After 50-60 min, the coagulated proteins were precipitated by centrifugation for 20 min at an acceleration of 6000g. As a rule, the supernatant had a golden color. Further purification of the virus suspension was effected by gel-chromatography on macroporous silicas. The choice of optimal support pore size was made based on a study of the virus distribution coefficient as a function of the ratio of average virus particle and pore sizes. As a result of the quasiequilibrium chromatography at an eluent flow rate of 0.1-1.0 cm/min, the distribution coefficient varied from zero at a pore diameter of 0.02 μm to 0.8 at a pore diameter of 0.2 μm. The data obtained, within experimental error, lie completely on the universal calibration curve for rigid colloidal particles. Porous glass of MPS-250GKh grade was used for preparative purification of the virus. The average pore diameter is 0.021 μm with a porosity of 1.2 cm^3/g and a narrow pore size distribution (deviation from the average diameter of not more than $\pm5\%$). The glass was modified by vinylpyrrolidone compounds, or analogously modified porous glass SMP-1M-200 was used for elimination of virus particle adsorption to the silica. A 0.01 M Tris-HCl buffer at pH 7.2 was used as eluent. A virus sample of volume corresponding to 0.2-0.1 of the total support volume on the column was placed on 0.9 × 50 or 2.6 × 45 cm columns. The eluent flow rate was from 1 to 5 cm/min. The virus concentration in the chromatographic fractions was determined by ELISA [13]. The chromatograms (Fig. 9) show that the virus particles elute from the columns first, separating from the smaller impurities, the main mass of which is found in the second peak. The activity of virus detected in the second (impurity) peak is apparently determined by the so-called soluble antigen, subunit fragments of destroyed virions. The yield of virus from the column was practically quantitative. Fractions of the first (virus) peak were selected for further studies. This peak corresponded to maximum virus activity.

2. CRYSTALLIZATION OF SPHERICAL VIRIONS

2.1. Basic Principles of Crystallization

The crystalline state of biosystems is a thermodynamically equilibrated state of the solid and a subject of thermodynamics and statistical physics. The minimum free energy of the system corresponds to this, the most stable state. Thus, the required condition for existence of the biocrystal can be represented by the expression $E = (U - \epsilon T)$, where E is the free and U the internal energy of the system, ϵ is the entropy, and T is the tem-

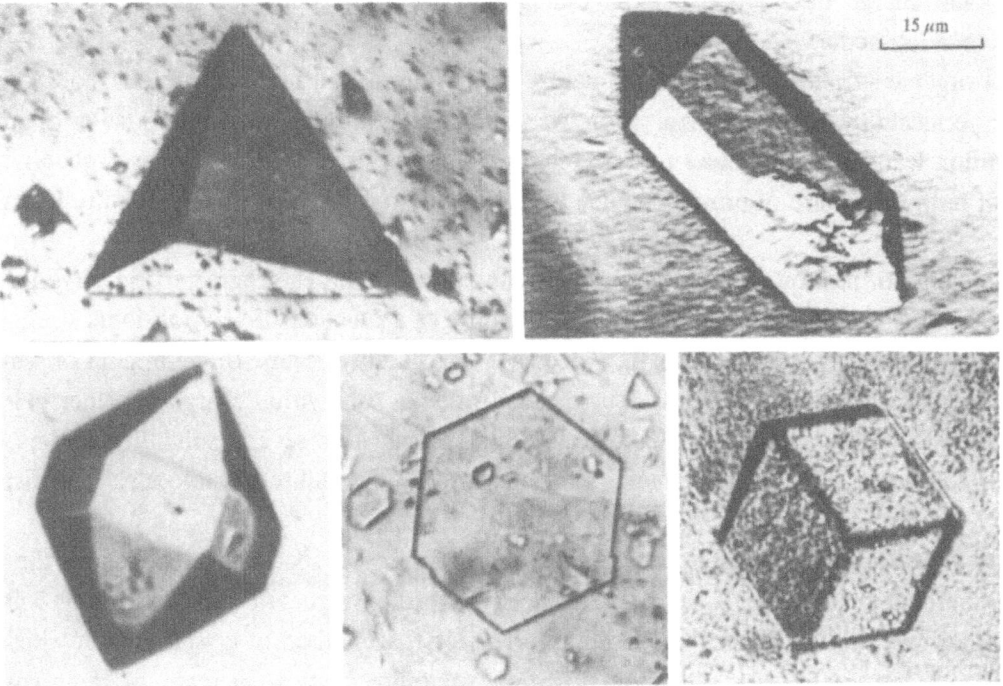

Fig. 10. Carnation mottle virus crystals.

perature. It should be immediately pointed out that the biocrystal is the stable state of the two-phase system crystal—mother liquor. Outside the mother liquor the biocrystal does not exist as a functional object. If crystallization can be modeled for "simple" structural units, then in the case of macromolecules, and even more so for their aggregates, the solution of the similar problem has not been realized (at least at present only data on nucleation and growth of biocrystals have been gathered).

Crystallization of biomolecules belongs more to the domain of art than science. However, there are general principles, approaches, and basic considerations governing such factors as solution ionic strength, the effect of pH on the medium, the temperature, and the presence of organic solvents, initiators, and polyethyleneglycol (PEG). The accumulated experience on crystallization is presented and generalized in a number of papers [14-23]. Among these, the most informative are [20, 23]. Growth of virus crystals suitable for x-ray structural analysis follows mainly the same principles as proteins [14-23], but adjustments should be made for the high symmetry of the capsid which dominates crystallization of viruses.

Crystallization can formally be divided into two steps: creation of conditions for formation of single nuclei through a supersaturated state of virions in local microregions of the virus suspension and subsequent maintainence of the virion concentration required for growth of perfect single crystals. Therefore, studies of the solubility of a given specimen as a function of these factors should precede work on crystallization.

An increase of precipitant concentration at low solution ionic strength values leads to increased solubility of virions (a salting-in effect). Apparently, this can be explained by the increased interaction of the biomolecules with water molecules. Further increase of the concentration of ions lowers solubility (salting-out effect). This results from preferential bonding of water molecules by precipitant ions. The solubility as a function of ionic strength can be expressed in analytical form as $\log S = \beta - K_S(I/2)$. Here $I/2$ is the ionic strength in moles per liter of water, S is the solubility of protein in grams per liter of water, and β and K_S are constants. The β value in the range of biocrystal growth conditions is mainly a characteristic of the proteinaceous surface and practically does not change with a change of buffer solution. The parameter K_S characterizes the slope of the descending portion of the solubility curve and is determined principally by the salt solution. Depending on the specific ionic charge distribution and its radius, the hydrating ability of various inorganic ions exhibit different precipitating activity on virions.

Use of organic solvents can change the nature of the interaction between protein molecules by changing the dielectric permeability of the medium. Moreover, addition to proteinaceous solutions of organic solvents at low concentration leads in several cases to a decrease of the number of nuclei. This is probably a consequence of a change of both the nature of protein globule hydration and the specific bonding of solvent molecules by the virion.

The solubility of macromolecules depends substantially on pH. The characteristic curve for virus suspensions shows that here the effect is opposite to the solubility as a function of solution ionic strength. In fact, at the beginning of the process (increasing pH), a decrease of solubility occurs, the minimum of which is attained at the isoelectric point which is characterized by the null value of total virion charge. Further pH increase leads to increased virus solubility. Individual biological specimens can have several solubility minima depending on pH at various solution ionic strength values. This is caused by the ability of molecules to redistribute surface charge.

The nature of changing temperature on solubility of biological specimens depends on the solution ionic strength value. The solubility coefficient for the majority of protein structures at low solution ionic strengths is positive. It is negative at high ionic strength values. This or that method of crystal growth which creates local supersaturation of viruses by slow reduction of virion solubility or slow increase of their concentration in solution is chosen through knowledge of the solubility as a function of the various factors. This is necessary for formation of single crystal nuclei and their subsequent slow growth. Without a doubt, each of the crystallization methods is in fact based on the cooperative influence of various factors that change solubility (concentration of virions). However, one or at least a small number of them dominate nonetheless.

The principal methods of spherical virus crystallization are vapor diffusion of solvent and equilibrium dialysis through a semipermeable membrane. Crystallization is carried out exclusively in a micromethod variant when the solution volume is 5-50 μl. The virion concentration in this case usually varies from 1 to 10%.

The vapor diffusion method is used when searching for crystallization conditions and during growth of crystals suitable for x-ray structural analysis. Studies of solubility as a function of various factors, such as precipitant type, its concentration, solution ionic strength and pH, the presence of various additives, etc. (all these parameters refer primarily to the equilibrium solution) can be carried out without changing the initial amount of protein.

The equilibrium dialysis method (dialysis of precipitant) is used in the same range as vapor diffusion. In our opinion, this method is most effective in the first step of crystal growth — the search for growth conditions for perfect single crystals and study of their solubility (concentration) as a function of various factors.

Methods based on creation of supersaturation through increased solution concentration of virus particles are promising methods for virion crystallization. Among these, electrodiffusion and centrifugation, including centrifugation in a density gradient [19, 21], are especially effective methods. They provide the possibility to carry out virion crystallization at low initial concentrations since in this case the requirements for purity and specimen homogeneity are much lower. The high symmetry of the capsid, its significant linear dimensions (hundreds of Ångstroms), and the large virion mass (millions of daltons) not only determine the crystallization conditions of the icosahedral virions but also impart characteristics to the virus crystals. Inclusion of a significant amount of mother liquor (up to 75% of the unit cell volume) is foremost among the features of virus crystals (Fig. 10). For proteins, this value is not more than 30-50%. This substantially lowers their stability, hindering (even in comparison to proteins) work with them.

Considering a virus suspension as a colloidal solution is apparently a good approximation to permit the nucleation and growth mechanisms of virus crystals to be ascertained to a certain degree. In fact, a virus suspension is not much different from a second class colloidal system ($S + L$). Here, the dispersing medium is an electrolyte (L) characterized by ionic strength, pH, and type of ions. The dispersed phase is the virus par-

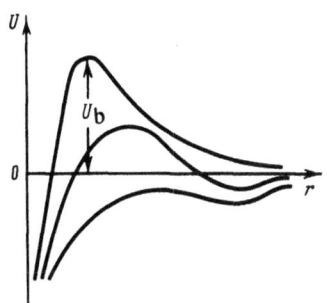

Fig. 11. Interaction energy between spherical colloidal particles.

ticle (S). Thus, crystallization of virions from aqueous solutions is predetermined by the nature of the interaction between the virus particles through the electric double layer which surrounds the virion. The basic components of the interactive forces are the electrical repulsive forces and the intermolecular attractive forces. In this case, the interactive potential energy U between virions has the form $U = U_a + U_e$, where U_a and U_e are the components of potential energy of intervirion interaction (attraction and electrical repulsion, respectively). Thus, it follows that at $|U_e| > |U_a|$ the virus suspension is stable and crystallization is possible in this range, whereas at $|U_e| < |U_a|$ the virions aggregate and form an amorphous precipitate. Both the path of the $U_e(r)$ and $U_a(r)$ curves and their value are determined by the state of the system. For given virions, they are determined to a large degree by the dispersing medium, i.e., by the type of ions and their quantity. These set the extent of each of the layers and the amount of their charge by forming electrical layers.

Figure 11 shows the $U(r)$ curves. Each of these is characterized by the parameter U_b (the value of the potential repulsion barrier). At large U_b values, the colloidal system is stable because only separate virions which have a sufficient amount of kinetic energy are capable of overcoming the potential repulsion barrier. This is a region of latent coagulation when only a small number of aggregates can be formed. In this region, nucleation and growth of isolated single crystals is possible. A decrease of the potential barrier value (Fig. 11) increases the quantity of aggregations which, in turn, leads to formation of a large number of nuclei (rapid crystallization). A transition of the potential barrier into the attraction region ($U < 0$) induces rapid aggregation which leads to conglomeration of virions and formation of an amorphous precipitate of them.

As already mentioned, the nature of the potential curve is determined by its components U_a and U_e which, in turn, are regulated by the electrolyte for a given type of virion. A decrease of the potential barrier value is possible due to a decrease of U_e through an increased concentration of ions. The latter leads to redistribution of charges in the electrical double layer of the virion, decreasing the thickness of its outer layer. The marked decrease in thickness of the layer leads to a negative value of the potential barrier, i.e., to a transition to the region of rapid formation of an amorphous precipitate of virions. It should be kept in mind that multiply charged counterions (electrolyte ions which have a charge opposite to that of the virion charge) contribute greatly to this process. In this case, aggregation is inversely proportional to the counterion charge to the sixth power. Thus, the ratio of aggregation threshholds (lowest value of electrolyte concentration which causes coagulation) for singly, doubly, and triply charged ions can be expressed by the ratio 1:0.016:0.0014 according to Deryagin and Landau.

A factor that stabilizes the virus suspension is also the adsorbed layer that forms around the virion due to solvation by the dispersing phase. Such layers, which decrease the mutual attractive forces of the virions, increase the value of the repulsive potential barrier and prevent approach of virions. Surface-active substances or high-molecular-weight compounds are effective stabilizing additives that prevent conglomeration of virions.

Thus, the choice of crystallization conditions is based on the nature of virion—salt and solution—virion interaction.

2.2. Types of Crystallization Arrangements

We will examine in greater detail the crystallization arrangements used for growth of virus single crystals. *Solvent Diffusion Method.* A siliconized layer on a glass plate is usually used as support for a protein suspension. We were forced to avoid this material for two reasons: 1) the drops of virus suspension of 40-60 μl volume, the optimal volume for growth of virus crystals of 2.0-3.0 mm size, spread out over the support surface during the growth period (4-6 weeks) and greatly decreased the crystal height and 2) the crystal of the size indicated above grew into the siliconized layer during growth and could not be isolated from the crystallizing solution.

Therefore, a cell consisting of two molded Teflon® rings between which a hydrophobic film was stretched was preferred. Both Teflon® (~15-25 μm thickness) and paraffin (Serva) films were tried. The surface of the Teflon® film becomes very dusty due to intense charging. Avoidance of this shortcoming requires special conditions that are not always possible. The most suitable material turned out to be the paraffin film (Serva) which was to a large degree free of these shortcomings. The dialysis cell was placed into a Petri dish with a polished upper rim for a hermetic seal of the glass plate. As a rule, the specimens prepared from different isolations nevertheless had somewhat different abilities to crystallize. Therefore, the growth conditions of virus single crystals chosen for specimens of one isolation were somewhat different from those for another. This, in turn, requires provision for fine-tuning the crystallization volume (for example, a change of precipitant concentration, different buffer, change of its pH, ionic strength, etc.) without losing the hermetic seal. This was done through an opening in the lid (4-5 mm diameter) which was covered with a piece of adhesive tape. With this arrangement we were able to grow carnation mottle virus crystals of sizes up to 3.0-3.7 mm with a high degree of reliability.

The second is a variation of the vapor diffusion method, the hanging drop method. This form of crystallization is especially effective for choosing the crystallization conditions, i.e., for constructing the phase diagram. A glass beaker of 20 mm height and internal diameter 15-20 mm has an opening of 3-4 mm diameter in the side at a height equal to half the beaker height. A cover (siliconized) glass with vacuum grease hermetically seals the top of the beaker. Then, a drop of the virus suspension (not more than 5-15 μl) and the equilibrating solution are placed on the cover glass through the side opening of the container. After this, the side opening is closed with transparent tape. If it is necessary to change the components of the virus suspension and (or) the equilibrating solution, the operation is carried out through the side opening with a syringe (practically without loss of hermeticity of the container). Paraffin film can also be used successfully instead of the cover glass.

Diffusion of Precipitant through a Membrane. This method is sufficiently suitable in the following modification. The container for the countersolution is a penicillin dish or glass cylinder of internal diameter 13-15 mm and a height of 50 mm. One end of the crystallization tube (glass cylinders of height 60 mm and internal diameter 2.6 and 3.5 mm, pipettes of 1 and 2 ml, respectively) is widened slightly. The crystallization cell is arranged as follows. A specially treated piece of the dialysis material is placed over the widened end of the tube. The siliconized tube (internal diameter of which is 1-2 mm) is gently pulled through it, acting as a channel for feeding the equilibrating solution onto the dialysis membrane. After adding 20-40 μl of virus suspension through the upper opening of the glass container and covering it with paraffin film, it is lowered to the required depth into the container with the precipitant solution. Continuous adjustment of the height of the crystallization cell relative to the precipitant solution is also accomplished with a siliconized hose attached to the tube and fixed in the neck of the container with countersolution.

A similar arrangement is uniquely convenient both to choosing the crystallization conditions and for growth of crystals suitable for x-ray structural analysis. The crystallization cell in this case can be used repeatedly with the same portion of virus suspension (with variation of various countersolution parameters). In conclusion, it should be noted that the crystals grown by diffusion of the precipitant solution are more ideal.

The method of crystallization and growth of crystals in thin-walled (0.01 mm) glass capillaries, which are used for data collection in x-ray structural analysis of biocrystals, deserves special mention. Several microliters of virus suspension solution are placed in the capillary. Both ends of the capillary are sealed with a material that permits slow evaporation of solvent. The rate of evaporation is regulated by varying the type of stopper. The procedure for growth of crystals proceeds as follows. A flat portion is preliminarily formed on the capillary wall using a small hot metal plate. A seed is mounted on the surface and covered with 15-20 μl of virus suspension. Then, one of the ends of the capillary is sealed with paraffin, the other with vacuum grease. Every second day the virus suspension is replaced with a fresh portion. The crystal growth is stopped when its faces reach the opposite wall of the glass capillary. The method is widely used for growth of satellite virus crystals of tobacco necrosis virus. The stability of the crystals can be raised considerably with this method, increasing their lifetime under the x-ray beam by a minimum of four times.

2.3. Growth Conditions for Some Virus Single Crystals

1. Satellite Virus of Tobacco Necrosis Virus. Single crystals of this virus are grown in thin-walled capillaries sealed with a material semipermeable to the vapors. Crystals grow up to 0.1 × 0.3 × 0.4 mm. These are then grown further to 0.2 × 0.6 × 1.0 mm by changing the mother liquor each week for 5-10 weeks. The virion concentration is 10-12 mg/ml (or 7-8 mg/ml and 0.4% P/V PEG-6000) with 1 mmole Mg^{2+} in sodium phosphate buffer (50 mmole, pH 6.5).

2. Tomato Bushy Stunt Virus. These were grown by addition of ammonium sulfate to an aqueous solution of virion at a concentration of 30 mg/ml. The final concentration of ammonium sulfate was 0.5 M. Crystals up to 0.3-0.5 mm were grown during several weeks at 4°C.

3. Southern Bean Mosaic Virus. Three types of these were grown. The first type grew in dialysis cells at a virion concentration of 20 mg/ml and 0.95 moles of ammonium sulfate as precipitant. Crystals of the second and third types were grown by equilibrium dialysis against 0.015 M ammonium sulfate at 10-11% virion concentration. The time for growth of crystals of the second and third types was several days.

4. Broad Bean Mottle Virus. These grew over several days upon addition of half-saturated ammonium sulfate or sodium citrate solution to a virus suspension with a virion concentration of 10 mg/ml at pH 6.5 and 25°C.

5. Tymovirus. Single crystals were grown by vapor diffusion. The virus suspension contained from 10 to 40 mg/ml virion and 4% P/V PEG 6000. The equilibrating medium contained 8% PEG 6000.

6. Belladonna Mosaic Virus. Crystals were grown (both for the upper and lower components) by vapor diffusion. The dialysis cell was filled with a virion suspension at a concentration of 25 mg/ml for the upper component (12 mg/ml for the lower) and 3.6% (P/P) PEG 6000 (2.5% for the lower) in 0.075 M (0.05 M for the lower) potassium phosphate buffer at pH 4.2 (4.5 for the lower). The equilibrating salt solution contained 10% (P/P) PEG 6000 in the same buffer at 0.1 M (0.05 M for the lower). Crystals grew over 2-7 days for the upper component (3-4 weeks for the lower).

7. Cowpea Chlorotic Mottle Virus. Single crystals were grown by vapor diffusion. The virus at a concentration of 50 mg/ml and 1.3% PEG 6000 was placed in 0.075 M acetate buffer at pH 4.2. The equilibrating salt solution consisted of 4% PEG 6000 in 0.1 M acetate buffer at pH 4.0. Crystals up to 1.0-2.5 mm grew after several days.

8. Cowpea Mosaic Virus. Single crystals were grown by vapor diffusion (a modification of the hanging drop). The virion at a concentration of 20 mg/ml was placed in 0.3 M citrate buffer at pH 4.9. The equilibrating solution was 0.6 M citrate buffer. Addition of ammonium sulfate to the equilibrating solution permitted crystals of 1.0 × 1.0 mm to be grown.

9. Rhinovirus. Single crystals were grown by the hanging drop method at room temperature. The initial virion concentration was 5 mg/ml with 0.125% (P/V) or 0.25% (P/V) PEG 6000 and 0.01 M CaCl$_2$ in 0.01 M Tris buffer at pH 7.1. The equilibrating solution contained 0.25% (P/V) or 0.5% (P/V) PEG 8000, respectively, and 0.02 M CaCl$_2$ in 0.01 M Tris-HCl buffer at pH 7.2. Crystals up to 0.9 mm grew over a period of a week.

10. Carnation Mottle Virus. Single crystals were grown by equilibrium dialysis against 20-40% saturated ammonium sulfate solution in 0.1 M Tris-Mal/NaOH buffer at pH 5.9-6.2. The virus suspension had a concentration of 20-40 mg/ml in the same buffer at 0.01 M and pH 6.5-6.8. Crystals of tetrahedral shape grew up to 3.5 mm after 4-6 weeks.

11. Bacteriophage MS2 Virus. Single crystals were grown by vapor diffusion. A 20-μl drop of virus suspension at a 1% virion concentration in 0.2 M sodium phosphate buffer with 1.5% PEG 6000 and 0.02% NaN$_3$ at pH 7.4 was suspended from the internal surface of a plastic Petri dish. The equilibrating solution was 0.4 M sodium phosphate buffer at pH 7.4.

REFERENCES

1.   V. M. Kolikov and B. V. Mchedlishvili, *Chromatography of Biopolymers on Macroporous Silicas* [in Russian], Nauka, Leningrad (1986).

2.   B. I. Venzel' and S. P. Zhdanov, "Kinetics of dimensional growth of the borate-phase zone in sodium borosilicate glasses," *Fiz. Khim. Stekla*, **1**, 122-127 (1975).

3.   B. G. Belen'kii and V. G. Mal'tsev, "Aqueous high-efficiency exclusion chromatography of polymers, problems and prospects," in: *Applied Chromatography* [in Russian], Nauka, Moscow (1984), pp. 54-64.

4.   R. C. Collins and W. Haller, "Protein-sodium dodecylsulfate complexes," *Anal. Biochem.*, **54**, 47-53 (1973).

5.   W. Haller, "A single equation relating molecular weight, pore-size and elution coefficient in controlled pore glass chromatography on proteins," *J. Chromatogr.*, **85**, 129-131 (1973).

6.   W. Haller, "Critical permeation size of dextran molecules," *Macromolecules*, **10**, 83-86 (1977).

7.   B. G. Belen'kii and D. Z. Vilenchik, *Chromatography of Polymers* [in Russian], Khimiya, Moscow (1978).

8.   M. B. Tennikov, V. V. Nesterov, and T. D. Anan'eva, "Estimation of the porosity and average pore size by passing a solution of luminescent substance over sorbent granules," *Kolloidn. Zh.*, **41**, 301-307 (1979).

9.   H. Determann, *Gel Chromatography*, Springer-Verlag, New York (1968).

10. J. J. Kirkland, "High-speed liquid chromatography with controlled surface porosity supports," *J. Chromatogr. Sci.*, **7**, 7-12 (1969).

11. K. Valegard, T. Unge, J. Montelius, and B. Strandberg, "Purification, crystallization and preliminary x-ray data of the bacteriophage MS2," *J. Mol. Biol.*, **190**, 587-591 (1986).

12. L. B. Él'bert, I. V. Krasil'nikov, B. V. Mchedlishvili, et al., "Chromatography of tick-borne encephalitis virus inactivated by formalin on macroporous silicas," *Vopr. Virusol.*, No. 1, 72-75 (1981).

13. A. Voller et al., "Enzyme immunoassays in diagnostic medicine," *Bull. W. H. O.*, **53**, 55-65 (1976).

14. M. Dixon and E. Webb, *Enzymes*, Academic Press, New York (1964).

15. V. R. Melik-Adamyan and T. I. Zhilyaeva, "Crystallization of proteins and transfer ribonucleic acids," in: *Molecular Biology*, Vol. 2 [in Russian], VINITI, Moscow (1973), pp. 7-54.

16. A. McPherson, "Crystallization of proteins from polyethylene glycol," *J. Biol. Chem.*, **251**, 6300-6303 (1976).

17. M. Zeppezauer et al., "Microdiffusion cells for the growth of single protein crystals by means of equilibrium dialysis," *Arch. Biochem. Biophys.*, **126**, 564-573 (1968).

18. T. Blundel and L. Johnson, *Protein Crystallography*, Academic Press, New York (1976).

19. R. W. G. Wyckoff and R. B. Corey, "The ultracentrifugal crystallization of tobacco mosaic virus protein," *Science*, **84**, 513-515 (1936).

20. A. McPherson, *Preparation and Analysis of Protein Crystals*, Wiley, New York (1982).

21. Chang-chen Chin et al., "Crystallization of human placental estradiol 17 β-dehydrogenase," *J. Biol. Chem.*, **251**, 3700-3705 (1976).

22. F. R. Salemme, "A free interface diffusion technique for the crystallization of proteins for x-ray crystallography," *Arch. Biochem. Biophys.*, **151**, 533-539 (1972).

23. R. S. Feigelson (ed.), "Protein crystal growth," *J. Cryst. Growth*, **76**, 527-726 (1986).

Part V

EXPERIMENTAL METHODS
FOR STUDYING CRYSTALLIZATION PROCESSES

STUDY OF MELT STRUCTURE AND CRYSTALLIZATION PROCESSES
BY HIGH-TEMPERATURE RAMAN SPECTROSCOPY

Yu. K. Voron'ko, A. B. Kudryavtsev,

V. V. Osiko, and A. A. Sobol'

INTRODUCTION

Recently, much attention has been paid to the influence of melt structure on its crystallization. The existence in the melt of stably bonded groups of atoms or ions with a definite structure has been considered a basic concept in this regard. It is assumed that this structure can be transformed with a change of melt temperature, and it is suggested that similarity or difference in structure of separate melt and crystalline fragments should play a deciding role in crystallization of a particular species [1-5]. For example, crystallization of metastable phases from melts of aluminum and gallium garnets and Al_2O_3 [1-5] were explained based on such assumptions.

Naturally, understanding of the nature of crystallization processes relative to melt structure cannot be achieved without development of experimental methods for studying the melt structure itself. Unfortunately, such a method as x-ray diffraction, although effective for studying the structure of crystalline substances, does not always give unambiguous results in the case of melts. The principal drawback of this method is that the radial distribution function which characterizes the probability of finding a pair of atoms at a given distance reduces the spatial distribution of atoms to smooth one-dimensional functions [6, 7]. In this respect, the local features of melt structures are averaged. The x-ray diffraction methods are very useful for melts with simple structure, for example for monatomic substances [7], but the information from them decreases substantially for studies of melts of multicomponent compounds.

A convenient and simple method for studying melt structures, as well as phase transitions in crystals, is Raman spectroscopy (RS). However, until recently its use was limited to substances with melting points up to 1000-1200°C [8, 9]. Thus, melts of a broad class of high-melting materials which have both scientific and practical interest, in particular melts of a number of complicated oxides, have been outside the limits of RS.

It should be noted that statements on melt structures have frequently been made based on studies of RS or IR spectra of glasses [10]. Such an approach is valid only when there is evidence that the structure of the glass and melt is identical. Moreover, not all substances can be prepared as a glass even using ultrarapid quenching methods. Therefore, unambiguous conclusions about the structure of substances in a melt can be made based only on RS of their melts.

We were able to broaden substantially the temperature range in which Raman spectra can reliably be recorded (up to 1750°C). Thus, a number of questions related to the structure of melts of high-melting compounds and their crystallization processes can be resolved.

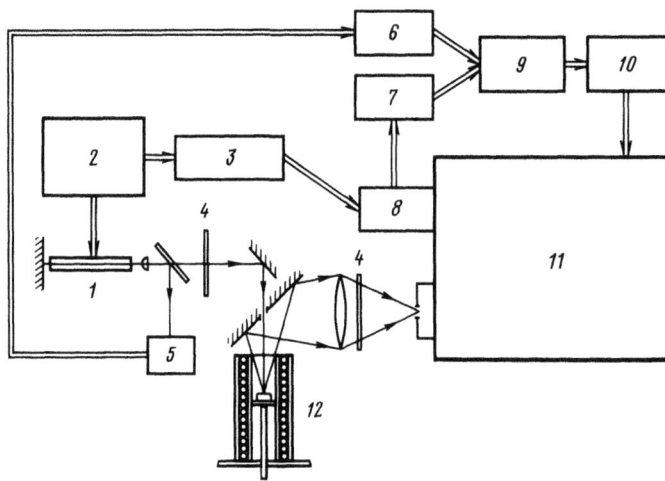

Fig. 1. Apparatus for Raman studies at high temperatures. 1) Copper vapor laser;
2) laser feed block; 3) modulator; 4) polarizer and analyzer; 5) reference photomultiplier; 6, 7) pulse count shaper; 8) recording photomultiplier; 9) coincidence
circuit; 10) treatment of experimental data and spectrometer control; 11) Spex–
Ramalog double monochromator; 12) resistance furnace.

In the present article, we present a review of studies concerned with studying melts of simple and complicated oxides based on phosphates, gallates, germanates, niobates, and tungstates by high-temperature RS, as well
as a study of the crystallization mechanism of these substances in relation to their melt structures.

1. METHOD FOR STUDYING RAMAN SPECTRA OF SUBSTANCES
AT HIGH TEMPERATURES IN THE CRYSTALLINE AND MELT STATES

The principal barriers which hinder high-temperature studies of materials by RS are problems of spectral
recording at these temperatures and the necessity to construct special heated cells. The intensity of thermal
emission from the sample and the heated apparatus in the visible spectral region increases with increasing
temperature. Excitation and recording of Raman spectra is usually carried out in this region. Thus, Raman
spectra should be recorded with a constant background due to thermal emission. The studies carried out in [8,
9] indicate that solid substances up to ~1300°C can be studied using traditional methods of RS. However, an
increase of excitation intensity to a power on the order of 100 kW is required for expanding the temperature
range above this limit. This is possible with use of a laser operating in a pulsed mode. In this case, the
temperature range for recording Raman spectra, taking the fluctuations used in [11] into account, can be
extended to ~2200°C. Experiments on the recording of Raman spectra of substances in crystals and melts up
to temperatures of ~1750°C were performed using the apparatus shown in Fig. 1. This value is limited by the
limiting temperature of the heater [12]. A copper vapor laser operating in pulsed mode at a frequency of 10
kHz was used as the excitation source. The pulse power was ~80-100 kW at a pulse length of 8 nsec. Use of
a laser with a high pulse power permits a satisfactory signal/noise ratio to be obtained at high temperatures.
The high pulse rate facilitates use of a photon counting procedure with a short accumulation time (0.5-1 sec),
which greatly reduces the time required to record the Raman spectrum. In our experiments, this time did not
exceed that required for recording the same spectra at room temperature.

A feature of the apparatus depicted in Fig. 1 was simultaneous use of two signal gating circuits. The
first circuit closed the recording part of the apparatus in the absence of an excitatory laser pulse. Due to this,
a large part of the thermal background intensity was not recorded by the photon counting circuit. The width of
the time window of the first gating circuit Δt_1 was ~10 nsec. The second gating circuit was used for prevention

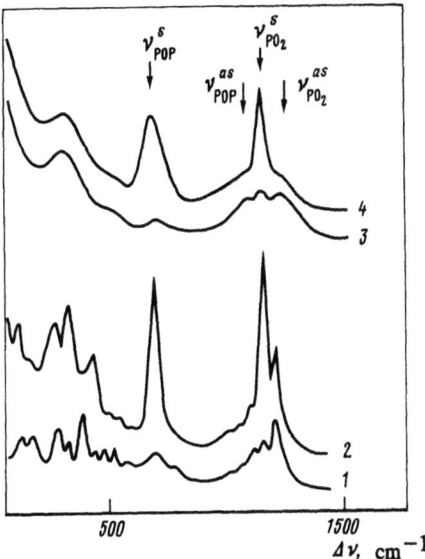

Fig. 2. Raman spectra of the binary phosphate LiLa(PO₃)₄:
1, 2) in the crystal at $T = 1020°C$ and 3, 4) in the melt at $T = 1260°C$ at various scattering geometries. 1) (ZX); 2) (ZZ); 3) VH; and 4) VV. Arrows show characteristic vibrations of the catenated phosphorus–oxygen anion.

of recording photomultiplier current saturation due to increased intensity of extraneous illumination of the photomultiplier photocathode by thermal emission at temperatures above 1500°C. Modulation of the photomultiplier with electrodes was used for this. The time window of the second gating circuit Δt_2 was ~150 nsec. Since $\Delta t_1 < \Delta t_2$, extraneous pulses caused by transitional processes occurring as a result of detection of the modulating signal by the photomultiplier electrodes did not affect the recording part of the circuit.

Figure 1 shows the heater cell designed for recording Raman spectra at a scattering geometry under 180°. This geometry is suitable for studying Raman spectra of melts since difficulties related to use of optical windows that are high-melting and chemically stable to the melt are obviated. In this case, excitation of Raman spectra and recording of scattered radiation is carried out through the upper edge of the melt. This same geometry readily permits study of the change of Raman spectra upon melting and crystallization of the material. An alloy of Pt–30% Rh, operating up to 1750°C, was used in the heater cell. The melts were studied in Pt or Pt–10% Rh crucibles. The crucible diameters were 10 mm and the length 30 mm. The weight of the samples was 0.1-2 g. Stabilization and regulation of temperature was maintained to an accuracy of 0.5°C.

2. DETERMINATION OF THE STRUCTURE OF COMPLICATED OXYGEN-CONTAINING COMPLEXES IN MELTS BY RS

Use of RS as one method of vibrational spectroscopy for studying the structure of substances is predicated on the close connection of vibrational spectra to the crystal lattice. Substances having structural units that are complexes with stable covalent bonds are of special interest. These substances include so-called inorganic polymers and silicates and their analogs that contain [MO₄] tetrahedra in a complicated anionic motif. In view of the stability of the bonding inside such a complex, its structure can be preserved upon a transition from the crystalline to the melt state. The Raman spectrum, which corresponds to internal vibrations of the [MO₄] complex, serves as an indicator of structural changes in the substance both in the crystalline and in the melt state.

Table 1. Characteristic Vibrational Frequencies of Isolated [MO₄] Tetrahedra in the Melt (M = Ga, Ge, P, Nb, and W)

Type of tetrahedral complex	Starting crystalline compound	Melting point, °C	Melt Raman line frequency, cm^{-1}	Raman line assignment
$[GaO_4]^{5-}$	Li_5GaO_4	1250	640 (P)	$\nu_1(A_1)$
			566	$\nu_3(F_2)$
$[GeO_4]^{4-}$	Li_4GeO_4	1320	741 (P)	$\nu_1(A_1)$
			692	$\nu_3(F_2)$
			469	$\nu_4(F_2)$
			295	$\nu_2(E)$
$[PO_4]^{3-}$	Cs_3PO_4	1300	976	$\nu_3(F_2)$
			871 (P)	$\nu_1(A_1)$
			535	$\nu_4(F_2)$
			410	$\nu_2(E)$
$[NbO_4]^{3-}$	Li_3NbO_4	1420	808 (P)	$\nu_1(A_1)$
	$LaNbO_4$	1720	720	$\nu_3(F_2)$
			270–440	$\nu_2(E); \nu_4(F_2)$
$[WO_4]^{2-}$	Na_2WO_4	900	918 (P)	$\nu_1(A_1)$
	$CaWO_4$	1600	820	$\nu_3(F_2)$
			324	$\nu_2(E); \nu_4(F_2)$

Note. P is a polarized band.

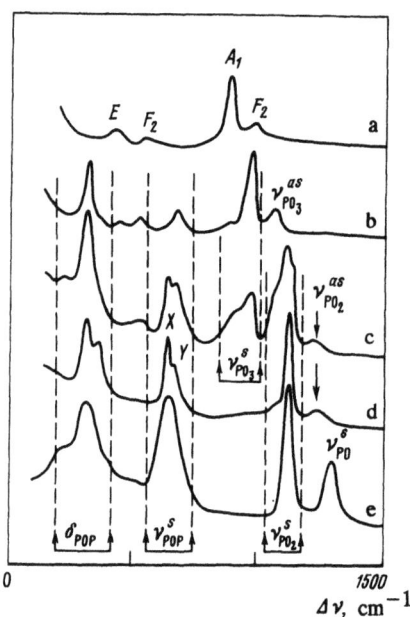

Fig. 3. Raman spectra of cesium phosphate melts at $T = 1300°C$ with anions consisting of various numbers of [PO₄] tetrahedra (R is the Cs_2O/P_2O_5 mole ratio). Isolated [PO₄] tetrahedra, $R = 3$ (a), pyrophosphate anion [P₂O₇], $R = 3.5$ (b), a short chain, $R = 7/5$ (c), an infinite chain $[PO_3]_\infty$, $R = 1$ (d), and an anion with branching, $R = 1/2$ (e). Arrows show characteristic vibrations of phosphorus–oxygen anions.

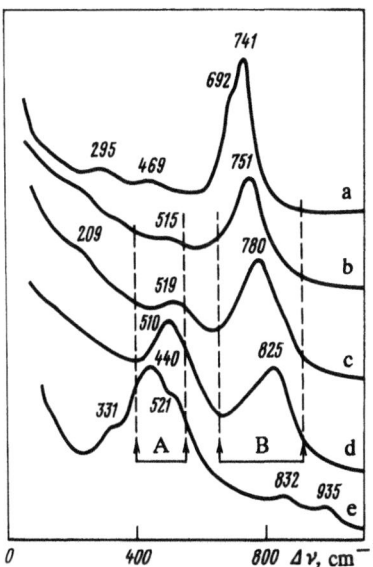

Fig. 4. Raman spectra of Li_2O–GeO_2 system melts consisting of model anions with various degrees of $[GeO_4]$ tetrahedra condensation. Isolated $[GeO_4]$ tetrahedra (a), pyrogermanate anion $[Ge_2O_7]$ (b), $[GeO_3]_\infty$ chains (c), $[Ge_2O_5]_\infty^\infty$ layers (d), and the $[GeO_2]_{\infty,\infty}^\infty$ backbone (e). A) $\tilde{\nu}_{GeOGe}^s$ vibration; B) $\tilde{\nu}_{GeO_3, GeO_2, GeO}^s$ vibration.

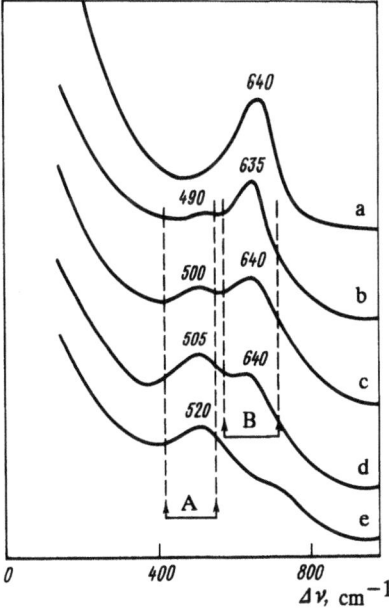

Fig. 5. Raman spectra of the $5Li_2O \cdot Ga_2O_3$ melt (a) and melts of the CaO–Ga_2O_3 system corresponding to model anions with various degrees of $[GaO_4]$ tetrahedra condensation ($T = 1250$-$1600°C$). Isolated $[GaO_4]$ tetrahedra (a), $[GaO_3]_\infty$ chains (b), model anions intermediate between a layer and backbone (c, d), and the $[GaO_3]_{\infty,\infty}^\infty$ backbone (e). A) $\tilde{\nu}_{GaOGa}^s$ vibration; B) $\tilde{\nu}_{GaO_3, GaO_2, GaO}^s$ vibration.

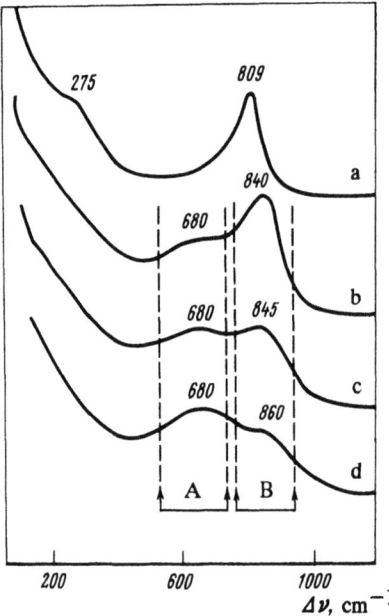

Fig. 6. Raman spectra of $Li_2O-Nb_2O_5$ melts corresponding to model anions with various degrees of $[NbO_4]$ tetrahedra condensation (T = 1260-1520°C). Isolated $[NbO_4]$ tetrahedra (a), $[NbO_3]_\infty$ chains (b), model anions intermediate between a chain and layer (c), and the $[Nb_2O_5]_{\infty,\infty}^\infty$ layer (d). A) ν_{NbONb}^s vibration; B) $\nu_{NbO_3, NbO_2, NbO}^s$ vibration.

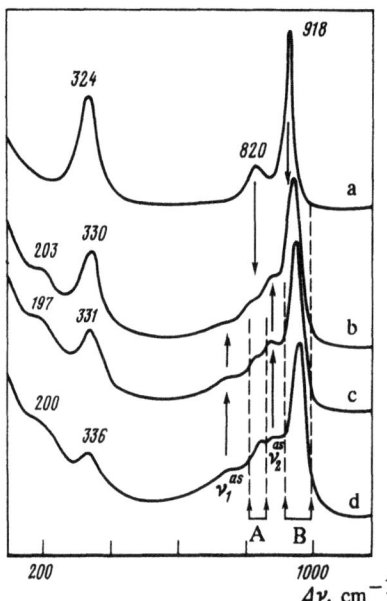

Fig. 7. Raman spectra of Na_2O-WO_3 melts corresponding to model anions with various degrees of $[WO_4]$ tetrahedra condensation (T = 900-1450°C). Isolated $[WO_4]$ tetrahedra (a), model anions intermediate between the ortho and pyroanion (b), pyrotungstate anion $[W_2O_7]$ (c), and short-chain $[W_4O_{13}]$ (d). A) ν_{WOW}^s vibration; B) ν_{WO_3, WO_2}^s vibration.

Assignments of RS lines corresponding to internal vibrations of isolated and catenated oxygen-containing [MO$_4$] complexes for various cations (M) — Si, Ge, and P — in the crystals are given in [13, 14]. The spectrum consists of lines that occur in the high-frequency region, in contrast to the external vibration spectrum that is located in the low-frequency region. The internal vibration spectrum of a tetrahedral complex remains constant if the structure in the melt is preserved upon transition from the crystalline to the melt state. Figure 2 shows a comparison of the Raman spectra of a crystal and melt of the complex polyphosphate LiLa(PO$_3$)$_4$. The Raman spectrum of the melt appears as a somewhat broadened crystal spectrum. Based on [14], groups of lines that correspond to internal vibrations of the phosphorus—oxygen motif can be separated in the Raman spectrum of the crystalline polyphosphate. It is important to note that the basic features of the crystalline Raman spectrum persist upon transition to the melt. These features include the position of high-frequency lines, the ratio of intensities between them, and the polarization of internal vibration lines with observation at defined scattering geometries (Fig. 2). The low-frequency lines, which are caused by external vibrations, disappear in the Raman spectra upon melting as a result of the loss of the translational symmetry of the crystal lattice. According to some published x-ray structural data, chains of catenated [PO]$_4$ complexes form from phosphorus—oxygen tetrahedra in LiLa(PO$_3$)$_4$ [15]. Given the coincidence of the Raman spectra of the crystal and melt of LiLa(PO$_3$)$_4$ and knowing the structure of the phosphorus—oxygen anion from x-ray structural data, the spectrum of this catenated anion can also be identified in the melt. The Raman spectra of simple and complex tetrahedral [PO$_4$] anions in the melt with the isolated, planar, or three-dimensional [PO$_4$] tetrahedral structure can be classified based on a similar principle.

These collections of standard Raman spectra are not difficult to obtain for substances that contain not only the [PO$_4$] group but also other [MO$_4$] complexes with M = Ga, Ge, Nb, etc. These standard Raman spectra can be used for studying the changes of melt structure itself of complex anions upon crystallization or fusion not only of materials with a single type of [MO$_4$] group but also with mixed anionic motifs.

We will now examine examples of selected standard Raman spectra for melts with different [MO$_4$] tetrahedral structure for M = P, Ga, Ge, Nb, and W.

2.1. Melts Containing Isolated [MO$_4$] Tetrahedra

Isolated tetrahedral complexes in an isotropic medium are known to have T_d symmetry [16]. This implies the presence of four vibrational modes which are active in the Raman spectra: ν_1 (A_1), ν_2 (E), ν_3 (F_2), and ν_4 (F_2). Since the fully symmetric vibration ν_1 (A_1) is characterized by a diagonal scattering tensor, the Raman line corresponding to this vibration even with an arbitrary orientation of the [MO$_4$] complex in an isotropic medium should be markedly polarized in contrast to lines of other vibrational modes ν_{2-4} [17]. The Raman spectra of isolated [MO$_4$] complexes in the melt can conveniently be identified by studying melts of ortho-compounds with the general formula Me$_{8-x}^y$(MO$_4$)$_y$, where M = P^{5+}, Ga^{3+}, Ge^{4+}, Nb^{5+}, and W^{6+}, and Me = alkali, alkaline earth, and rare earth metal ions, where x and y are the valence of M and Me cations, respectively. X-ray structural analysis has shown that the [MO$_4$] tetrahedra are isolated in the crystal lattice of these compounds [18-22]. In addition, Raman lines in these substances are also identified as internal vibrations of [MO$_4$] tetrahedra ν_{1-4} [13, 14, 23, 24].

A correspondence between the Raman spectra of [MO$_4$] complex internal vibrations in the crystal and melt is observed upon fusion of the ortho-compounds listed above, as for the case of LiLa(PO$_3$)$_4$ (Fig. 2). Thus, Raman lines of isolated tetrahedral complexes in the melt can be identified. The frequencies of these lines are given in Table 1. Figures 3-7 show the Raman spectra of isolated [MO$_4$] tetrahedra in the melt. Features of these spectra agree with the theory. Thus, the four ν_{1-4} lines are observed for Cs$_3$PO$_4$ and Li$_4$GeO$_4$. The most intense of these, ν_1 (A_1), is strongly polarized (Figs. 3 and 4a). The smaller number of lines in the spectra of LaNbO$_4$, Na$_2$WO$_4$, and CaWO$_4$ melts is caused by accidental degeneracy of the ν_2 (E)

and ν_4 (F_2) vibrations [23] (Figs. 6 and 7a). The presence of only the most intense polarized line ν_1 (A_1) for lithium orthogallate is related to the very weak Raman spectral intensity of the melt of this compound (Fig. 5a).

2.2. Melts with Condensed Tetrahedral Groups

Isolated tetrahedral groups are not the sole example of stable complexes in the melt. Anionic motifs in crystalline silicates and their analogs are known to have different structures depending on the catenating ability of the [MO$_4$] tetrahedra [25]. A variety of substances with various degrees of [MO$_4$] tetrahedra condensation can be prepared depending on the mole ratio R of Me$_2$O$_y$/M$_2$O$_x$. The degree of condensation is conveniently expressed by $n = (1/2)(x + Ry)$, which characterizes the number of oxygen atoms that occur on one metal atom. For an isolated tetrahedron, $n = 4$. A decrease in the range $n = 4\text{-}3$ leads to formation of chains of various lengths from joined [MO$_4$] tetrahedra vertices. The value $n = 3$ corresponds to an infinite chain. Further, layered complexes, in which the [MO$_4$] tetrahedra have three shared vertices, are formed at $n = 3\text{-}2.5$. Finally, a three-dimensional backbone in which tetrahedra are joined at all four vertices are formed at $n = 2.5\text{-}2$.

It should be noted that n_{max} for a certain M cation depends on its valence. Thus, $n_{max} = 3$ is possible for tungstates ($x = 6$). This corresponds to the chain model. An $n_{max} = 2.5$, characteristic of the layer model, occurs for phosphates and niobates ($x = 5$). The backbone model with $n_{max} = 2$ is found for germanates ($x = 4$). Since catenation is possible only with a maximum of the four tetrahedral vertices, the $n_{max} = 1.5$ for gallates ($x = 3$) implies the existence of a part of the Ga^{3+} ions outside the tetrahedral backbone. This part of the Ga^{3+} ions will fulfill the function of the main cations.

At first glance, the presence of a large number of atoms in the complex condensed anion would seem to make interpretation of its Raman spectrum impossible. However, it turns out that groups of atoms can be separated in the complicated complex. The vibrations of these atoms are determined by the different force constants. Such an approach is called the characteristic vibration method [13, 14]. Based on this method, the so-called bridging MOM bond at the catenation point of tetrahedra and the terminal MO$_3$ and intermediate MO$_2$ and MO groups, in which oxygen atoms are bonded to only one M cation, can be separated. Thus, one bridging and two terminal MO$_3$ groups occur at $n = 3.5$ (two joined tetrahedra). Vibrations of the intermediate MO$_2$ groups arise in longer chains. When contact occurs between three tetrahedral vertices and the planar network is formed, vibrations of the MO group arise. The vibrational spectrum of the backbone structure is interpreted based on the bridging MOM bond vibrations.

Crystalline structures in which the anionic motif consists of various degrees of condensation of tetrahedral [MO$_4$] groups and the melts of which can be used for studying Raman spectra of complicated complexes are not hard to find for germanates, phosphates, and gallates. The Raman spectra of these melts are shown in Figs. 3-7. Characteristic vibrations of complicated [MO$_4$] complexes in the melt were identified based on comparison of these spectra to Raman spectra of the starting crystalline materials. Here, it is important to note the presence of two groups of intense lines at medium and high frequencies. These correspond to the symmetric stretching vibrations (ν^s) of bridging bonds (A) and vibrations of the terminal and intermediate groups (B) (Figs. 3-7). The lines $\nu^s_{MO_3, MO_2, MO}$ (B) and ν^s_{MOM} (A) are also polarized in the Raman spectra of melts, similar to the ν_1 (A_1) vibration of isolated [MO$_4$] tetrahedra. This substantially helps in their identification.

Tables 1 and 2 give values for the frequencies of characteristic vibrations and their identification for melts with various degrees of [MO$_4$] tetrahedra condensation. The data of Tables 1 and 2 and Figs. 3-7 show that the Raman spectra reflect well the effect of increasing degree of polycondensation. An increase in relative line intensity for the bridging ν^s_{MOM} bond in comparison with the line intensity of terminal and intermediate groups $\nu^s_{MO_3, MO_2, MO}$ is observed. A change of the nature of [MO$_4$] vertex contacts also leads to a redistribution of intensities in the high-frequency group of lines between $\nu^s_{MO_3}$, $\nu^s_{MO_2}$, and ν^s_{MO}. The ν^s_{MO} lines have the highest frequency and appear for the highly condensed anionic motif. This effect is shown well in Fig. 3, where spectra of phosphate melts are given. These spectra have the narrowest line width. Lines of the symmetric

Table 2. Characteristic Vibrational Frequencies of Condensed Tetrahedral [MO_n] Groups (M = Ga, Ge, P, Nb, and W) at $n < 4$ in Melts

Type of complex [MO_n] anion, n	Melt composition	Melting point, °C	Melt Raman line frequency, cm^{-1}	Raman line assignment
[GaO_n]; 3–2	$CaO-Ga_2O_3$	1300–1600	630–650 (P)	$\nu^s_{GaO_3}, \nu^s_{GaO_2}, \nu^s_{GaO}$
			490–520 (P)	ν^s_{GaOGa}
			280–320	δ (*)
			765	ν^{as}(*)
[GeO_n]: 3.5–2	Li_2O-GeO_2	1150–1300	935	ν^{as}(*)
			750–830 (P)	$\nu^s_{GeO_3}, \nu^s_{GeO_2}, \nu^s_{GeO}$
			500–520 (P)	ν^s_{GeOGe}
			200–330	δ (*)
[PO_n]; 3.5–2.8	$Cs_2O-P_2O_5$	1300	1300 (P)	ν^s_{PO}
			1230–1260	$\nu^{as}_{PO_2}$
			1080–1130 (P)	$\nu^s_{PO_2}$
			1060	$\nu^{as}_{PO_3}$
			910–970 (P)	$\nu^s_{PO_3}$
			620–670	ν^s_{POP}
			200–510	δ (*)
[NbO_n]; 3–2.5	$Li_2O-Nb_2O_5$	1260–1520	815–870 (P)	$\nu^s_{NbO_3}, \nu^s_{NbO_2}, \nu^s_{NbO}$
			670–690 (P)	ν^s_{NbONb}
			650–860	ν^{as}(*)
			110–320	δ (*)
[WO_n]; 3.5–3.25	Na_2O-WO_3	900–1450	930–950 (P)	$\nu^s_{WO_3}, \nu^s_{WO_2}$
			865–875	$\nu^{as}_{WO_3}, \nu^{as}_{WO_2}$
			820–830 (P)	ν^s_{WOW}
			740–760	ν^{as}_{WOW}
			180–340	δ (*)

Note. ν^s is the symmetric vibration, ν^{as} is the asymmetric vibration, δ is the deformation vibration, (P) is a polarized band; (*) assignment of this vibration to a definite group is not clear (terminal, intermediate, bridging).

$\nu^s_{MO_3}$ vibrations cannot be separated from the $\nu^s_{MO_2}$ and ν^s_{MO} vibrations in Raman spectra of melts of other compounds because of the intense broadening. Therefore, the increased intensity of the $\nu^s_{MO_2}$ or ν^s_{MO} lines with increasing degree of condensation leads to a shift to high frequency of the whole nonuniform broadened polarized band B which belongs to vibrations of terminal and intermediate groups (Figs. 4-7).

It should be noted that the features of the Raman spectral changes with increased degree of polycondensation in melts of phosphates, germanates, and gallates are the same as those for melts of niobates and tungstates although for the latter, crystallines states with condensed [NbO_4] and [WO_4] tetrahedra are not known. This fact indicates that these complicated tetrahedral Nb and W complexes exist in the melt and that their characteristic vibrations can be identified in a manner similar to that for tetrahedral complexes of P, Ge, and Ga (Table 2).

3. THE CORRELATION OF ANIONIC STRUCTURE
IN THE CRYSTAL AND MELT

Based on the standard Raman spectra of complicated complexes with various melt structures given above, the rearrangement processes of these complexes upon fusion or the change in the amount of melt superheating of various materials can be studied using high-temperature RS. The following features observed during such studies can be identified.

3.1. Fusion of $Me_k[MO_n]$ Compounds

Here, Me is one or several main cations for which ionic bonding is predominant. The M cations in the crystal form an anionic motif of isolated or catenated $[MO_4]$ tetrahedra with predominantly covalent bonding. Compounds of this type are several phosphates, gallates, niobates, germanates, and tungstates of alkali, alkaline earth, and rare earth metals. Their melts were studied by high-temperature RS [11, 26-33]. Fusion of these compounds leads to persistence in the melt of the tetrahedral environment of the M cation and the same degree of anionic motif condensation as was seen in the crystal if a change of the mole ratio of oxides $Me_2O_y/M_2O_x = R$ does not occur upon fusion. This conclusion followed from an analysis of Raman spectra in which the relative intensity of ν_{MOM}^s and $\nu_{MO_3, MO_2, MO}^s$ vibrations was constant upon transition of the substance from the crystalline to melt state. The same effect was observed during superheating of melts of this type of compound if the value of R did not change. This indicated the absence of a superheating effect on the nature of $[MO_4]$ tetrahedra condensation for the studied materials.

A change of complex anion form was also detected upon fusion or a change of melt temperature based on Raman spectra of melts. In lanthanum polyphosphate $La(PO_3)_3$ of crystalline modification type I, the La ion has eight-fold oxygen coordination which is caused by a unique form of polyphosphate chains. Upon fusion, as the Raman spectra show, this form is not retained and a configuration identical to the form of polyphosphate chain in the polyphosphates with sixfold coordination of the rare earth ion is adopted [28].

An interesting effect was recorded upon fusion of polyphosphates of the alkali metals K, Rb, and Cs [11, 27]. Here, a splitting of the vibration related to the bridging bond ν_{POP}^s is observed in the Raman spectrum of the melt. It splits into narrow X and broad Y components for the melts (Fig. 3d). With increasing melt temperature, the intensity redistributes into the X line. This effect is related to the increased degree of ordering of the polyphosphate chain with increasing melt temperature due to dissociative processes [11, 27].

3.2. Fusion of $Me_k(MO_3)_m$ Compounds

In these compounds, the M cation has six-fold oxygen coordination although in other structures it can form tetrahedral complexes. Examples of $Me_k(MO_3)_m$ compounds are $LiNbO_3$ and $SmGaO_3$ in which the niobium and gallium form a backbone of catenated vertices of oxygen octahedra. In contrast to the fusion of substances that contain only tetrahedral $[MO_4]$ complexes, where a correspondence between the Raman spectra of the crystal and melt is observed, fusion of $Me_k(MO_3)_m$ compounds in the case examined leads to complete transformation of the vibrational spectra (Fig. 8). Comparison of the Raman spectra of $LiNbO_3$ and $SmGaO_3$ melts with the standard spectra, which illustrate the various cases of formation of complicated complexes from $[NbO_4]$ or $[GaO_4]$ tetrahedra, shows that a change of Nb and Ga coordination and formation of complexes from catenated $[NbO_4]$ and $[GaO_4]$ tetrahedra occur upon fusion of $LiNbO_3$ or $SmGaO_3$.

3.3. $Me_kM_l(MO_4)_mO_r$ Compounds

Here, the same M cation occupies two positions, tetrahedral and octahedral, in the crystal lattice. The sphene $CaGe(GeO_4)O$ and garnet $Sm_3Ga_2(GaO_4)_3$ have such structure. In these compounds, the $[GeO_4]$ and

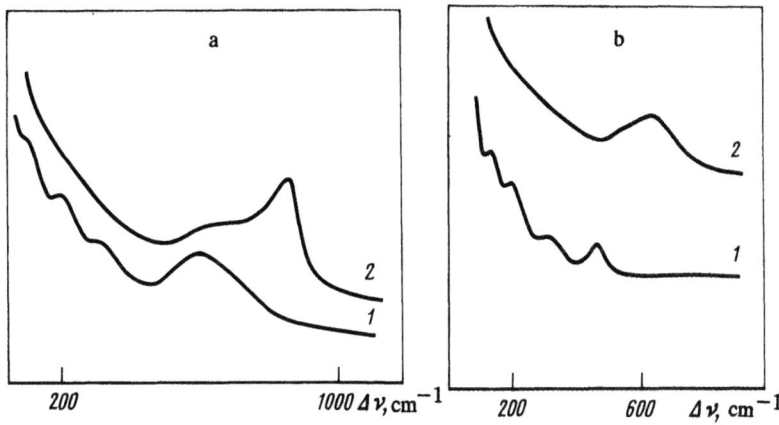

Fig. 8. Raman spectra of paraphases of lithium niobate (a) and samarium gallate (b): 1) before fusion and 2) in the melt. a: 1) $T = 1250$; 2) $T = 1270$; b: 1) $T = 1540$; 2) $T = 1570°C$.

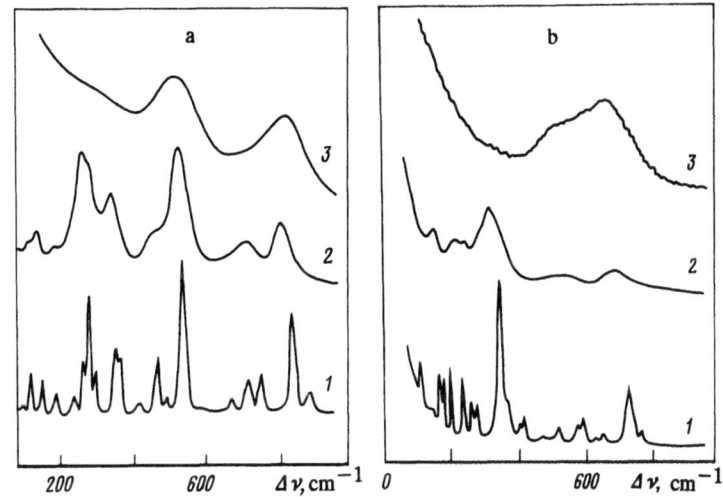

Fig. 9. Raman spectra of $CaGe(GeO_4)O$ (a) and $Sm_3Ga_2(GaO_4)_3$ (b): 1, 2) in the crystal and 3) in the melt. a: 1) 20; 2) 1200; 3) 1250; b: 1) 20; 2) 1640; 3) 1680-1720°C.

[GaO_4] tetrahedra are isolated from each other. In addition, the presence of contact between the [MO_4] tetrahedra and the octahedral gallium or germanium, which form partially covalent bonds to oxygen in contrast to the Me metals, leads to a significant perturbation of the internal vibration of the [GeO_4] and [GaO_4] complexes in the $Me_kM_l(MO_4)_mO_r$ lattice in comparison to the ortho-compounds $M_k(MO_4)_m$ where the M cation enters only the tetrahedral sites of the lattice. A drastic change of Raman spectra is observed upon fusion of $CaGe(GeO_4)O$ and $Sm_3Ga_2(GaO_4)_3$. This indicates a rearrangement of the anionic complexes of these compounds (Fig. 9). Instead of one polarized line, which should be observed in Raman spectra of melts with isolated [MO_4] complexes, two strong polarized lines are observed in the spectra of $CaGe(GeO_4)O$ and $Sm_3Ga(GaO_4)_3$ melts (Fig. 9). Comparison of the spectra in Fig. 9 with the standard Raman spectra for anions composed of [GeO_4] or [GaO_4] tetrahedra (Figs. 4 and 5) suggests that the ν^s line of bridging bonds and the ν^s line of intermediate groups are observed in the calcium digermanate and samarium gallium garnet melts. These attest to the formation in these melts of condensed anions from catenated tetrahedral groups. Such an effect is possible only due to incorporation into the anionic tetrahedral motif of the melt of Ge or Ga cations which

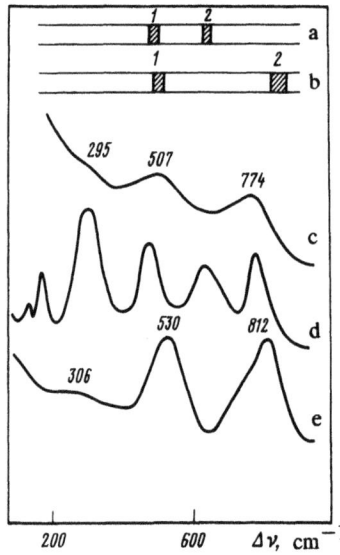

Fig. 10. Vibrational frequencies of: 1) bridging bonds ν^s_{MOM} and 2) intermediate groups $\nu^s_{MO_2}$ for Ga (a), Ge (b), and the Raman spectrum of $Ca_3Ga_2Ge_3O_{12}$ garnet in the melt at 1400°C (c), before fusion at 1350°C (d), and as a glass at 20°C (e).

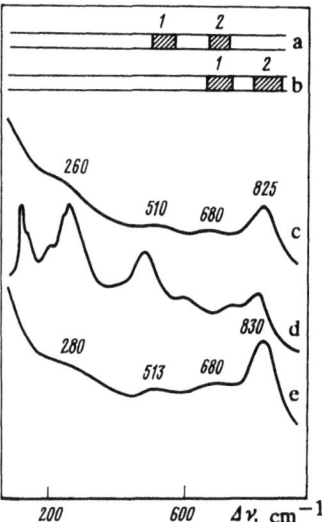

Fig. 11. Vibrational frequencies of: 1) bridging bonds ν^s_{MOM} and 2) intermediate groups $\nu^s_{MO_2}$ for Ga (a), Nb (b), and the Raman spectrum of $Ca_3Ga_{3.5}Nb_{1.5}O_{12}$ garnet in the melt at 1470-1600°C (c), before fusion at 1450°C (d), and as a glass at 20°C (e).

have octahedral coordination in the crystal. Then, the anion with the layer structure $[Ge_2O_5]^\infty_\infty$ is formed in the melt of $CaGe(GeO_4)O$ whereas a complicated anion which has a structure intermediate between the layer and backbone is formed in the $Sm_3Ga_2(GaO_4)_3$ melt.

$Me_k M_l^2 (M^1 O_4)_m$ Compounds

These belong to the case examined above in which the octahedral sublattice is formed by the M^2 cation and the tetrahedral by M^1. Examples of these compounds are the complex garnets $Ca_3 Ga_2 (GeO_4)_3$ and $Ca_3 (Nb_{1.5} Ga_{0.5})(GaO_4)_3$. A rearrangement of the anionic motif occurs upon their fusion. A mixed motif of catenated $[GaO_4]$ and $[GeO_4]$ tetrahedra in the first case and $[GaO_4]$ and $[NbO_4]$ tetrahedra in the second is formed in the melt [33, 34]. This rearrangement shows up well upon comparison of Raman spectra of the crystals and melts of these garnets (Figs. 10 and 11 show the frequency ranges at which characteristic vibrations of the $[GaO_4]$, $[GeO_4]$, and $[NbO_4]$ complexes occur). The ν^s frequencies of the bridging GaOGa and GeOGe bonds are seen to be identical whereas the $\nu^s_{GeO_2}$ frequency is markedly higher than $\nu^s_{GaO_2}$. In this respect, one line in the bridging frequency range and two in the intermediate group region are expected in the Raman spectra of the $Ca_3 Ga_2 (GeO_4)_3$ melt. In fact, splitting of the $\nu^s_{GaO_2}$ and $\nu^s_{GeO_2}$ lines does not occur and one ν^s vibration with intermediate frequency is observed (Fig. 10).

A different situation occurs in the $Ca_3 (Nb_{1.5} Ga_{0.5})(GaO_4)_3$ melt where a splitting of both the bridging and intermediate group vibrations of the $[GaO_4]$ and $[NbO_4]$ tetrahedra is observed (Fig. 11). Apparently, this condition is related to formation in the mixed anionic motif of aggregates of $[GaO_4]$ and $[NbO_4]$ tetrahedra in the $Ca_3 Nb_{1.5} Ga_{3.5} O_{12}$ garnet melt whereas the anionic motif consists of randomly distributed $[GaO_4]$ and $[GeO_4]$ tetrahedra in the $Ca_3 Ga_2 (GeO_4)_3$ melt.

Comparison of the Raman spectra shown in Figs. 9-11 shows that the spectrum of the $Ca_3 (Nb_{1.5} Ga_{0.5})(GaO_4)_3$ melt differs substantially from the Raman spectra of the other garnet melts due mainly to the more intense line, which corresponds to $[NbO_4]$ vibrations. As demonstrated in [34], this is caused by a scattering probability for $[NbO_4]$ group vibrations which is almost an order of magnitude larger than that for $[GaO_4]$ complexes.

The features given in the preceding sections indicate that several cations, P, Ga, Ge, Nb, and W, which tend toward formation of tetrahedral oxygen groups due to the predominance of covalent bonds, have a tendency to form anionic motifs in the melt which consist of $[MO_4]$ tetrahedra. If these cations had an environment with a different coordination in the crystal, then its change to tetrahedral occurs upon fusion.

It is important to note that changes in the Raman spectra of the studied compounds, which would indicate a change of the degree of condensation of the tetrahedrally constructed motifs or their rearrangement with formation of anions with another principal structure, were not observed upon superheating of the melts up to 300°C above the melting point. This conclusion refutes the hypothesis proposed in [2-4] on the occurrence in melts of rare earth gallium garnets of a structural change of the gallium—oxygen complexes with formation of an octahedral gallium environment upon superheating of the melt by 60-70°C above the melting point.

4. STUDY OF MELT CRYSTALLIZATION PROCESSES BY HIGH-TEMPERATURE RAMAN SPECTROSCOPY

High-temperature RS was also used by us for studying crystallization of substances from a melt. Phase transitions could be recorded and phase composition could be determined directly upon melt cooling as a result of the simplicity and informative nature of this method.

The most interesting substances were those for which a mismatch of the anionic motifs in the crystal and melt was observed. In our studies, crystallization was achieved by cooling crucibles with the melt. This corresponded to conditions of homogeneous nucleation. In this case, the rate of formation of crystalline nuclei (I) that are capable of further growth can be represented according to the data of [35] as

$$I = NA \exp\left[(-\Delta\Phi - \Delta\Phi_D - \Delta\Phi_s)/KT\right]; \tag{1}$$

here $\Delta\Phi$ is the potential barrier for formation of a nucleus of the critical size r^*;

$$\Delta\Phi = \frac{4\pi r^3}{3} \Delta\varphi + 4\pi r^2 \delta, \quad r^* = -\frac{2\delta}{\Delta\varphi}, \tag{2}$$

$\Delta\varphi$ is the change of melt thermal potential upon formation of a unit volume of a definite crystalline phase ($\Delta\varphi < 0$, if $T < T_c$, where T_c is the melt crystallization temperature); δ is the specific surface energy of the nucleus; A is a constant; N is the number of sites of possible crystalline phase nucleation, and $\Delta\Phi_D$ and $\Delta\Phi_s$ are the activation energies of diffusion and structural arrangement at the crystalline nucleus and melt interface, respectively.

Equation (1) shows that a structural mismatch of complexes in the crystal and melt increases the value of the barrier $\Delta\Phi_s$. Moreover, according to [35], this mismatch should also lead to a higher value in (1) for the $\Delta\Phi_s$ in comparison to the case of similar structure of the crystal and melt. As a result, crystallization of these melts should lead to separation of metastable phases. Furthermore, cooling of melts can also lead to such effects as their significant overcooling or glassification.

We will now examine further examples of studies of crystallization processes for materials in which a difference in structure of isolated fragments of the crystalline and melt states is found.

4.1. Lithium Niobate LiNbO$_3$

The anionic motif in a LiNbO$_3$ melt was shown to consist of [NbO$_4$] tetrahedra catenated through vertices. However, the crystal lattice of lithium niobate is constructed of [NbO$_6$] octahedra. In spite of this, the lithium niobate melt can be overcooled by 40-50°C. It glassifies with difficulty even under ultrahigh quenching conditions [37]. The single crystallization product of this melt is LiNbO$_3$ with octahedral niobium—oxygen complexes.

This example shows that the conditions of homogeneous nucleation of crystalline LiNbO$_3$ are easily fulfilled in this case even with a mismatch between the melt and crystal structure.

4.2. Lanthanum Polyphosphate La(PO$_3$)$_3$ and Calcium–Gallium–Germanium Garnet

Cooling melts of these substances under quasi-stationary conditions leads to crystallization of equilibrium structures. The anionic motifs of these differ from their structure in the melt.

Moreover, cooling of the same melts under conditions which prevent rearrangement of their anionic motifs, for example with rapid quenching at a rate of 10^3-10^4 deg/sec, leads to crystallization of metastable phases or to melt glassification [28, 33]. Glass is formed at a quenching rate of 10^4 deg/sec. The same anionic motif is found in this as in the melt (Fig. 10) [28, 33]. Lanthanum polyphosphate crystallizes in a structural modification type C which has not been identified earlier for this substance at lower quench rates. This modification has six-fold oxygen coordination for the lanthanum ion. The reason for this is the similarity of the polyphosphate chain form in the type C structure to the configuration of this chain in the melt [28].

Rapid cooling of the Ca$_3$Ga$_2$(GeO$_4$)$_3$ garnet melt at a rate of 10^3 deg/sec leads to its decomposition according to the formula

$$7Ca_3Ga_2(GeO_4)_3 \rightarrow 3Ca_2Ga_2GeO_7 + 4Ca_3Ga_2Ge_4O_{14} + Ca_3Ge_2O_7.$$

In this case, mainly Ca$_2$Ga$_2$GeO$_7$ and Ca$_3$Ga$_2$Ge$_4$O$_{14}$ crystallize from the melt. The anionic motif of these is constructed of catenated [GaO$_4$] and [GeO$_4$] tetrahedra. Just such a structure for the anionic motif, as was demonstrated above, is characteristic of the Ca$_3$Ga$_2$(GeO$_4$)$_3$ melt. These examples show that the metastable crystalline forms can compete with equilibrium forms in nucleation processes if the structure of the anionic motifs of these metastable structures is similar to the melt structure. In the examples given, this competition is

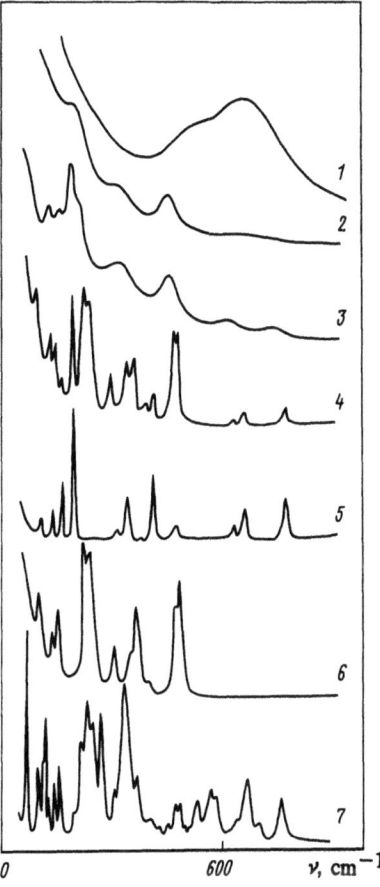

Fig. 12. Raman spectra of a $3Sm_2O_3 \cdot 5Ga_2O_3$ melt heated to 1720°C at various stages of cooling. 1) 1440°C, melt; 2) 1400°C, $SmGaO_3$ + melt; 3) 1290°C, $3SmGaO_3$ + β-Ga_2O_3; 4) 20°C. Raman spectrum at T = 20°C of polycrystals of: 5) β-Ga_2O_3; 6) $SmGaO_3$; 7) $Sm_5Ga_{11}O_{24}$.

successful only for nonstationary crystallization conditions when nucleation occurs in the melt with high viscosity. A similar effect, but for low-temperature crystallization of silicate glasses, was observed in [35], where the nucleation of a certain type of structures depended on the correspondence of the crystalline and glassy structures.

4.3. Compounds with Garnet Structure, $Sm_3Ga_2(GaO_4)_3$ and $Ca_3Nb_{1.5}Ga_{3.5}O_{12}$

The mismatch of the structure of complex anions in the melt and solid is important for these compounds, as for the calcium—germanium—gallium garnet in the case examined above. However, compounds with a tetrahedral anionic motif in the crystal do not occur in the Sm_2O_3–Ga_2O_3 system. Nevertheless, this anionic structure does exist in the melt. This phenomenon can be explained [32] by the unique behavior of melts in the Sm_2O_3–Ga_2O_3 system upon cooling.

Melts in the Sm_2O_3–Ga_2O_3 system were shown in [2-4, 32] to be capable of significant overcooling to 200-300°C if they are previously superheated to 60-70°C above the melting point. As shown above, such superheating does not lead to a change of melt structure, and its role in the subsequent effects which occur upon cooling is related to elimination of crystallization centers of the garnet phase [32].

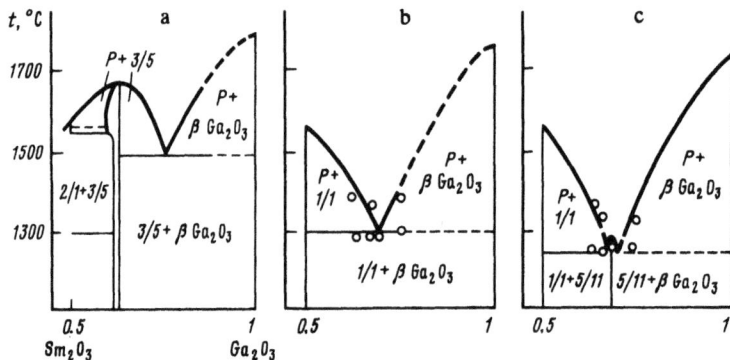

Fig. 13. Equilibrium (a) [4] and nonequilibrium (b, c) phase diagrams of the Sm_2O_3–Ga_2O_3 system.

Crystallization of a previously superheated melt of $Sm_3Ga_2(GaO_4)_3$ was detected earlier to take place in two steps [2, 3]. It was proposed that crystallization of $SmGaO_3$ occurs in the first step at a temperature of T_1. Part of the substance remains in the melt state. Further cooling at temperature T_2 leads to eutectic decomposition of the remaining melt. Crystallization of the metastable phases can be represented by the equation

$$Sm_3Ga_2(GaO_4)_3 \rightarrow 3SmGaO_3 + \beta\text{-}Ga_2O_3.$$

The nature of phase transitions during cooling of melts in the Sm_2O_3–Ga_2O_3 system was followed for the first time using high-temperature RS [32]. Results of these studies for samarium–gallium garnet are shown in Fig. 12. The Raman spectra indicated that the melt superheated to 1720°C with subsequent cooling up to temperature T_1 retained the tetrahedral structure of the gallium–oxygen complexes (Fig. 12, curve 1). In the Raman spectrum at a temperature $T_2 < T < T_1$, lines of crystalline $SmGaO_3$ and melt are recorded (Fig. 12, curve 2). This confirms the hypothesis proposed in [2, 3]. With subsequent cooling at a temperature $T < T_2$, lines of polycrystalline β-Ga_2O_3 are observed in the Raman spectrum in addition to lines of $SmGaO_3$ (Fig. 12, curve 3).

Equilibrium crystallization of Sm_2O_3–Ga_2O_3 melts in the range 50-75 mole % Ga_2O_3 were also studied using RS in [32]. The composition of a melt with 68.8 mole % Ga_2O_3 was determined. This experienced only eutectic decomposition upon cooling. β-Ga_2O_3, and not $SmGaO_3$, crystallized in the first step with shift of this composition toward excess gallium oxide, in contrast to the $Sm_3Ga_2(GaO_4)_3$ melt (62.5 mole % Ga_2O_3). Based on these results, we constructed a nonequilibrium phase diagram for the Sm_2O_3–Ga_2O_3 system (Fig. 13b) in contrast to the equilibrium one (Fig. 13a). We also noticed that this diagram is not unique. The nature of transitions in the first cooling step is preserved with slow stepwise cooling of the same melts. However, eutectic decomposition at the second step does not occur. The composition with 68.8 mole % Ga_2O_3 under these conditions is completely crystallized as the compound $Sm_3Ga_{11}O_{24}$, which was not known earlier. The Raman spectrum of this compound at room temperature is shown in Fig. 12, curve 7, where the Raman spectra of crystalline phases known for the Sm_2O_3–Ga_2O_3 system are shown for comparison.

A second nonequilibrium phase diagram for the Sm_2O_3–Ga_2O_3 system was constructed based on our results. A eutectic decomposition point was absent in this diagram, in contrast to the first (Fig. 13c).

Such a unique behavior of melts in the Sm_2O_3–Ga_2O_3 with 50-75 mole % Ga_2O_3 above all attests to the difficulty of crystallizing a garnet phase under homogeneous nucleation conditions. A role played by the large activation energy for structural rearrangement $\Delta\Phi_s$, which is caused by the structural mismatch of the melt anionic motifs and the garnet structure, is also likely. Furthermore, compounds in the Sm_2O_3–Ga_2O_3 system do not exist for which the crystal lattice is constructed only of catenated $[GaO_4]$ tetrahedra. The $SmGaO_3$

structure is constructed of $[GaO_6]$ octahedra [4], whereas β-Ga_2O_3 is composed of $[GaO_6]$ octahedra joined by edges and tetrahedral chains of $[GaO_4]$ groups [36]. Thus, neither $SmGaO_3$ nor β-Ga_2O_3 should *a priori* have an advantage over $Sm_3Ga_2(GaO_4)_3$ from the viewpoint of a more facile rearrangement of the anionic motif during crystallization. Apparently, this explains the large degree of supercooling (up to 300°C) of melts in the Sm_2O_3–Ga_2O_3 system. The fact that crystallization of $SmGaO_3$, β-Ga_2O_3, or $Sm_5Ga_{11}O_{24}$ occurs nonetheless at a definite temperature indicates fulfillment of nucleation conditions for these structural forms upon attainment of the required amount of supercooling. Apparently, a role is played not only by the quantity $\Delta\Phi_s$, but also by the driving force of the transition $\Delta\varphi$ in formulas (1) and (2), which have values at a definite amount of supercooling that guarantee fulfillment of nucleation conditions only for the nongarnet phases.

The peculiar crystallization of the melt previously heated above the melting point was also observed for the calcium–niobium–gallium garnet. The same two stages of phase transitions, but with different final products [14], are observed during slow cooling of a $Ca_3Nb_{1.5}Ga_{3.5}O_{12}$ garnet melt, similar to $Sm_3Ga_2(GaO_4)_3$:

$$Ca_3Nb_{1.5}Ga_{3.5}O_{12} \xrightarrow{T_1} Ca_2NbGaO_6 + \text{melt} \xrightarrow{T_2} Ca_2NbGaO_6 + CaGa_4O_7 + \text{unknown phases.}$$

An interesting feature of $Ca_3Nb_{1.5}Ga_{3.5}O_{12}$ garnet is that it can be prepared as a glass by rapid cooling of the melt, in contrast to $Sm_3Ga_2(GaO_4)_3$. The same tetrahedral anionic motif as in the melt of calcium–niobium–gallium garnet is retained in the glass, as demonstrated by results of RS studies (Fig. 11).

The examples given for studying crystallization processes of such substances as $LiNbO_3$, $Sm_3Ga_2(GaO_4)_3$, and $Ca_3Nb_{1.5}Ga_{3.5}O_{12}$ are also interesting from the viewpoint of possible melt rearrangement in the temperature region directly preceding its crystallization. Since the structure of the anionic groups in melts of these compounds differs greatly from their structure in the crystalline state, the occurrence of a melt structural change in the precrystallization temperature region should be readily recorded using RS. However, the effect of such a rearrangement was not detected by us for any of the studied compounds. Apparently, rearrangement of the anionic motif, if it occurs, has a fluxional nature and does not affect a significant volume of the melt of these compounds.

5. CONCLUSIONS

Study of the influence of identical or different structures of the melt and crystal fragments on the crystallization process under conditions similar to homogeneous nucleation showed the following.

Such effects as glassification and melt supercooling, as well as its crystallization into metastable structures, was observed to be related to the mismatch of the melt and crystal structures. However, the existence of such a mismatch itself is not a required condition for existence of the anomalies. In each separate case, the influence of the mismatch appears differently.

Cases of equilibrium phase crystallizations are possible, despite the difference of their anionic motifs in the crystal and melt. In other examples, such a difference creates more favorable conditions for nucleation of metastable structures that have crystal lattice fragments similar to the melt.

Third, the mismatch of anionic motifs in the melt and crystal hinders nucleation of equilibrium structures. Under these conditions, metastable forms with a lattice structure far from the melt structure are preferably crystallized. It should be noted that the examples included are correct for homogeneous nucleation conditions. It is possible that similarity or difference in melt and crystal structure also affects the growth of crystalline faces or the formation of structural defects during synthesis of single crystals from the melt. However, this question requires special studies.

REFERENCES

1. J. L. Caslavsky and D. J. Viechnicki, "Melting behavior and metastability of yttrium aluminum garnet (YAG) and $YAlO_3$ determined by optical differential analysis," *J. Mater. Sci.*, 15, No. 7, 1709-1718 (1980).

2. M. A. DiGiuseppe and S. L. Soled, "Perovskite transformations in the Sm_2O_3–Ga_2O_3 and Gd_2O_3–Ga_2O_3 systems," *J. Solid State Chem.*, **30**, No. 2, 203-208 (1979).

3. M. A. DiGiuseppe, S. L. Soled, W. M. Wenner, and J. E. Macur, "Phase diagram relationships of the garnet–perovskite transformation in the Gd_2O_3–Ga_2O_3 and Sm_2O_3–Ga_2O_3 systems," *J. Cryst. Growth*, **49**, No. 4, 746-748 (1980).

4. J. Nicolas, J. Coutures, J. P. Coutures, and B. Boudot, "Sm_2O_3–Ga_2O_3 and Gd_2O_3–Ga_2O_3 phase diagrams," *J. Solid State Chem.*, **52**, No. 1, 101-113 (1984).

5. A. Nukui, H. Tagai, H. Morikawa, and S. I. Iwai, "Structural conformation and solidification of molten alumina," *J. Am. Ceram. Soc.*, **59**, No. 11/12, 534-536 (1976).

6. A. I. Kitaigorodskii, *X-Ray Structural Analysis of Finely Crystalline and Amorphous Solids* [in Russian], Gos. Izd. Tekh. Teor. Lit., Moscow (1952).

7. V. A. Vatolin and É. A. Pastukhov, *Diffraction Studies of the Structure of High-Temperature Melts* [in Russian], Nauka, Moscow (1980).

8. D. W. James, "Vibrational spectra of molten salts," in: *Structure of Molten Salts* [Russian translation], Mir, Moscow (1966), pp. 398-425.

9. M. Ishigame and T. Sakurai, "Temperature dependence of the Raman spectra of ZrO_2," *J. Am. Ceram. Soc.*, **60**, No. 78, 367-369 (1977).

10. P. McMillan, "Structural studies of silicate glasses and melts – applications and limitations of Raman spectroscopy," *Am. Miner.*, **69**, No. 7/8, 622-644 (1984).

11. A. B. Kudryavtsev and A. A. Sobol', "Detuning from thermal radiation during the study of Raman effect spectra at temperatures to 1950 K," *Kratk. Soobshch. Fiz.*, No. 1, 17-22 (1984).

12. A. F. Banishev, Yu. K. Voron'ko, A. B. Kudryavtsev, V. V. Osiko, and A. A. Sobol', "Raman spectroscopy at high temperatures," in: *Spectroscopy of Crystals* [in Russian], Nauka, Leningrad (1985), pp. 243-252.

13. A. N. Lazarev, *Vibrational Spectra and Structure of Silicates* [in Russian], Nauka, Leningrad (1968).

14. A. N. Lazarev, A. P. Mirgorodskii, and N. S. Ignat'ev, *Vibrational Spectra of Complex Oxides* [in Russian], Nauka, Leningrad (1975).

15. K. K. Palkina, "Crystal chemistry of rare earth element phosphates," *Izv. Akad. Nauk SSSR, Neorg. Mater.*, **18**, No. 9, 1413-1428 (1982).

16. K. Nakamoto, *Infrared Spectra of Inorganic and Coordination Compounds* [Russian translation], Mir, Moscow (1966).

17. T. Gilson and P. Hendra, *Laser Raman Spectroscopy*, Wiley, New York (1970).

18. W. H. Zachariasen, "The crystal structure of the normal orthophosphates of barium and strontium," *Acta Crystallogr.*, **1**, No. 5, 263-265 (1948).

19. H. Vollenkle and A. Wittmann, "Die Kristallstruktur von Li_4GeO_4," *Z. Kristallogr.*, **128**, No. 1/2, 66-71 (1969).

20. F. Stewner and R. Hoppe, "Zur Kristallstruktur von β-Li_5GaO_4," *Z. Anorg. Allg. Chem.*, **381**, No. 2, 140-148 (1975).

21. G. R. Wilkinson, "Raman spectra of ionic, covalent, and metallic crystals," in: *Application of Raman Spectra* [Russian translation], Mir, Moscow (1977), pp. 408-578.

22. S. Tsunekawa and H. Takei, "Twinning structure of ferroelastic $LaNbO_4$ and $NdNbO_4$ crystals," *Phys. Status Solidi A*, **50**, No. 2, 695-702 (1978).

23. S. P. S. Porto and J. F. Scott, "Raman spectra of $CaWO_4$, $SrWO_4$, $CaMoO_4$, and $SrMoO_4$," *Phys. Rev.*, **157**, No. 3, 716-719 (1967).

24. G. V. Anan'eva, A. M. Korovkin, A. B. Kudryavtsev, et al., "Fergusonite–scheelite phase transition and Raman spectra of a $LaNbO_4$ crystal and the solid solution $(CaWO_4)_{1-x}(LaNbO_4)_x$," *Fiz. Tverd. Tela*, **23**, No. 4, 1079-1086 (1981).

25. D. Yu. Pushcharovskii, *Structural Mineralogy of Silicates and Their Synthetic Analogs* [in Russian], Nedra, Moscow (1986).

26. A. F. Banishev, Yu. K. Voron'ko, V. V. Osiko, and A. A. Sobol', "Raman scattering in melts of alkali metal phosphates," *Dokl. Akad. Nauk SSSR*, **274**, No. 1, 559-561 (1984).

27. A. F. Banishev, Yu. K. Voron'ko, V. V. Osiko, and A. A. Sobol', "Study of Raman spectra of metaphosphates of alkali metals at high temperatures," *Kratk. Soobshch. Fiz.*, No. 6, 27-30 (1984).

28. G. M. Balagina, A. F. Banishev, Yu. K. Voron'ko, et al., "Study of phase transitions in a number of polyphosphates of rare earth metals $Ln(PO_3)_3$ by Raman scattering," *Izv. Akad. Nauk SSSR, Neorg. Mater.*, **21**, No. 5, 712-720 (1985).

29. A. F. Banishev, N. V. Vinogradova, Yu. K. Voron'ko, A. A. Sobol', and N. N. Chudinova, "Study of thermal transitions of binary rare earth element and alkali metal phosphates of composition $M^I Ln(PO_3)_4$ by Raman scattering," *Izv. Akad. Nauk SSSR, Neorg. Mater.*, **23**, No. 2, 292-297 (1987).

30. A. F. Banishev, Yu. K. Voron'ko, A. B. Kudryavtsev, V. V. Osiko, and A. A. Sobol', "High-temperature studies of Raman scattering spectra of calcium tungstate in crystalline and melt states," *Kristallografiya*, **27**, No. 3, 618-620 (1982).

31. Yu. K. Voron'ko, A. B. Kudryavtsev, V. V. Osiko, et al., "Raman spectroscopy of melts of Li_2O–Nb_2O_5," *Kratk. Soobshch. Fiz.*, No. 2, 34-36 (1987).

32. Yu. K. Voron'ko, A. B. Kudryavtsev, V. V. Osiko, et al., "Study of crystallization of superheated melts in the Sm_2O_3–Ga_2O_3 system by Raman scattering," *Dokl. Akad. Nauk SSSR*, **298**, No. 1, 87-91 (1988).

33. Yu. K. Voron'ko, A. B. Kudryavtsev, V. V. Osiko, et al., "Study of the structure of crystalline and fused germanates by Raman scattering," *Dokl. Akad. Nauk SSSR*, **283**, No. 6, 1333-1336 (1985).

34. Yu. K. Voron'ko, A. B. Kudryavtsev, N. A. Es'kov, V. V. Osiko, et al., "Raman scattering in crystals and melts of calcium–niobium–gallium garnet," *Dokl. Akad. Nauk SSSR*, **298**, No. 3, 604-607 (1988).

35. V. N. Filipovich, "Special features of the crystallization of glasses during the formation of glass–ceramics," in: *Structural Transitions in Glasses at Elevated Temperatures* [in Russian], Nauka, Moscow–Leningrad (1965), pp. 30-43.

36. D. Dohy and G. Lucazeau, "Raman spectra and valence force field of single-crystalline β-Ga$_2$O$_3$," *J. Solid State Chem.*, **45**, No. 2, 180-192 (1982).

37. K. Nassau, C. A. Wang, and M. Grasso, "Quenched metastable glassy and crystalline phases in the system lithium–sodium–potassium metaniobate–tantalate," *J. Am. Ceram. Soc.*, **62**, No. 9/10, 503-510 (1979).

ACOUSTIC EMISSION STUDY OF
CRYSTALLIZATION FROM THE MELT.
A PROBLEM OF IDENTIFICATION

É. L. Lube and A. T. Zlatkin

1. STATEMENT OF THE PROBLEM

Compilation of data on the actual structure of a crystal is one of the most important problems of basic and applied research. The data are necessary for establishment of relationships between the actual structure of the crystal, the ultimate parameters of crystallization, and the parameters that determine the growth conditions, optimization of the process for preparing crystals of given perfection, and definition of the crystal quality indices which are related to its further use. At present, the actual structure of a crystal in the majority of cases is studied upon completion of the growth process. The methods that are used are nondestructive and require large expenditures of effort and time in sample preparation (cutting, polishing, etching, etc.). Since in these cases the actual structure of the crystal represents a montage of influences in the course of the crystallization process, the correlation between the actual structure and the growth conditions is not always unambiguous. In view of the peculiarities of crystallization — spatial nonuniformity of the distribution parameters, primarily temperature and concentration, their variability, and the effect of unregulated and uncontrolled factors — determination of the structural perfection of a crystal directly during growth is desirable. Operational monitoring of responses of the crystal to a change of growth conditions can with good reliability establish a correlation between them and can provide information on the dynamics of the process. Such monitoring also resolves concerns about the immediacy of crystal characterization.

Until recently, x-ray topography was the sole *in situ* method for studying the actual structure of a crystal during growth from the melt [1]. A silicon crystal of 0.8-1 mm thickness was placed in a miniature furnace which was built in the carriage of a Lang topographic camera. An x-ray tube with a rotating molybdenum anode (focal point 0.5×10 mm, potential 60 kV, current 0.5 A) was used as the source in order to guarantee the necessary quickness of topogram recording. A topogram of 9×13 mm size was recorded at a resolution of 25 μm by a vidicon sensitive to x-rays. The picture was accumulated during movement of the carriage with the crystal (from 3 to 10 sec).

It should be noted that the x-ray diffraction method, which has a high sensitivity and spatial resolution, is not applicable for studying growth processes of large crystals, especially at high temperature, due to a number of limitations in principle and construction. Among these, we single out the limit to crystal thickness in a transmission topograph, the presence in the path of x-rays of strongly absorbing tungsten and molybdenum heat shields and crucible walls, and the large distances between the source, the studied crystal, and the x-ray detector (since the growing crystal is located in the heater cavity which is surrounded by heat shields).

Analysis of alternative physical methods for studying the actual structure of a crystal directly during growth showed that acoustic emission, which was developed by us for application to the growth experiment [2-6], is at present the only possibility. This method is feasible over a wide crystallization temperature range and for growth of single crystals of any size. Use of the method requires minimal changes in construction of the crystallization apparatus and does not disturb the growth of crystals.

2. MECHANISM OF ACOUSTIC EMISSION

Acoustic emission (AE) is commonly defined as emission of elastic waves by a material due to a dynamic local rearrangement of its structure. Acoustic emission is connected with effects that accompany elastic and plastic deformations, phase transitions with a volume change, and generation and propagation of cracks. It may also be caused by detachment of dislocation loops of critical size from the points of attachment [7-10], by multiplication of dislocations using a Frank–Read source [8, 11], by micrononuniformities of the plastic flow [12], by development of slipbands [12, 13], and by twinning [13-15].

The energy that determines the dynamics of a separate dislocation is too small for generating an AE signal that can be recorded. It was suggested in [9] that a coherent elastic wave is generated by 10^5-10^6 segments of a dislocation line which are incorporated into an avalanche through the stimulating effect of the traveling acoustic wave that is formed at the origin of one or several dislocation segments. This hypothesis was proposed from experiments on plastic deformation of single crystals of LiF and KCl.

The relation between the AE signals and the formation of slipbands as a result of the avalanche of dislocations was found during deformation of Cu, Mg, and Fe single crystals [12]. An average deformation $\epsilon = 8 \cdot 10^{-7}$ for the Mg crystal and $5 \cdot 10^{-7}$ for the copper crystal was found to correspond to one pulse. Electron microscopic investigation of the dislocation structure showed that about $2 \cdot 10^4$ dislocation sources participate in formation of a slipband and generation of an AE pulse.

The energy of AE signals caused by twinning is much larger than for slip deformation. The AE signals are generated upon both onset and disappearance of an elastic twin [13, 15]. Low-energy signals which correspond to development of slipbands [13] are recorded during deformation of Zn single crystals between signals caused by formation of twins.

The energy of AE signals during formation of cracks is several orders of magnitude larger than during the effects examined above. In spite of the large number of theoretical and experimental studies, an unambiguous correlation between the parameters of the AE signals and the crack formation processes has not at present been obtained.

The examined AE mechanisms, which are related to rearrangement of the actual structure of the material upon its deformation, can be taken separately or in combination as an adequate model of sources in the crystal during growth and cooling. Thermal stresses in the crystal cause the same effects as mechanical loadings: movement of dislocation pile-ups, slippage, twinning, and crack formation. Nonuniform distribution of components in the crystal, inclusion of secondary phases and bubbles, and polymorphic transformations of the crystal lattice upon cooling also lead to formation of stresses and dislocations. In turn, these stresses in the crystal are tied to the crystallization parameters: temperature distribution, melt composition, crystallization rate, crystal cooling rate, etc.

3. MEASUREMENT AND ANALYSIS OF ACOUSTIC EMISSION SIGNALS

The elastic wave arising as a result of local deformation or disruption of a solid is propagated in its bulk at the speed of sound and can be transformed into an electrical signal using a piezotransformer that has acoustic

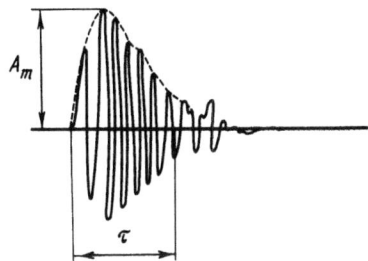

Fig. 1. AE transformer output signal.

contact with the studied specimen. An example of the transformer output signal is shown in Fig. 1. A signal of this type, which has a burst form, is considered an explosive AE in contrast to a continuous AE which appears as high-frequency noise with an amplitude which is substantially smaller than signals of the first type. The continuous emission is assigned to incoherent movement of dislocations as a result of impedance during interaction with each other or with other defects. The explosive emission is associated with an avalanche of detachments from points of attachment [16, 17]. Since AE is the result of discrete effects, the division into explosive and continuous is arbitrary. In keeping with the idea of [18], that the continuous emission is the superposition of a large number of explosive AE pulses of small amplitude, we note that the quasicontinuous emission at separated time intervals can have a significant amplitude. For example, the AE registered by us from a large number of needlelike crystals, which were precipitated on a single crystal grown by a gas-flame method (so-called whiskers), was the superposition of a large number of pulses of explosive emission.

Several theoretical models [11, 20, 21], which do not have strict experimental confirmation, have been proposed for description of the shape of the primary AE signal.

The AE elastic wave on the path from the source to the detector undergoes attenuation and phase distortions. Reflected waves, which can overlap the primary signal, arise at the interfaces. In the end, the pulse parameters at the piczotransformer output depend not only on the source characteristics but also on the mechanical properties of the material and the form and sizes of the studied specimen. These distortions of the original AE signal hinder identification, i.e., the relation of the signal and the plastic deformation or destruction caused by it. Difficulties arise during acoustic emission studies of crystallization from the melt, especially at high temperatures. The dimensions, shape, and temperature of a crystal change during growth. Since the temperature of the piezotransformer should not exceed the Curie point (300°C), the piezotransformer is joined to the crystal, at a temperature near 2000°C, by an acoustic line. The elastic characteristics of the sonoconductor can change. Mechanical contact with the crystal is not always reproducible. It should also be noted that sources of AE at sites of the crystallization chamber exist besides the multitude of AE mechanisms in the growing crystal. These sources are susceptible to thermal stresses. Since these sites are acoustically related to the crystal, an additional problem of AE signal selection arises.

The shape, amplitude, and frequency of the AE signal are analyzed in detail, taking into account the above considerations, using a computer that searches for the most informative parameters of the actual AE signal from the growing crystal, i.e., signals that are largely related to these parameters [4]. In this case, the maximal amplitude, the duration, time, and rate of ascent and descent of amplitude, the number of carrier pulses, their average and instantaneous frequency, the number of sign changes of the first derivative of amplitude envelopes, and the integrals, squares, and dispersion of amplitudes are determined. The time of appearance is recorded with each AE signal received. In further experiments, the number of AE signal parameters was reduced. The most informative parameters were chosen as the maximal amplitude, the signal duration, the product of the square of the maximal amplitude and the duration, the average frequency of effective pulses, the time interval between signals, and the frequency distribution of the collected signals.

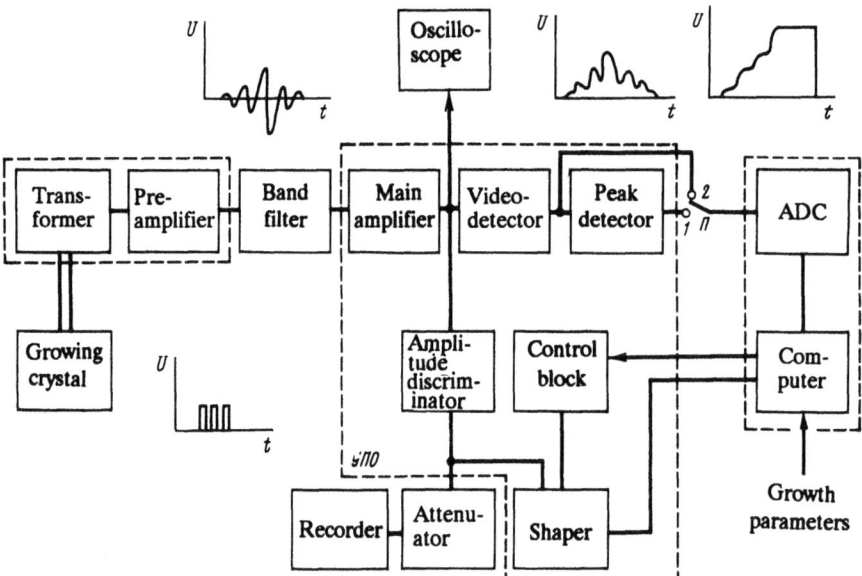

Fig. 2. Diagram of the automatic system for AE studies.

The following experimental features were considered for development of a system for measurement and analysis of AE signals [4]: 1) linear and rotational movements of the crystal during growth which hinder connection to the AE signal detector; 2) an extension of the crystallization process to several days, which increases demands on the reliability of the system and methods of data processing; 3) exposure of the piezo-electric transformer with the preamplifier for the whole process to the active high-temperature zone and a strong variable electromagnetic field inside the hermetic crystallization apparatus, which raises requirements for reliability and protection from the effects of temperature and interference. A diagram of the system is shown in Fig. 2. The AE pick-up, a shielded water-cooled mechanism which contains a disk piezotransformer from lead zirconate-titanate (PZT) ceramic and a low-noise preamplifier inside, is connected to the growing crystal through an acoustic line. The preamplifier specifications are: multiplication factor 100, band pass 0.3-1000 kHz, and mean square noise value at the output up to 2.5 μV. Input is taken from galvanic batteries for minimization of interference. The switchable band-pass filter is an array of 16 three-section high- and low-pass LC-filters (from 30 kHz in steps of 1 octave up to 1500 kHz) with an amplitude—frequency decay curvature characteristic in the barrier band of not less than 60 dB/octave. Use of passive means for suppression of electromagnetic interferences (shielding, input from batteries, galvanic bypasses, and filters) permitted their reduction to a level of 0.1 pulse/min AE activity. Further suppression of interference was accomplished by data processing programs for signal treatment on the computer. Parameters that are effective for discrimination of interference are the ascent and descent rate of the amplitude, the duration of the AE signal, and the AE count rate. Use of this assortment of data allows the effect of electromagnetic interference on the AE measurements to be excluded almost completely.

Passing the band switch filter, the signal encounters the preliminary treatment module (PTM). The main amplifier output signal is monitored on a storage oscilloscope. The videodetector separates the signal envelope, which is fed into the analog-digital converter (ADC) of the computer with switch P in position 2. This digitizes the envelop in steps of 60 μsec. The measured amplitude values are saved in a file on the magnetic disk. The amplitude discriminator forms square-wave pulses from the sinusoidal waves from AE electrical signals which effectively surpass a given threshhold. The first of the formed pulses triggers the computer timer for determination of the signal duration and initiates counting of these pulses by the number-pulse signal input module

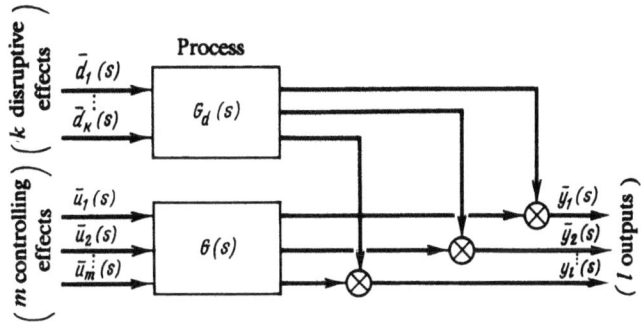

Fig. 3. Crystallization as a multidimensional linear system.

(NPSIM) of the computer. A drop of the trailing envelope front below the discriminator threshhold determines the end of the AE signal and stops the timer and the NPSIM. In the mode that corresponds to position 2 of switch P, the NPSIM is sampled at time intervals that are equal to the digitization step of the envelopes. The readings obtained are stored in the computer memory. In the mode corresponding to position 1 of switch P, the computer transfers the PTM control block to the input of the peak detector after receiving the first effective pulse formed by the amplitude discriminator. The peak detector separates and stores in analog form the maximal amplitude of the initial signal. At the end of the AE signal, this amplitude is measured using the ADC and saved in the computer memory, where the timer and NPSIM values, the AE signal duration and total AE count during the signal, are also recorded. The PTM control block is placed in the initial state by a computer command which also erases the peak detector memory and sets the NPSIM to zero. The PTM parameters are the following: signal range at the main amplifier input 1.5-200 mV, determination error of maximal amplitude and AE signal envelope amplitudes less than 10%, and deadtime between two events of 0.25 msec. The attenuator and recorder measure and record in analog form the AE count rate.

A wide band pass filter of 60-1000 kHz and an amplitude discriminator threshhold corresponding to a signal level at the preamplifier input of 5-10 μV are used in the absence of *a priori* information on the AE parameters (which is related to the crystallization features mentioned above). Mechanical noise from drives and cooling water is eliminated. The principal selection of signals is made by mathematical treatment on the computer.

As noted above, one factor that distorts AE signals is the overlap of reflected signals with the main signal. An adaptive duration shaping block is added to the measurement system for exclusion of this effect [5]. This block chooses the main pulse of maximal amplitude from a packet of overlapping signals. Depending on the value, it limits the duration of the main signal, suppressing reflection in the tail.

4. METHODS OF IDENTIFICATION

We now study the relationship between the actual structure of the crystal and the parameters of crystallization, assuming for simplicity that the process is a linear multidimensional system with averaged parameters (Fig. 3). The system has m control inputs, k disruption inputs, and l outputs, where

$$\overline{y}(s) = \overline{G}(s)\,\overline{u}(s) + \overline{G}_d(s)\,\overline{d}(s),$$

where $\overline{y}(s)$, $\overline{u}(s)$, and $\overline{d}(s)$ are vectors and $\overline{G}(s)$ and $\overline{G}_d(s)$ are matrix components of the equalities:

$$\overline{d}(s) = \begin{bmatrix} \overline{d}_1(s) \\ \vdots \\ \overline{d}_k(s) \end{bmatrix}, \quad \overline{u}(s) = \begin{bmatrix} \overline{u}_1(s) \\ \vdots \\ \overline{u}_m(s) \end{bmatrix}, \quad \overline{y}(s) = \begin{bmatrix} \overline{y}_1(s) \\ \vdots \\ \overline{y}_l(s) \end{bmatrix}, \quad \overline{G}(s) = \begin{vmatrix} g_{11}(s) \dots g_{1m}(s) \\ \dots \dots \dots \dots \dots \\ g_{l1}(s) \dots g_{lm}(s) \end{vmatrix}, \quad \overline{G}_d(s) = \begin{bmatrix} g_{11_d}(s) \dots g_{1k_d}(s) \\ \dots \dots \dots \dots \dots \\ g_{l1_d}(s) \dots g_{lk_d}(s) \end{bmatrix},$$

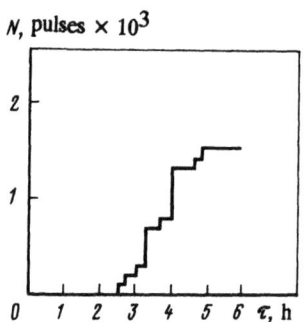

Fig. 4. Total AE count as a function of time during development of two cracks along cleavage planes in a leucosapphire crystal.

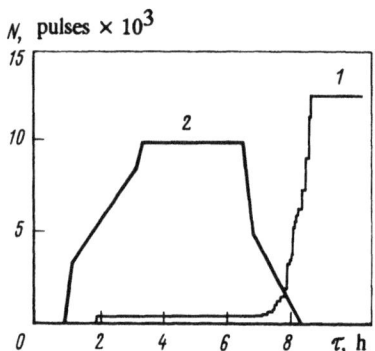

Fig. 5. 1) Total AE count (1) and change of heater potential (2) during thermal shock of a leucosapphire crystal.

$\overline{y}(s)$, $\overline{u}(s)$, and $\overline{d}(s)$ are Laplace transforms of the output $\overline{y}(t)$, controlling $\overline{u}(t)$, and controlled disruption $\overline{d}(t)$; $g(s)$ and $g_d(s)$ are transfer functions for control and disruption, respectively. In this case, we take each of the l outputs in correspondence with a certain mechanism for emission generation: detachment of dislocation pile-ups from the attachment points, development of slipbands, twinning, and crack formation; or as reasons for their generation: nonuniform distribution of crystal components, inclusion of secondary phases and bubbles, and polymorphic transformations of the crystal lattice. The disruptive effects on the k inputs should be monitored. They can arise from random deviations of process parameters or can be created specifically, for example, by a change of heater power, rate of extrusion, etc. However, we do not in fact have a direct measurement of $y(t)_l$ in this sense and should determine them from AE signal measurements.

Solution of the problem of AE signal identification involves not single signals but their collection. This is related to the search for parameters that are most sensitive to the nature of the source and to the gathering of data on the dynamics of processes that disrupt these signals. In agreement with accepted terminology [16], we examine the principal parameters of the AE signal ensemble: the number of pulses N_Σ is the number of recorded pulses of discrete AE for the observation period; the total count N is the number of recorded AE pulses that surpass a given discriminator level for the observation period; the activity $\dot{N}_\Sigma$ and the count rate $\dot{N}$ are the ratios of the number of AE pulses and the total AE count to the observation period, respectively; the amplitude distribution of AE signals is the distribution of AE pulse amplitudes for the observation period; and the AE spectral density as a random process describes the total frequency structure of the process [17].

The information gained from the last parameter explains its connection to the rate of the mechanism generating the AE signals. This should facilitate its identification. However, quantitative estimates of this parameter as a function of source parameters are not completely correct. Computer calculation of the energy spectrum of the AE signals [22] is rather complicated. Determination of other parameters of the ensemble of AE signals is simpler but does not always resolve the identification problem, especially for simultaneous operation of several mechanisms and substantial dependence of the signal parameters on the excitation conditions.

A relatively simple identification problem is the analysis of the total count N as a function of time τ during growth of a leucosapphire crystal [3]. Acoustic emission pulses are completely absent during growth of a nonblock crystal with a small dislocation density while crystallizing and for a significant part of the cooling cycle. The two jumps in the distribution, shown in Fig. 4, between time points 3 and 5 h are identified as practically instantaneous development of two parallel cracks along the rhombohedral planes. After the 5-h time point, pulses did not appear up to full cooling of the crystal. For verification of the identification, the crystal was again heated and cooled. In this case, another crack formed, also along the rhombohedral plane. Figure 5 shows the total AE count for this experiment as a function of time and the heater potential. The maximal potential corresponds to a crystal temperature of about 1900°C.

In more complicated cases, use of multiparametric distributions was required [4]. The examples below give the distribution histogram for the number of AE signals according to average effective frequency (columns F, kHz, and M), average values of the maximal amplitude (A, mV) and duration (τ, msec), total AE signal count (N) falling in a given frequency range, the number of signals with low and high average effective frequency (in the examples given, <80 kHz and >80 kHz, respectively), their average effective amplitudes, durations, and frequencies (A, τ, and F), the number of signals for which the amplitude exceeds the upper limit of the ADC ($A > 50$ mV), the number of AE signals for a measurement period (total M), and the average effective frequency for all signals (F_{av}).

Example 1

F, kHz	M	A, mV	τ, msec	N
50–60	1	6.30	.71	39
60–70	2	6.30	.71	97
70–80	1	6.43	.50	38
80–90	1	24.16	1.67	138
90–100	1	15.77	1.61	154
100–110	5	16.81	1.58	706
110–120	7	21.67	1.58	1285
120–130	4	30.28	2.47	1219
130–140	3	24.16	1.78	697
140–150	2	21.29	1.19	346
150–160	1	17.53	1.16	185

< 80 kHz	> 80 kHz	$A > 50$ mV
$M = 4$	$M = 24$	$M = 7$
$A = 6.37$ mV	$A = 22.53$ mV	(25,0%)
$\tau = .60$ msec	$\tau = .72$ msec	$\tau = 2.81$ msec
$F = 60.61$ kHz	$F = 119.75$ kHz	$F = 120.50$ kHz

Total $M = 28$ $F_{av} = 117.20$ kHz

Example 2

F, kHz	M	A, mV	τ, msec	N
10–20	1	4.90	.52	48
20–30	7	4.60	.46	84
30–40	11	4.24	.58	218
40–50	13	4.30	.49	289
50–60	13	4.23	.54	393
60–70	26	4.75	.72	1188
70–80	24	7.72	1.50	2694
80–90	26	6.48	1.21	2653
90–100	28	5.79	.69	1834
100–110	22	6.03	.74	1726
110–120	17	5.41	.63	1250
120–130	13	4.74	.60	966
130–140	9	4.34	.62	741
140–150	1	4.32	.65	95
160–170	1	4.03	.45	77

< 80 kHz

$M = 95$

$A = 5.30$ mV

$\tau = .85$ msec

$F = 60.61$ kHz

> 80 kHz

$M = 117$

$A = 5.68$ mV

$\tau = .79$ msec

$F = 100.95$ kHz

$A > 50$ mV

$M = 3$

(1.4%)

$\tau = 10.33$ msec

$F = 78.88$ kHz

Total $M = 212$ $F_{av} = 82.10$ kHz

Example 3

< 80 kHz

$M = 752$

$A = 52.25$ mV

$\tau = 11.44$ msec

$F = 54.25$ kHz

> 80 kHz

$M = 270$

$A = 125.67$ mV

$\tau = 20.95$ msec

$F = 91.95$ kHz

$A > 900$ mV

$M = 7$

(.7%)

$\tau = 46.24$ msec

$F = 40.77$ kHz

Total $M = 1022$ $F_{av} = 69.20$ kHz

A small scatter of signals according to average effective frequency, a high average effective frequency for all signals, a small number of signals with $F < 80$ kHz, and a large signal amplitude are evident in Example 1.

In Example 2, the total number of signals is large. Of these, 70% lie in a broad frequency band. The fraction of signals with $F < 80$ kHz is large. The signal amplitude is lower. In Example 3, the amplitudes and durations of signals are very large. In this case these indicators are sufficient for determination of the class. Therefore, distribution of signals according to average frequency was not done. The first class was identified as formation of microcracks in the seed crystal, the second was separation of a secondary phase in the crystal, and the third was development of cracks in the crystal bulk upon cooling.

The variety of mechanisms for generation of AE in a growing crystal, the correspondingly large range of AE signal parameters, and the change in conditions for generation and detection of AE signals, which affects the absolute values of several signal parameters, required development of an identification method best suited to measurements of the conditions. This was also related to the intention of reducing the number of modelling experiments on crystal growth necessary for an automatic identification system.

An identification method having sufficient flexibility and utility was developed by us using the principles of pattern recognition [23-24]. In the classical statement of the problem of pattern recognition, it is assumed that the discriminated objects are given a collection of indicators from which the actual indicators or properties should be extracted. Further, an algorithm is synthesized which defines for each newly entering object to which

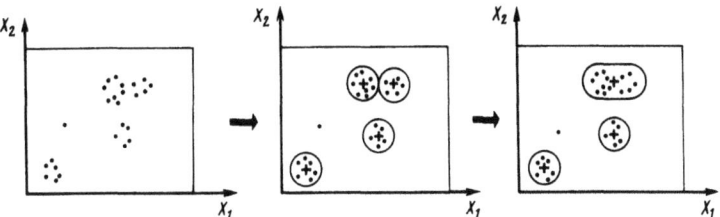

Fig. 6. Working principle of the classification algorithm.

of the final number of classes it belongs. Here, class means that category which is defined by a number of properties common to all its elements.

The procedure for pattern recognition consists of the following stages: representation of the discriminated object as a vector $\bar{X} = (X_1, X_2, ..., X_n)'$, the components of which are measured characteristics X_i; a separation of characteristic indicators and a lowering of the dimensionality of the indicator vectors; and construction of decisive classification procedures.

The main approaches to automatic classification as a function of completeness of *a priori* class description are:

Principle of Enumeration of Class Members. A discrimination process through comparison with standard shapes that are stored in the system memory is proposed.

Principle of Commonality of Properties. Shapes belonging to a certain class have a number of common properties or indicators that reflect similarities of these samples.

Principle of Clustering. Shapes are viewed as points in an *n*-dimensional Euclidean space. The distance scale in this space is set. A compact region, called a cluster, in which points belonging to the same class are grouped, corresponds to each class. Overlap of clusters is possible due to incompleteness of available information and noise distortions of measured sample characteristics.

Of the classification principles examined, we chose the clustering principle, which satisfies better the features stated above for measurement of AE signals during crystallization from the melt. Separation into groups of signals from several different AE sources is equivalent to the search for clusters in vector space, the components of which are measured characteristics of the AE signals. With this, we do not discard any information either on the proposed cluster centers or on their number.

We developed an algorithm for cluster revelation which operates on a minimal distance criterion. The expression below was chosen as the sample metric space

$$D(\bar{X}_i, \bar{X}_j) = \left[\sum_{k=1}^{n} \alpha_k \left(\frac{X_{ki} - X_{kj}}{X_{ki} + X_{kj}} \right)^2 \right]^{1/2},$$

where X_{ki} and X_{kj} are the kth components of the vectors $\bar{X}_i$ and $\bar{X}_j$ and α_k are assigned weights of components that reflect the different information levels of the AE signal characteristics.

The working principle of the algorithm is illustrated schematically by the succession of diagrams (Fig. 6) for the two-dimensional case (in reality, the number of vector components of the shape is larger than two). The algorithm is a succession of iterative procedures.

The first procedure carries out the search for similar shapes in the initial group. The metric $D(\bar{X}_i, \bar{X}_j)$ is calculated for each X_i, X_j pair according to the formula given above. If D_{ij} does not exceed some threshold, then $\bar{X}_i$ and $\bar{X}_j$ are joined into a cluster S_m with center $\bar{Z}_m = (1/2)(\bar{X}_i + \bar{X}_j)$. In the case where $\bar{X}_i$ (or $\bar{X}_j$) have already formed a cluster S_p with center $\bar{Z}_p$ in an earlier iteration, then $D(\bar{X}_i, \bar{Z}_p)$ (or $D(\bar{X}_j, \bar{Z}_p)$) is calculated. If this distance is less than the threshold, then the object $\bar{X}_i$ (or $\bar{X}_j$) is included in the cluster S_p and the

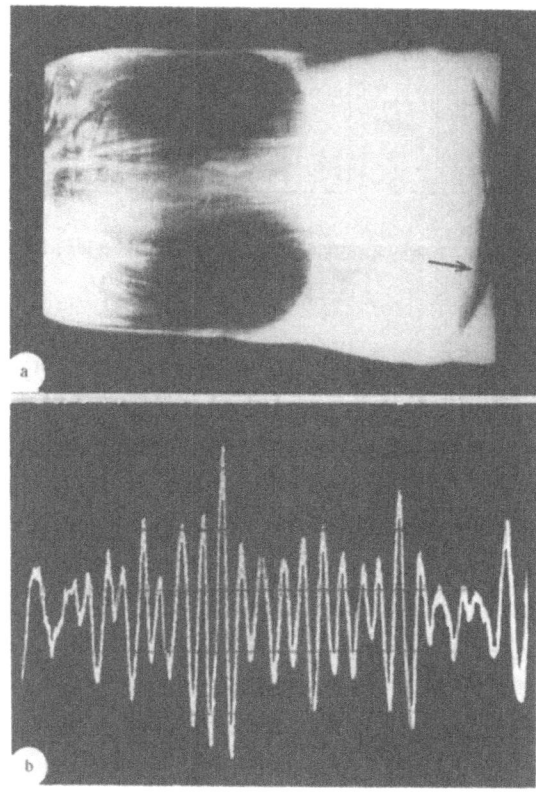

Fig. 7. Nonfusion flaw in a ruby crystal (shown by an arrow in a) and the AE signal corresponding to this effect (for b the scale is 10 μsec/division).

center is readjusted:

$$\overline{Z}_{P'} = \frac{1}{N_P + 1} (N_P \overline{Z}_P + \overline{X}_i),$$

where N_p is the number of objects in the cluster S_p. The result of the first procedure is a preliminary separation of the initial group into r clusters, for each of which the center $\overline{Z}$ and several scattering parameters of fields around it are calculated.

The second procedure for each object X calculates its distance from all cluster centers found in the first procedure and determines the minimal $D_{min} = \min_{S_k} D(\overline{X}_i, \overline{Z}_k)$. In the case where $\overline{X}_i \in S_m$ and D_{min} is reached for this same cluster S_m, the classification remains unchanged or else a transfer of $\overline{X}_i$ from S_m into the cluster S_p occurs, for which the distance is minimal. In the latter case, the centers of $\overline{Z}_m$ and $\overline{Z}_p$ are corrected. This procedure works iteratively, in which the initial group is sorted into different directions until a stable classification is attained, i.e., until transfer of objects from cluster to cluster ceases. After this, several parameters which characterize the quality of the classification are calculated: the number of clusters, the number of objects in the cluster, the mean square value of the scatter of objects around the center in the cluster, cross distances between centers, the designation of clusters nearest each other, the number of unclassified objects, and several others.

The third procedure based on this information performs pairwise fusion of clusters if the distance between their centers is less than the threshold. Here, several clusters may be eliminated if the number of objects in them is too small and the members of such clusters are included in neighboring clusters.

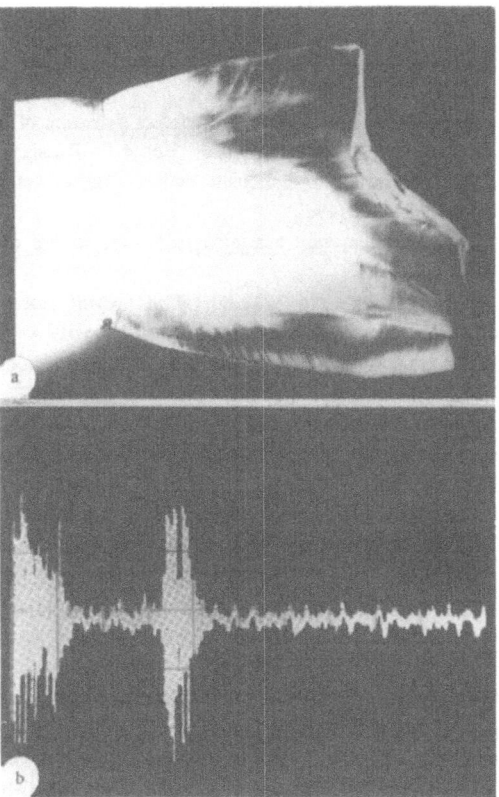

Fig. 8. Slipbands in a ruby crystal in polarized light (a) and the typical AE signal (b) that accompanies their development (for b the scale is 10 μsec/division).

The automatic system of pattern recognition operates in a dialog mode. This enables the investigator to correct operatively the classification process when the weight of indicators for the best separation of classes must be changed and the clusters must be joined or subdivided.

The result of clusterization is the separation of the whole group of AE signals into a certain number of classes, in each of which the signals are arranged in succession according to acquisition time. This provides a picture of the dynamics of the ensemble of mechanisms that cause AE. With sufficient control of the constructive and disruptive effects on the process (Fig. 3), this picture, as the l outputs of the process, can be related to these effects.

For automatic identification of AE signals and their sources, the system was tested on model experiments in which disruptions that cause primarily formation in the crystal of a certain type of actual structural defects, were created. Such an approach is required due to the absence of methods for direct observation of the actual structure of large crystals during growth in a melt.

Figure 7 shows an example of a nonfusion flaw in a ruby crystal formed in a model of the gas-flame growth method upon abrupt increase of mixture supply. Below is shown the AE signal corresponding to the effect of nonfusion flaw. The automatic identification system eventually detected lack of fusion during crystallization in a crystal of ~1 mm size.

Figure 8 shows a test example of discrimination of slipbands. Disruption was created by changing the temperature program. For the AE signals corresponding to this effect, a small duration and high effective pulse frequency are characteristic.

REFERENCES

1. J.-I. Chikawa, "Technique for the video display of x-ray topographic images and its application to the study of crystal growth," *J. Cryst. Growth*, **24/25**, No. 1, 61-68 (1974).

2. É. L. Lube, Kh. S. Bagdasarov, E. A. Fedorov, et al., "Automatic defect detection of large crystals during growth at high temperatures," in: Abstracts of the VIth Int. Conf. on Growth of Crystals, Vol. 4 [in Russian], Moscow (1980), pp. 212-213.

3. É. L. Lube, Kh. S. Bagdasarov, E. A. Fedorov, et al., "Acoustic emission defect detection of large crystals during growth at high temperatures," *Kristallografiya*, **27**, No. 3, 584-587 (1982).

4. É. L. Lube and A. T. Zlatkin, "Automatic system with a minicomputer for studies of high-temperature crystallization by acoustic emission," *Defektoskopiya*, No. 10, 7-13 (1985).

5. A. T. Zlatkin and É. L. Lube, Inventors' certificate No. 1250941 (USSR), *Byull. Izobret.*, No. 30 (1986).

6. É. L. Lube, Kh. S. Bagdasarov, and E. A. Fedorov, Inventors' certificate No. 859490 (USSR), *Byull. Izobret.*, No. 25 (1982).

7. C. A. Tatro and R. G. Liptai, "Acoustic emission from crystalline substances," in: Proc. Symp. Phys. Nondestructive Testing, Boston (1962), pp. 145-158.

8. R. T. Sedgwick, "Acoustic emission from single crystals of LiF and KCl," *J. Appl. Phys.*, **39**, No. 3, 1728-1740 (1968).

9. D. R. James and S. H. Carpenter, "Relationship between acoustic emission and dislocation kinetics in crystalline solids," *J. Appl. Phys.*, **42**, No. 12, 4685-4697 (1971).

10. O. V. Gusev, *Acoustic Emission upon Deformation of Single Crystals of High-Melting Metals* [in Russian], Nauka, Moscow (1982).

11. V. D. Natsik and K. A. Chishko, "Dynamics and acoustic emission of a Frank–Read dislocation source. II. Formation of a dislocation pile-up," Preprint Physicotech. Inst. for Low Temperatures, Acad. Sci. USSR, Khar'kov (1976).

12. R. M. Fisher and L. S. Lally, "Microplasticity detected by an acoustic technique," *Can. J. Phys.*, **45**, No. 2, 1147-1159 (1967).

13. B. H. Schofield, "Research on the sources and characteristics of acoustic emission," in: Acoustic Emission: ASTM Spec. Tech. Publ. 505, Baltimore (1972), pp. 11-19.

14. V. S. Boiko, R. I. Garber, V. F. Kivshik, and L. F. Krivenko, "Synchronous recording of dislocation shifts and the acoustic emission generated by them," *Fiz. Tverd. Tela*, **17**, No. 5, 1541-1543 (1975).

15. V. S. Boiko, R. I. Garber, L. F. Krivenko, and S. S. Krivulya, "Acoustic emission of twinning dislocations during their escape from a crystal," *Fiz. Tverd. Tela*, **11**, No. 12, 3624-3626 (1969).

16. V. A. Greshnikov and Yu. B. Drobot, *Acoustic Emission* [in Russian], Izd. Standartov, Moscow (1976).

17. V. M. Baranov and K. I. Molodtsov, *Acoustic Emission Instruments for Atomic Energy* [in Russian], Atomizdat, Moscow (1980).

18. A. A. Pollock, "Acoustic emission," *Engineering*, **209**, No. 5433, 639-642 (1970).

19. R. W. B. Stephens and A. A. Pollock, "Waveforms and frequency spectra of acoustic emission," *J. Acoust. Soc. Am.*, **50**, No. 3, 903-910 (1971).

20. R. G. Liptai, D. O. Harris, R. B. Engle, and C. A. Tatro, "Acoustic emission techniques in materials research," *Int. J. Nondestr. Test.*, **3**, No. 3, 215-275 (1971).

21. V. I. Ivanov, "On possible forms of acoustic emission signals," *Defektoskopiya*, No. 5, 99-101 (1979).

22. B. Woodward, "Identification of acoustic emission source mechanisms by energy spectrum analysis," *Ultrasonics*, No. 11, 249-255 (1976).

23. J. Tou and R. Gonzales, *Pattern Recognition Principles*, Addison-Wesley, Reading, Massachusetts (1975).

24. K. Fukunaga, *Pattern Recognition*, Academic Press, New York (1972).

ELECTRON MICROSCOPY
IN STUDIES OF DEFECT FORMATION PROCESSES
IN SEMICONDUCTING CRYSTALS

S. K. Maksimov

The present article is based on a review presented by the author at the IVth All-Union Conference on Crystal Growth (Tsakhkadzor, Armenian SSR, 1985 [1]), which has been revised and substantially expanded through inclusion of materials of the Symposium on High-Resolution Electron Microscopy (Tempe, Arizona, 1985 [2]) and of the XIth International Congress on Electron Microscopy (Kyoto, Japan, 1986 [3]). A translation of the monograph by J. Spence, *Experimental High-Resolution Electron Microscopy* [4], appeared in Russian in 1986. Thus, discussion of the prospects and limitations of this technique were somewhat settled. More attention then could be directed to analytical electron microscopy, primarily to methods based on loss spectra of inelastically scattered electrons. Unfortunately, these methods (as well as methods which use convergent beam electron diffraction) are practically not used in domestic studies and did not receive wide coverage.

Transmission electron microscopy (TEM) is used for studying the principal aspects of crystallization. The atomic structure of objects that are forming [2, 3], the structure of surfaces and interfaces [2, 3] during homo- and heteroepitaxial crystallization or formation of dielectric layers [2, 3], the presence and characteristics of composition fluctuations, short-range and long-range orders, and segregations [2, 3], and the features of the distribution and mechanisms of defect generation of the crystal structure and their role in crystallization processes [2, 3] are all studied using TEM.

The whole arsenal of modern electron microscopic methods is used in these studies.

If the required information level can be guaranteed using microphotography with a resolution greater than 1 nm during identification of surface or volume defects, study of their distribution, and analysis of interfaces and composition fluctuations, then TEM with diffraction contrast is used. A required resolution of 1.0-2.0 nm necessitates a high-resolving variant of similar microphotography, weak beam imaging (WBI) [5]. Information at an atomic level of resolution (for example, on the structure of dislocation nuclei, structure of heterojunctions, etc.) must be obtained using microphotography with direct resolution of the crystal lattice (HREM) [2, 3]. Additional structural information and identification of phases can be gathered using small-angle diffraction (SAD) or convergent beam diffraction (CBED) [2, 3, 6]. High-resolution electron microscopy contrasts with analytical electron microscopy (AEM) in which chemical information is obtained along with electron-optical data. The chemical composition can be studied by energy-dispersive x-ray microanalysis (EDX), electron energy loss spectroscopy (EELS), and Auger electron spectroscopy (AES) [2, 3]. The best spatial resolution of the analytical methods is achieved in instruments for which a scanning transmission electron microscopy (STEM) [2, 3] procedure can be employed in a primary or accessory mode.*

Much of the success that is associated with use of electron microscopy for answering various problems of the physics and structure of materials has been attained due to HREM methods. However, the apparent obviousness and simplicity of interpretation in methods based on direct resolution of the crystal lattice are deceptive. The links between the structure of an object and the intensity distributions in its images are far from direct [2-4, 7-9]. Therefore, significant difficulties arise even in the interpretation of images of defect-free crystals [4, 7-10]. These can be partially overcome by invoking additional information obtained by integral methods, for example CBED methods [6, 11]. However, the main course of interpretation is related to construction of theoretical photomicrographs that correspond to various possible structural models, foil thicknesses, diffraction conditions, defocusings, and illumination conditions [2-4]. Quantitative comparison of experimental microphotographs to these parameters is also possible [2-4]. Similar procedures require significant expenditures of machine time. It demands incorporation of a theory of information transfer in distorted systems [12, 13], a theory of elastic and inelastic electron scattering [12], and, finally, solution of the problems of three-dimensional object reconstruction from two-dimensional projections [4, 8]. Methods for independent and precise determination of such experimental conditions as foil thickness, diffraction conditions, defocusing, etc. [2-4] are very important for unambiguous interpretation.

The situation is even more complicated for defect imaging due to obscuring of multiwave scattering processes by strongly distorted regions near the center of the imperfection [14]. The very same contrast effects can be explained based on several alternative models of imperfection. We will examine the results of [15] as an example.

In [15], two types of images corresponding to defects located in the {111} planes were observed during a study of precipitation in a supersaturated solid solution of GaAs(Te). An extraplanar region appeared in some photomicrographs and was absent in others. The two types of images were identified with two types of defects which correspond to different stages of precipitation based on indirect reasons. The images without the extraplanar region were recorded for segregations of dopant atoms whereas those which had these regions were taken by separation of the Ga_2Te_3 phase. Images of both types of defects were commensurate with the foil thickness of ≈ 20 nm. The disappearance in the photomicrographs of traces of the extraplanar region can also be related to the formation conditions of the image [4, 7].

However, a third explanation for these results may be proposed. Both types of images represent the same imperfection, which is similar in geometry to a Frank interstitial loop. In this case, defects which have images with the extra semiplane traverse the foil from above to below whereas images of the second type represent imperfections which are trapped individually in the foil and are localized either at the upper or lower surface.

Even if the two types of images are found to reflect an actually existent defect structure geometry as a result of analysis of image features formed under various conditions, there remain two alternative models. In the first, the two types of images reflect differences in the nature of the defects. In the second, their position in the crystal differs.

Theoretical analysis confirmed the possibility of explaining the images by both models and showed that a choice between them is impossible from photomicrographs with direct resolution of the lattice. Therefore, additional information obtained either from photomicrographs with low resolution or using integral methods [4, 11, 12] is especially important in HREM studies.

---

*All abbreviations used here and further are found widely in the literature, for example in [2, 3].

The complicated nature of scattered wave amplitude variations, the presence of phase breaks with changes of crystal thickness [16, 18], and the variety of phase change effects with variations of excitation uncertainties [17] necessitate the development of methods for studying complex amplitudes [19]. Moreover, invoking phase information for identification of observed objects can be useful in those cases where information contained in the intensity distributions (which is determined from the square of the modulus of the complex amplitude) is insufficient for definition of the nature of the object [19-21]. Phase information is also necessary for studying those parameters that cannot be studied by traditional methods, for example, the average internal crystal potential or potential variations near a defect [22]. Development of analytical methods for phase information is one of the directions in which modern electron microscopy is being improved [19-24]. Two approaches exist for incorporation of phase information. One is based on use of defocused electron photomicrography [19-22]. The other is holography [23, 24].

The second method seems more promising. Already, holograms with an intensity oscillation period of ≈ 0.1 nm have been obtained [27]. However, improvement of the instrument is necessary for successful application of the holographic technique. Instruments with increased illumination coherency [due to use of instruments with cold (field) cathodes] and microscopes equipped with Mollenstadt biprisms are needed [23-25].

Methods based on defocused microphotographs are being successfully developed on standard instruments with average resolution [19-21]. Theoretically, defocused photomicrographs permit reduction of the complex amplitude distribution [12, 26]. However, the complex amplitude reduction procedure from defocused photomicrographs entails such accuracy in intensity measurements and such expenditures of machine time that a shift in this direction is substantially limited at this time.

Use of defocused photomicrographs with low and average resolution for identification of imperfections is more likely. Methods based on defocused photomicrographs successfully complement traditional ones, especially in those cases when use of the latter is hindered or impossible. Pores of unit nanometer size [20] and atomic surface steps [16] are revealed and identified using effects which arise during defocusing. The nature of small point defect clusters [21] and tetrahedra of stacking faults localized in the center of the foil [19] are defined.

With respect to the use of defocused photomicrographs for identification of the nature of defects, we will mention another method which is based on defocusing. This is the so-called $2^1/_2$-D method [29]. Results obtained with this method should be treated with caution, as pointed out in [30, 31].

Neither the production nor the interpretation of electron photomicrographs of crystals is possible without inclusion of diffraction information [6, 11]. The SAD method is the main source of additional information in low- and average-resolution electron microscopy. The electron beam is focused in the image plane during production of the SAD picture and the specimen is illuminated by a beam with very small angular dispersion (theoretically parallel). The angular dispersion of the illuminating beam determines the angular resolution of the method for a perfect crystal. The spatial resolution depends on a selective diaphragm that limits the region of diffraction picture formation. As a result of aberrations, the spatial resolution in SAD cannot be better than tenths of a micrometer.

Upon changing to HREM methods, the information that is contained in SAD pictures turns out to be insufficient. They are insensitive to deviations of crystal axes of the object from the optical axis of the instrument (on which the HREM image depends substantially), do not permit the instrument adjustment to be monitored, do not have the required spatial resolution, do not provide information on the symmetry of the object in the direction of electron travel, etc. Therefore, procedures in which the electron beam is focused in the plane of the object are used in place of SAD. (A similar focusing is also exceedingly important in AEM methods.) Diffraction pictures from regions of units to tenths of nanometers can be obtained. Two main procedures are possible in this case. If the impinging beam remains parallel, then nano-probe diffraction with coherent illumination is involved. If a convergent electron beam impinges on the sample, then diffraction pictures in convergent beams (CBED) are formed. The information derived from coherent nano-probe electron patterns corresponds to information of SAD pictures (only spatial resolution is improved).

The informational possibilities of CBED are much greater, and it has been used widely [6, 11, 30-37]. From the viewpoint of electron beam dispersion, three types of CBED pictures can be formed. If half of this angle is smaller than the Bragg angle ($\theta_s/2 < \theta_B$), then electron patterns for which separate reflections do not overlap are formed. For $\theta_s/2 \approx \theta_B$, overlap of reflections is observed. Finally, Kossel patterns are formed at sufficiently large dispersion angles.

The most important parameter that determines the CBED features is the coherence of the incident beam. If the relation $x_a < 0.2R_a$ is fulfilled between the transverse coherence width x_a in the plane of the condensing diaphragm, which limits the probe size, and the radius of this diaphragm, then the probe can be considered partially coherent. At $x_a \geq R_a$, the probe is completely coherent. In practice, CBED is formed under conditions of coherent illumination in instruments with a field emitter. In microscopes with thermal emissive illumination, the CBED pictures can be calculated using an incoherent illumination approximation [11].

For incoherent CBED, the probe diameter is defined by the formula

$$d^2 = d_g^2 + d_d^2 + d_s^2 + d_c^2 , \tag{1}$$

where d_g is the Gaussian spot diameter and d_d, d_s, and d_c are broadenings caused by diffraction effects and spherical and chromatic aberrations, respectively. In the coherent case, expression (1) gives only a rough estimate of the probe dimension. Computer calculations which consider the illumination parameters are required for its estimate. For coherent illumination, large intensity oscillations within the probe and extended intensity tails are characteristic. The most compact probe is observed near the "dark focus" upon defocusing

$$\Delta f = -0.44(C_s \lambda)^{1/2}.$$

The coherency also determines the intensity distribution in CBED pictures that are formed under conditions of reflection overlap or in CBED that corresponds to portions which contain defects [6, 11].

Studies of probe parameters can play an important role in adjustments. Theoretical intensity distributions are observed only for a properly adjusted instrument. For example, misalignment of the illumination axis and the objective lens or improper positioning of the condensing diaphragm can distort the shape of the probe in the plane of the object [6]. Correct adjustment is especially important for the probe shape in HREM where other methods do not ensure the proper precision [6].

Important sources of information in CBED are reflections which correspond to high order Laué zones (HOLZ).

In particular, the HOLZ and Kossel lines in HOLZ pictures from portions which contain a dislocation, for which the condition $gb \neq 0$ is fulfilled, are split. Those for which $gb = 0$ remain unsplit. Analysis of HOLZ-line splitting allows the spatial orientation of the Burgers vector to be determined from one picture [6]. A method which also permits the sign of the shift vector to be defined using HOLZ lines was developed for stacking faults [30]. CBED can also be used for determination of the nature of distortions and the magnitude of deformations which are observed around a coherent precipitate and pores [32] or for layer junctions with different lattice constants [6]. A method is proposed in [31] which is based on HOLZ diffraction and permits the composition amplitude modulation in the GaAlAs—GaAs superlattice to be determined to an accuracy commensurate with that of EDX measurements. (For $Ga_{1-x}Al_xAs$—GaAs at $x \lesssim 0.3$, determination of the modulation parameters by other electron microscopic methods is hindered as a result of the similarity of lattice constants of the joined layers [33]).

Information on the bond lengths and angles in amorphous materials is obtained using small ($\lesssim 1.0$ nm) coherent probes for studying them. This same method can be used for studying dispersed objects. Study of

CBED symmetry with a change of accelerating potential facilitates investigation of the excitation conditions of Bloch waves, which are very sensitive to crystal structure parameters. HOLZ diffraction enables the unit cell constants to be monitored with an accuracy near that of x rays [35], including along the projection axis [36].

However, the principal area of CBED applicability remains the study of the symmetry of objects [6, 11, 37], as well as their spatial orientation. A combination of HREM and CBED methods can sharply increase the reliability of crystal structure interpretation of many complicated objects [37]. CBED studies which are carried out using a probe of 0.2-0.3 nm diameter can play an important role in the interpretation of complicated structures. This can reveal the structure of separate portions of large unit cells [11].

Analytical electron microscopic methods and modern diffraction methods are both carried out using focused and adjusted electron beams [11]. Three main methods are used in AEM. These are ADX, EELS, and AES. However, we will discuss only the EELS method in the present article. First, the other methods have been described completely and in detail in Russian (including monographs) whereas similar information for EELS is not available. Second, the other methods are practically not used in domestic practice and claims of its prospects are irrelevant. Third, EELS is especially important in microscopic studies. This is the single analytical method that can be used in an instrument operating in a HREM mode. Moreover, energy filtering of electrons is beginning to be used for production of high-quality HREM [38] where this method reduces the effect of chromatic aberration and is suitable for studies on thicker samples [38].

The distinguishing features of EELS are its exceptional spatial and energy resolution. The spatial resolution approaches the level reached in HREM. The energy resolution in serial instruments reaches 0.7 eV. (Reports of attainment of an energy resolution ≈ 0.1 eV have appeared [38].) Further improvement of spatial and energy resolution in serial instruments is connected with the change to instruments with field emission which ensures attainment of energy resolution (without use of monochromators) at the 0.25-0.3 eV level. This is sufficient for solution of the majority of problems.

Characteristic losses are divided into two categories: <50 eV and >50 eV. The first are related to valence electrons. The second involve inner electron shells.

The following areas of EELS application are pointed out.

1. Elemental Analysis. It is especially effective for revealing and studying the distributions of light and very light elements [38-43], for example Li, B, and O. An atomic fraction of 10^{-4} is the characteristic detection limit in many cases. Calculated for those volumes which are usually studied in EELS, this means that the method can detect tens or even individual atoms of one type or another.

2. Analysis of Electronic Structure. Especially large advantages in comparison with other methods for studying energy structure of solids are associated with the localized nature of EELS. The information can be obtained from volumes on the order of 10 nm^3. Thus, studies of characteristic effects which are caused, for example, by the presence of persistent defects can be carried out [38]. Moreover, this, this information can be correlated with structural information obtained through HREM [42].

3. Analysis of Absorption Spectrum Fine Structure in Regions: a) near the absorption edge (ELNES) (information on the valence states) and b) extended fine structure (EXELFS) (information on the nature of near neighbors) [38, 40, 42].

A number of problems must be solved for implementation of the very broad prospects of EELS. The most important of these concerns the destruction of the structure of the object as a result of the high-energy electron beams. A density of $\approx 10^8$ e/nm^2 is required for optimization of the signal/noise ratio. This is an order of magnitude above the maximally permissible dose for the majority of objects [39, 42]. The structure of even relatively stable materials can be destroyed at such large doses due to loss of light elements in such compounds as TiC and Fe_2O_3, by vaporization of surface layers of objects, by radiation-stimulated diffusion, etc. [38, 41].

A second important problem in EELS is connected with multiple scattering. Its negative role is different for different areas of EELS application. The most rigid demands for elimination of multiple scattering effects arise for analysis of electronic structure, which is carried out based on single scattering [38]. In other methods, multiple scattering leads to a rise of background level, degradation of the signal/noise level, and distortion of the absorption edge. Therefore, it must be considered when making the corresponding calculations [38, 44, 45]. A decrease of sample thickness weakens the useful signal. This effect is not even completely suppressed for samples of 2-3 nm thickness [38, 44, 45]. A more promising approach uses increased accelerating potential [46]. In this case, the probability of multiple scattering decreases and the effectiveness of electron collection increases (due to decreased scattering angles). The decrease of the ionization cross-section is compensated by the increase of electron source intensity [46]. The useful thickness of a sample doubles with a change of potential from 100 kV to 300 kV [46].

The third problem with EELS is the presence of a residual background in the absorption edge region which is observed after removal of multiple scattering effects [38]. Errors in the calculation of this background lead to substantial errors in interpretation of results [38-40].

Serious discrepancies in interpretation of EELS results are caused by inaccuracies in determination of effective cross-sections [38]. Only after their refinement can a quantitative standardless EELS method be developed which is based on first principles [39].

The required comparison and correlation of EELS results with various details of the structure found from electron-optical studies raises the question, "What determines the resolution and localization of signals in this method?" Even if EELS is built into a given microscope, a change of procedures causes changes in signal localization due to physical and instrumental reasons. Electron photomicrographs are formed by elastically scattered electrons and are determined by the spatial distribution of the scattering material. Inelastic scattering, which is related to the generation of excited states, is used in EELS. The differences in the nature of the signals lead to differences in their localization. Changes in the excitation levels of the lenses, which can lead to shifts in probe position, occur with a change of electron microscope operating regime. Moreover, the coincidence of signals in various regimes depends on the stability of the probe position with time.

A discussion of the parameters of individual instruments is beyond the scope of this article. Only physical limitations, as applied to STEM equipped with a field cathode [40], will be examined here.

The EELS sensitivity is tied to spatial resolution. In AEM, three different characteristics of this quantity are used: 1) d_0 is the minimal size of the illuminating probe. The minimal number of detected atoms is equal to the minimal detected concentration relative to the irradiated volume which, in turn, is proportional to the square of the probe radius. Therefore, the absolute detection limit is reached at the minimal probe radius; 2) d_p is the practical resolution which is determined by the attained intensity; and 3) d_c is the actual resolution which is determined by the permissible critical dose.

The electron distribution in the probe and its size are described by a combination of three functions: the point distribution function which characterizes the probe for a rigidly monochromatic point source; the polychromatic point distribution function which describes probe broadening related to chromatic aberrations; and the extended source distribution function which considers broadening of the probe as a result of its length. In [40], calculated probe parameters for various illuminating system regimes for field cathode instruments are given.

The electron beam experiences broadening upon passage through a sample due to elastic and inelastic scattering. This effect in EELS is limited by the collecting diaphragm with a small exit angle β. The broadening for thin samples is proportional to $t\beta$, where t is the sample thickness. For thick samples (t equal to tens of nanometers), the broadening is calculated on a computer [40].

The resolution d_p is defined by the expression [40]

$$d_p = \frac{3}{x} \left(\frac{BT}{S_T} \frac{1}{DP\eta} \right)^{1/2} ,$$

(2)

where x is the dopant concentration, S/B is the signal/noise ratio at $x = 100\%$, T/τ is the ratio of total count time to active time of one channel, D is the irradiation dose, and $P\eta$ is the product of quantum yield and light effectiveness. The value d_p for EELS can be estimated based on the following characteristic values of these quantities: $S/B = 1$, $P \approx 10^{-5}$; for modern instruments $T/\tau \approx 100$ and $\eta \approx 0.1$-0.5. Equation (2) allows one to make a number of conclusions. A dose increase of 10^2 is required for a 10-fold improvement in spatial resolution. A 10-fold improvement of resolution can be achieved with development of systems with detectors acting in parallel. Doses $\approx 10^8$ e/nm^2 are necessary for detection of composition fluctuations on a nanometer scale.

Equation (2) also characterizes the resolution at limiting irradiation doses. Thus, doses $\approx 10^8$ e/nm^2 are required according to (2) for studying SiO_2 particles of 1 nm size in silicon. However, the particles dissipate at these doses. This prevents quantitative measurements [42].

Scattering effects can result in excitation of atoms localized at distances L from the average trajectory of the electron beam in the crystal. Knowledge of the delocalization distance is necessary for comparison of EELS results with results of structural studies. The delocalization distance can be estimated from the formula [40]

$$L = \lambda E / \pi \Delta E, \tag{3}$$

where E is the primary beam energy and ΔE are the average losses.

According to (3), the characteristic amount of delocalization comprises several nanometers. This greatly surpasses the minimal probe size in EELS with field emission. Since to a first approximation $\lambda \approx E^{1/2}$, the precision of localization worsens with an increase in accelerating potential, i.e., the constraints for conditions which guarantee elevated sensitivity (resolution) and precise localization oppose each other. The precision of localization increases with an increase of the amount of characteristic losses, i.e., with an increase of atomic number of the analyzed dopant (if analogous absorption edges are compared). If a certain dopant corresponds to several absorption edges, then the best localization is achieved using the edge which corresponds to the most losses. The precise localization of defect states is worse than localization of dopant atom distributions.

More rigorous approaches for estimation of delocalization distances than those used in the derivation of (3) exist. However, experiments show that (3) gives sufficiently reasonable values and provides actual routes which permit improvement of this parameter [40].

The delocalization effect can in some cases distort the true picture. For example, a heterojunction with an abrupt composition change appears as a gradient according to EELS data (also for EDX data). (Delocalization in EDX is much more pronounced). However, the significance of this effect should not be overemphasized. Comparison of results from structural studies and EELS shows that large errors are possible as a result of "apparatus effects." These effects are changes of probe position with a change of working regime and its drift during measurements. It should also be noted that delocalization in images with diffraction contrast under conditions that are nearly dynamic has a larger value than in EELS. Only in WBI is it comparable or somewhat smaller than in EELS. The only type of photomicrographs for which the localization precision of an object is much higher than in AEM is HREM obtained with correct adjustment and formation conditions.

Apparently, computer treatment of EELS signals produces the true source distribution of these signals. A similar approach ensures not an intuitive but a valid comparison of HREM and AEM results.

An example of work in which various study methods (diffraction electron microscopy HREM, EELS, and EDX) are successfully combined is [42]. In this work, precipitates which decorate dislocation cores in Si and Ge grown by the Czochralski method are identified. The precipitates are shown to have a crystalline structure using a moiré pattern in [42]. HREM found that they are localized in that dislocation core region where expansion stresses dominate. This same technique allowed the size of the inclusions as a function of the Burgers vector of the decorated dislocation to be evaluated. Precipitates of maximal size are observed near dislocations with Burgers vector ⟨100⟩. Those of minimal size are Shockley dislocations. This is explained well by the dependence of the size of the decorating particles on the local elastic deformations that are related to the defect and that are proportional to the square of the Burgers vector. A small anomaly is observed for the complete dislocations with Burgers vector $1/2 \langle 110 \rangle$ and Frank dislocations. Inclusions near the Frank dislocations are large.

EDX showed that the precipitates cannot be treated as inclusions of metallic phases. Studies of the regions which contain inclusions by EELS with a 0.7 nm probe found that they are distinguished by a high oxygen content. More accurate quantitative studies of their composition were impossible. The inclusions dissipated during EELS measurements under the influence of the electron beam. A repeat investigation by HREM of samples studied earlier by EELS confirmed the disappearance of the precipitates.

At present, EELS is being rapidly improved. Methods of standardless quantitative analysis are under development [43]. New filtration methods for electrons which undergo multiple scattering are being devised [44]. Physical aspects of inelastic scattering processes caused by defects are being studied [45].

EELS has been shown to possess many unique capabilities. For example, it can be used for analysis of states that are related to individual defects, for investigations of nonvertical ($q \neq 0$) transitions, and for study of dopant atom distributions between nonequivalent positions [45]. However, implementation of these and other capabilities is a matter for the future due to difficulties in interpretation of results. At present, the fine structure of absorption jumps is unambiguously interpreted only for metals. Various models are suggested in the literature for explanation of observed spectra for substances with directed bonds (semiconductors, ionic crystals). Different treatments of EELS spectra are proposed even for such a relatively simple compound as MgO, for which a nonequivalence of absorption edges corresponding to Mg^{2+} and O^{2-} is observed [40].

In the concluding sections of the article, we will note some results that are related to the use of electron microscopic methods for studying defect structure of semiconducting materials.

Recently, studies of structural inhomogeneities caused by composition fluctuations have been thrust into the forefront, judging by the number of publications. Thus, objects with such technically introduced fluctuations as heterotransitions and superlattices [35, 47-49] and with spontaneously generated inhomogeneities [50, 51] are studied.

In [49], the following classification of heterostructures is introduced. These also have different problems which are solved using electron microscopy and are used in study methods. 1) Similar structures with a small lattice mismatch, 2) similar structures with a large mismatch, 3) different structures with a small mismatch, and 4) different structures with a large mismatch.

There are difficulties in studying the first type of composition that are related to discrimination of layers of different composition. The most studied composition of this type is $Ga_{1-x}Al_xAs$—GaAs at $x < 0.3$. A number of special techniques were developed for these studies. The sections on CBED mentioned some of these. Others involve excitation of 200 reflections [49]. However, the latter method does not reveal the heterojunction in foils of thickness $\lesssim 10$ nm. A thickness which is increased to ≥ 20 nm (when visualization of layers of different composition is facilitated) hinders interpretation of the images of the heterojunction itself. In [33], Al fluctuations near the junction in GaAlAs were revealed for a superlattice based on this composition, which was prepared by organometallic chemical vapor deposition. It is possible that these fluctuations have the same nature as the automodulated structures which were detected in [50]. Other compositions of the first type that have been studied by electron microscopy methods are $ZnSe_{1-x}S_x$ on (001) GaAs, GaP, Ge, and CdTe—(001)InSb.

Among the similar structures with a large lattice mismatch, the compositions GaAsP—GaAs and SiGe—Si, as well as compositions mentioned in the previous paragraph but with other component ratios, were studied. Identification of layers of various compositions in this case did not present problems and the main attention of the investigators was directed to the questions: What is the heterojunction structure? How do the mismatch stresses relax? How are the crystal structure defects distributed? In [52], the nature of the defect structure formed was shown to be determined not only by the lattice mismatch but also by the orientational relationships.

The results noted in [49] apparently confirm this position. An epitaxial layer of ZnS on GaAs with a 4% mismatch is completely disordered as a result of a high density of twins and stacking faults. However, the epitaxial layer for the composition (CdHg)Te—GaAs with a 15% mismatch of lattice constants is distinguished by a high degree of perfection. The junction is guaranteed by the presence of a quasiperiodic subjunction disloca-tion structure with an average distance between defects of 2.5 nm. With such a high density of imperfections, HREM is the sole method for observation and study of individual dislocations [49].

The different structures with a small mismatch include compositions of several metal silicides on Si. Much attention is devoted to the study of possible atomic configurations in the heterojunction plane. $NiSi_2$ and $CoSi_2$ have identical lattices with similar constants. The mismatch of lattices for the pair $NiSi_2$—Si is 0.4%. That for $CoSi_2$—Si is 1.2%. However, an atomic motif for the $NiSi_2$—Si heterojunction is preferred in which the Ni atom has 7 nearest neighbors. The metal atom for the $CoSi_2$—Si heterojunction has 5 nearest neighbors [48, 49]. The reasons for such differences in the structure of a single type of heterojunction is not clear at present. The $NiSi_2$—(001)Si heterojunction has been confirmed to be formed by the {111} and (001) facets.

Problems caused by the nonplanar nature of the junction often arise during studies of silicon—silicide compositions by HREM methods. In this case, the information can be obtained using either photomicrography with diffraction contrast or diffraction methods.

The heterocomposition Si—sapphire has been studied in most detail for the different structural types with large mismatch. Problems which arise during these studies and the methods used for this case are analogous to those which were described for similarly mismatched compositions [48].

Epitaxial growth has been demonstrated to be related to spontaneous composition fluctuations [50, 51, 54] for very different epitaxial layer growth technologies (gas and liquid epitaxy, molecular beams). HREM methods are not applicable for studies of this effect due to the three-dimensional nature of the fluctuations [51]. Therefore, studies are carried out using diffraction methods [54].

Use of methods based on dilatation contrast phenomena is promising in these studies. These can reveal composition variations which cause relative variations of the lattice constant $\Delta a/\langle a \rangle \geq 4 \cdot 10^{-5}$ [53]. Micropho-tography with moiré contrast is also used for studies of modulated structures. If the extent of the gradient layer is not a multiple of the extinction distance, then a moiré pattern does not arise. It can be determined whether the junction is gradational or abrupt in nature, and the extent of the gradient layer and the amplitude of the modulation composition can be determined by studying the appearance of moiré patterns upon changes of the inclination angle of the junction relative to the direction of the electron beam [53].

Two reasons for generation of modulated structures are implicated for semiconducting solutions upon epitaxial crystallization. These are nonlinear effects at the crystallization front [50] and effects which are similar to spinodal decomposition [51]. In the first case, modulation occurs along the growth direction and is distin-guished by regularity. Superlattices are spontaneously formed at constant growth conditions [50]. In the second, it is observed in the growth plane [51]. Both mechanisms can operate during liquid-phase heteroepitaxy. Variations of composition arise initially in the heterotransition plane. However, they change into fluctuations along the growth direction during the growth process [54].

Precipitation processes in supersaturated semiconducting solutions are distinguished by a significantly larger role for intrinsic point defects, in contrast to analogous processes which occur in metal solutions. The precipitation mechanism and the morphology of the phases which arise can change with a change of sign of the nonequilibrium intrinsic point defects which dominate in a local volume.

Structural information extracted from moiré patterns is analyzed in many works studying precipitation processes [42]. The method in which moiré patterns which correspond to various projection axes are compared is distingushed by elevated information levels and reliability. In this case, the structure of inclusions and the features of their coupling with the matrix, which are determined from the moiré patterns for the same projections and reflections, are checked by analogous images that correspond to different previously chosen projections and reflections.

High-voltage electron microscopy (HVEM) is being used more widely for studies of defects in semiconducting crystals and compositions [55, 56]. It has a number of advantages, especially for studying technical defects arise in semiconducting materials as a result of technological influences during production of solid state electronic devices [55, 56].

First, samples of thickness greatly exceeding the thickness of objects that are studied in instruments with 100 kV potential (in 1000 keV HVEM, Si foils of thickness up to 8 μm and GaAs up to 1.5 μm are studied) can be studied. The defect distribution in multilayered device compositions can be investigated [55, 56]. For example, the reasons for generation of dislocations upon formation of dielectric layers can be studied [55].

Second, *in situ* observations in high-voltage microscopes are possible. Processes which occur under industrial conditions, for example, generation of defects during high-temperature annealing, are modelled by these [55, 56].

REFERENCES

1. S. K. Maksimov, "New electron microscopic studies of thin defect formation in semiconducting crystals," in: Abstracts of Papers of the VIth All-Union Conf. on Growth of Crystals, Vol. 3 [in Russian], Izd. Akad. Nauk Arm. SSR, Erevan (1985), pp. 5-6.
2. P. R. Buseck (ed.), "High-resolution electron microscopy: Proc. Centennial Symp.," *Ultramicroscopy*, 18, 1-473 (1985).
3. T. Imura, S. Maruse, and T. Suzuki (eds.), "Electron microscopy 1986: Proc. XIth Int. Cong. Electron Microscopy, Kyoto (Japan)," *J. Electron Microsc.*, 35, suppl., 1-1792 (1986).
4. J. Spence, *Experimental High-Resolution Electron Microscopy*, Oxford Univ. Press, New York (1981).
5. D. J. H. Cockayne, *Weak Beam Electron Microscopic Methods* [Russian translation], Metallurgiya, Moscow (1984), pp. 50-76.
6. R. W. Carpenter and J. C. H. Spence, "Application of modern microdiffraction to materials science," *J. Microsc.* (Oxford), 136, 165-178 (1984).
7. G. R. Anstis and D. J. H. Cockayne, "Calculations and interpretation of high resolution electron microscopy images of lattice defects," *Acta Crystallogr. Sect. A: Cryst. Phys., Diffr., Theor. Gen. Crystallogr.*, 35, 511-524 (1979).
8. A. Howie, "Problems of interpretation in high resolution electron microscopy," *J. Microsc.*, 129, 239-251 (1983).
9. L. D. Marks, "Image localization," *Ultramicroscopy*, 18, 33-38 (1985).
10. P. G. Self, R. W. Glaisher, and A. E. C. Spargo, "Interpretation of high-resolution transmission electron micrographs," *Ultramicroscopy*, 18, 49-62 (1985).
11. J. M. Cowley, "High-resolution electron microscopy and microdiffraction," *Ultramicroscopy*, 18, 11-18 (1985).
12. J. Cowley, *Diffraction Physics*, Elsevier, New York (1981).
13. F. Lenz, "Transfer of image information in electron microscopy," in: *Electron Microscopy in Material Sciences*, Academic Press, New York (1971), pp. 541-568.
14. D. J. H. Cockayne, "Developments in diffraction contrast studies of defects," in: *Electron Microscopy 1982*, Vol. 2, OCCEM, Hamburg (1982), pp. 79-88.
15. S. K. Maksimov, M. Ziegler, I. I. Khodos, et al., "High resolution electron microscopic investigations of dislocations in deformed GaAs single crystals doped with Te," *Phys. Status Solidi A*, 84, 79-86 (1984).
16. N. I. Borgardt, V. L. Egorov, and S. K. Maksimov, "Use of dynamic effects in electron scattering to increase the sensitivity of methods for studying surface relief details," *Poverkhnost*, No. 4, 127-131 (1982).
17. S. K. Maksimov and N. I. Borgardt, "Changes in electron wave phases at the exit surface of a scattering crystal located near the reflection position," *Phys. Status Solidi B*, 119, 685-692 (1984).
18. S. Horiuchi, "Effect of the phase shift due to dynamic scattering on the contrast of crystal structure images," *Ultramicroscopy*, 10, 229-236 (1982).
19. S. K. Maksimov, N. I. Borgardt, V. N. Kukin, D. I. Piskunov, and I. I. Khodos, "Defocused electron microphotography of substances with strong pulse contrast," *Izv. Akad. Nauk SSSR, Ser. Fiz.*, 33, 1729-1734 (1984).
20. M. Ruhle and M. Wilkens, "Defocusing contrast of cavities. I. Theory," *Cryst. Lattice Defects*, 6, 129-140 (1975).
21. N. I. Borgardt, S. K. Maksimov, and D. I. Piskunov, "Phase information in transmission electron-microscope investigations of defects. Dynamic images with weak amplitude contrast," *Phys. Status Solidi A*, 86, 55-65 (1984).
22. M. Ruhle and S. L. Sass, "The detection of the change in mean inner potential at dislocations in grain boundaries in NiO," *Philos. Mag. A*, 49, 759-782 (1984).
23. K.-J. Hanszen, "Holography in electron microscopy," in: *Advances in Electronics and Electron Physics*, Vol. 59, Academic Press, N.Y. (1982), pp. 1-77
24. A. Tonomura, "Electron holography," *J. Electron. Microsc.*, 35, suppl., 9-14 (1986).
25. F. J. Franke, K.-H. Herrman, and H. Lichte, "Off-axis electron holography with digital object reconstruction," *J. Electron Miscrosc.*, 35, Suppl., 667-668 (1986).

26. S. N. Bezryadin, M. G. Bezryadina, and S. K. Maksimov, "Phase contrast in electron microscopic images formed by one reflection," *Dokl. Akad. Nauk SSSR*, **260**, 339-342 (1981).

27. J. B. Mitchel and W. L. Bell, "Characterization of point-defect clusters by $2^1/_2$-D TEM," *Acta Metall.*, **24**, 147-152 (1976).

28. D. I. Piskunov, "Are the results of $2^1/_2$-D electron microscopy method correct?" *Fiz. Tverd. Tela*, **27**, 3521-3525 (1985).

29. G. J. Hardy, M. L. Jenkins, and M. J. Whelan, "Determination of the nature of small stacking fault tetrahedra by $2^1/_2$-D electron microscopy," *J. Electron Microsc.*, **35**, suppl., 1199-1200 (1986).

30. M. Tanaka and T. Kaneyama, "CBED studies of imperfect crystals," *J. Electron Microsc.*, **35**, suppl., 203-205 (1986).

31. D. J. Eaglesham and C. J. Humphreys, "A new technique for microanalysis using CBED: model study of (Al/Ga)As," *J. Electron Microsc.*, **35**, suppl., 709-710 (1986).

32. D. Cherns and A. R. Preston, "Convergent beam diffraction studies of crystal defects," *J. Electron Microsc.*, **35**, suppl., 721-722 (1986).

33. C. J. Humphreys, "The interpretation of high resolution and convergent beam patterns of interfaces in semiconductors," *J. Electron Microsc.*, **85**, suppl., 105-108 (1986).

34. H. Matsuhata and J. W. Steeds, "Precise measurement of zone axis critical voltage from observation in zero layer and HOLZ CBED discs," *J. Electron Microsc.*, **85**, suppl., 697-698 (1986).

35. H. Matsumura, Y. Tomokiyo, N. Kuwano, M. Kominami, T. Okuyama, and K. Oki, "Applicability of kinematical treatment in determination of lattice parameters from a HOLZ pattern," *J. Electron Microsc.*, **85**, suppl., 707-708 (1986).

36. P. Liu and G. L. Dunlop, "A general relation between the radius of HOLZs and crystallographic parameters," *J. Electron Microsc.*, **85**, suppl., 703-704 (1986).

37. A. F. Moodie and H. J. Whitefield, "Combined CBED and HREM in the electron microscope," *Ultramicroscopy*, **13**, 265-268 (1984).

38. M. Isaakson, "Fundamentals, applications, problems and prospects for electron energy loss spectroscopy within the electron microscope," *J. Electron Microsc.*, **35**, suppl., 37-40 (1986).

39. P. R. Batson, "Prospects for high-resolution electron energy-loss experiments with the scanning transmission electron microscope," *Ultramicroscopy*, **18**, 125-130 (1985).

40. C. Colliex, "An illustrated review of various factors governing the high spatial resolution capabilities in EELS microanalysis," *Ultramicroscopy*, **18**, 131-150 (1985).

41. C. Colliex, "High spatial resolution electron energy loss spectroscopy," *J. Electron Microsc.*, **35**, suppl., 41-44 (1985).

42. A. Bourret and C. Colliex, "Combined HREM and STEM microanalysis on decorated dislocation cores," *Ultramicroscopy*, **9**, 183-187 (1982).

43. L. E. Thomas, "Light element analysis with electrons and x rays in high-resolution STEM," *Ultramicroscopy*, **18**, 173-184 (1985).

44. G. R. Bradley and P. C. Gibbons, "How to remove multiple scattering from core-excitation spectra. Non-zero scattering angles," *Ultramicroscopy*, **19**, 317-324 (1986).

45. J. C. H. Spence, "The structure sensitivity of electron energy-loss near edge structure (ELNES)," *Ultramicroscopy*, **18**, 165-172 (1985).

46. J. Bentley and P. Angelini, "Microanalysis by electron energy loss spectroscopy at 300 kV," *J. Electron Microsc.*, **35**, suppl., 521-522 (1986).

47. J. L. Hutchinson, "High-resolution electron microscopy in the study of semiconducting materials," *Ultramicroscopy*, **15**, 51-60 (1984).

48. J. L. Hutchinson, "Recent advances and applications of high resolution electron microscopy," *J. Microsc.*, **136**, 127-135 (1984).

49. J. L. Hutchinson, "The accommodation of mismatch epitaxial interfaces in semiconductors: some contributions from HREM," *Ultramicroscopy*, **18**, 349-354 (1985).

50. S. K. Maksimov and E. N. Nagdaev, "Phenomenon of the automodulation of the composition of GaAsP films," *Dokl. Akad. Nauk SSSR*, **245**, 1369-1372 (1979).

51. M. M. J. Treacy, J. M. Gibison, and A. Howie, "On elastic relaxation and long wavelength microstructures in spinodally decomposed $In_xGa_{1-x}As_yP_{1-y}$ epitaxial layers," *Philos. Mag. A*, **51**, 389-417 (1985).

52. A. K. Gutakovskii, O. P. Pchelyakov, and S. I. Stenin, "Possibilities of controlling the predominant type of mismatch dislocations during heteroepitaxy," *Kristallografiya*, **25**, 806-814 (1980).

53. S. K. Maksimov and E. N. Nagdaev, "Electron microscopy images of composition-modulated crystals. I. Theory," *Phys. Status Solidi A*, **68**, 645-652 (1981); "II. Dilatation contrast images," *ibid.*, **69**, 505-512 (1982); "III. Moiré fringe images," *ibid.*, **72**, 135-145 (1985).

54. S. K. Maksimov, L. A. Bondarenko, A. S. Petrov, and V. V. Kuznetsov, "Self-modulation phenomenon of the composition of epitaxial layers during crystallization from the liquid phase," *Fiz. Tverd. Tela*, **24**, 628-631 (1982).

55. J. Van Landuyt, J. Vanhellemont, H. Bender, and S. Amelinckx, "HVEM studies of processing induced defects in semiconductors," *J. Electron Microsc.*, **35**, suppl., 1059-1064 (1986).

56. J. Heydenreich, "HVEM studies of semiconductors," *J. Electron Microsc.*, **35**, suppl., 1047-1052 (1986).

Part VI

MATHEMATICAL MODELING AND REGULATION
OF CRYSTALLIZATION PROCESSES

COMPUTATIONAL EXPERIMENTATION
IN STUDIES OF CRYSTALLIZATION PROCESSES

A. A. Samarskii

INTRODUCTION

Computational experimentation as a powerful means for theoretical study has convincingly demonstrated its high effectiveness in various branches of science and technology, and primarily in physics. The ideological basis of the computational experiment is mathematical modeling. The methodological basis is the theory of computational algorithms. The technical basis is the modern computer.

In computational experimentation, problems concerned with crystallization processes present significant difficulties. These difficulties are related to the fact that computational–theoretical studies on this topic are not limited to a qualitative analysis of the effects but must make accurate recommendations for optimal construction and operation of equipment. This requires examination of a large number of interrelated processes, consideration of geometric details, and accurate description of the properties of the studied media, etc. All of this greatly complicates the mathematical model and hinders its investigation. Under such conditions, the construction of effective numerical methods that can produce the required results with the use of existing computational techniques and the development of programs enabling computational experiments to be carried out by technologists who are not themselves specialists in computational mathematics take on a special significance. The corresponding sets of programs (we will call them "engineering") can broaden the pool of active users and will aid the incorporation of computational experimentation technology into daily practice.

In the present article, a review of efforts on mathematical modeling of crystallization processes is given. These efforts were carried out under the direction of the author at the M. V. Keldysh Institute of Applied Mathematics of the Academy of Sciences of the USSR. In spite of the variety of technical processes studied in them, all mathematical models have a single basis, the Navier–Stokes equations. These equations describe the flow of a viscous noncompressed liquid (gas) [1]. Other factors considered are convective heat and mass transport, chemical reactions, phase transitions, turbulence, etc. More detailed data on the models are contained in the descriptions of actual computational experiments which are given below.

1. MATHEMATICAL MODELING OF
THE CZOCHRALSKI PROCESS

The surface of a melt occupying a crucible of radius R_c and height H_c ($G = |(r, z)| 0 < r < R_c;$ $0 < z < H_c$) contacts a crystal of radius R_{cr} which is rotating at an angular velocity Ω_{cr}. The angular velocity of crucible rotation Ω_c at the surface of the crucible Γ_c, crystal Γ_{cr}, free surface of the melt Γ_m, and the axis Γ_a set the thermal regime. The crucible can be placed in a magnetic field $B = (B_h(r, z), 0, B_z(r, z))$. A flux can be placed on the melt surface. Shields can be fixed in the melt and on its surface.

Complex flow, which is caused by thermal, forced, and thermal capillary convections, arises in the melt. The nature of the flow greatly affects the distribution of dopant in the melt. With the solution of the model construction problem in hand, questions on the existence and configuration of steady-state melt flow, on the behavior of isotherms in the subcrystal zone, on the distribution of dopant along the crystal radius, and, in the case of oxygen (in silicon), on its quantity in the crystal all become interesting. These factors also determine the mathematical model.

The mathematical model consists of solution in the region G with boundaries $\Gamma = \Gamma_h \cup \Gamma_{cr} \cup \Gamma_m \cup \Gamma_a$ of the Navier–Stokes equations with the Boussinesq approximation [1-3]. The corresponding statement of the problem in the presence of a magnetic field was given in [4]. Statements of the problem for the concentration of dopant are discussed in [3, 5, 6]. Differential equations in these works are stated in the variables $\omega = \omega^{(\varphi)}/r$, $\psi = \psi^{(\varphi)}r$, $M = u^{(\varphi)}r$, T, and c (or C_1') (see below), where $\omega^{(\varphi)}$, $\psi^{(\varphi)}$, and $u^{(\varphi)}$ are radial components of the vortex ($\omega = \text{rot } u$), current ($u = \text{rot } \psi$), and rate u. The quantity T is the deviation of temperature from the phase transition temperature. The concentration C of a nonvolatile dopant ($\partial C/\partial n = 0$, $(r, z) \in \Gamma_c \cup \Gamma_m$) is given as $C = \bar{C}(t)(1 + C_1')$, where $\bar{C}(t)$ is the average dopant concentration in the crucible [3, 5]. It was found that $C' = O(w_0)$, where w_0 is the rate of crystal growth, a minor parameter of the problem. The solution of C' was sought as a function in steps of w_0.

For the first approximation $C_1' \sim w_0$, a second-order boundary condition $D(\partial C_1'/\partial z)|_{\Gamma_{cr}} = (k_0 - 1)w_0$ is placed on Γ_{cr}. The source term $[R_{cr}^2(k_0 - 1)w_0]/R_c^2 H_c$ arises in the right part of the convective diffusion equation for C_1'. Here k_0 is the dopant equilibrium distribution coefficient and D is the diffusion coefficient. The problem for C_1' has a nonzero solution for each fixed H_c. The scatter of concentration along the radius $\delta c = (C_1'\,_{max} - C_1'\,_{min})/2$. For a volatile dopant $C|_{\Gamma_m} = 0$, $D(\partial C_1'/\partial z)|_{\Gamma_{cr}} = (k_0 - 1)w_0 C|_{\Gamma_{cr}}$ at the crystal surface, and the condition $C = C_s$, where C_s is the saturated concentration, occurs at the crucible surface for oxygen in silicon. On the axis, $Dr(\partial C/\partial r)|_{\Gamma_0} = 0$.

A similar condition is also placed on Γ_0 for temperature. The temperature (or heat exchange conditions with irradiation) is fixed in the remaining part of the interface in the simplest case. Natural conditions are placed on Γ for the rate u: the functions ψ, ω, M: $\psi|_{\Gamma} = 0$; $\partial\psi/\partial n|_{\Gamma_c \cup \Gamma_{cr}} = 0$; $\omega|_{\Gamma_m} = 0$; $\partial\omega/\partial r|_{\Gamma_0} = 0$; $M|_{\Gamma_{cr}} = \Omega_{cr}r^2$; $M|_{\Gamma_c} = \Omega_c r^2$; $\partial M/\partial z|_{\Gamma_m} = 0$; and $M|_{\Gamma_0} = 0$. If the surface tension coefficient α depends on T, then $\omega|_{\Gamma_m} = -1/_r(\partial\alpha/\partial T)(\partial T/\partial r)$. In the presence of a magnetic field, additional source terms appear in the equations for M and ω. Also, the equation for the electrical current function ψ_J, $\psi_J|_{\Gamma} = 0$, must be solved.

The Re, Gr, Ro, Pr, Sc, Ma, etc. numbers appear after solution of the equations in a dimensionless form.

The finite difference method was used for solution of these problems. This involves difference schemes of variable directions [3], implicit difference schemes [7], etc. Thus, the schemes in [7] were built on nine-point cross models with the approximation $O(h^2 + \tau)$ (h is the average grid step in spatial variables and τ is the time). The schemes, which conserve kinetic energy on the grid to an accuracy up to $O(h^3 + \tau)$, are practically monotonic. The ICCG method and direct methods (programs from the Yale University set) are used for solution of the grid equations. Calculations can be carried out separately. The method is transferable and effective. A program set was created.

The methods of [3, 7] were used to calculate a large number of problems [?, 9]. The nature of the melt flow (germanium, silicon, gallium arsenide) was derived for the various parameters Re, Ro, Gr, Ma, Pr, Sc and the geometric parameters R_{cr}, R_c, and H_c. Problems on liquid flow in a magnetic field [3] were solved. The oxygen distribution in a silicon melt was studied [8]. Several recommendations which have accelerated development of new technologies and have improved the construction of the thermal unit were made.

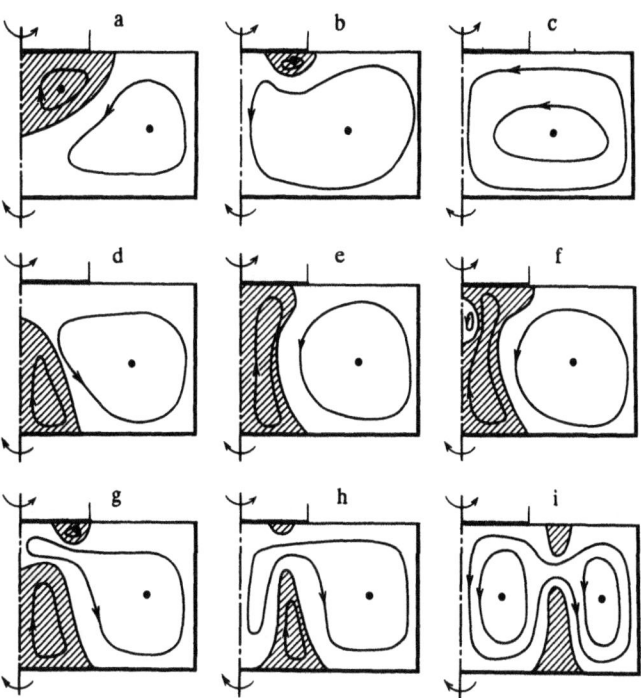

Fig. 1. Various pictures of melt flow.

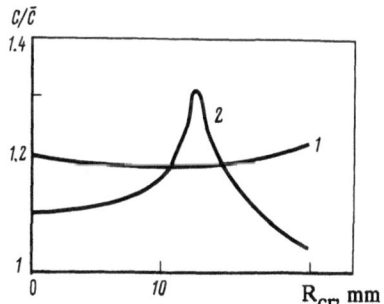

Fig. 2. Radial distribution of dopant at
the crystallization front.

We now give several examples of the calculations. Figure 1 shows the development of flow for an initially resting melt which is heated to a constant temperature $T = 0.5(T_{max} + T_{min})$. Thus, a steady-state flow with a subcrystal vortex is established with rapid crystal rotation ($\Omega_{cr} = 60$ rpm) (Fig. 1a). The flow is transformed (Fig. 1a-d) and becomes steady (Fig. 1d) at small Ω_{cr} ($\Omega = 20$ rpm). At a medium rate $\Omega_{cr} = 30$ rpm, the approach to the steady state is shown in Fig. 1e-h. The steady-state flow picture has the form of Fig. 1e for small loadings (small H_c). At small R_c, steady-state configurations like that shown in Fig. 1i are observed. Frequently (at large Gr/Re^2), when steady-state flow is completely absent, an oscillatory mode was observed in the calculations.

The decreased spreading of dopant along the single crystal radius at increased rotation rate Ω_c and the presence of a subcrystal vortex were confirmed by calculations. Figure 2 (curve 1) illustrates the gallium dopant distribution along the crystal radius in a germanium melt at $\Omega_{cr} = 60$ rpm. The spreading δC is not large ($\delta C \approx 4\%$). The spreading is much larger (Fig. 2, curve 2) for the flow shown in Fig. 1e.

The quantity of oxygen in a silicon single crystal depends substantially on the configuration of the melt flow. With the subcrystal vortex under a large part of the free surface of the component melt, the velocity

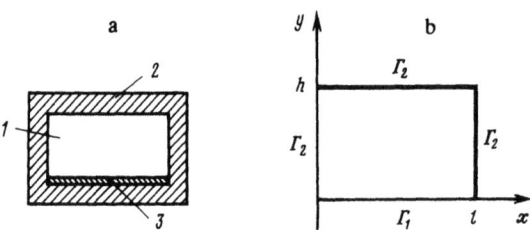

Fig. 3. Diagram of an experimental cell (a) and the calculation
region (b). 1) Melt; 2) graphite; 3) substrate.

$u_z < 0$. This leads to decreased oxygen vaporization from the melt surface and increased fraction of oxygen in the silicon single crystal.

The calculations showed that a steady-state gallium arsenide melt flow during growth of a single crystal ($R_{cr} \ll R_c$) can be achieved through a decrease in the value Gr/Re_c^2 (a decrease of ΔT and Gr) or an increase in Ω_c and $Re_c = R_c^2\Omega_c/\nu$. We examined cases of small R_c. Stabilization of flow can be attained by examining the flow in a magnetic field with $B_z \approx 0.2$ T. Results of calculations are given in [2, 3, 8, 9].

2. MATHEMATICAL MODELING OF THE PREPARATION OF $A_x^3B_{1-x}C^5$ STRUCTURES BY LIQUID-PHASE EPITAXY

We now examine epitaxial crystallization of $A_x^3B_{1-x}^3C^5$ compounds from a limited melt volume (Fig. 3a). The melt exists as a dilute solution of A and C components in solvent B at growth temperatures which are usually used in practice. At present, convective mass transport in the melt is well known to have a substantial influence on the conditions of epitaxial layers in several regimes, for example, when the melt thickness exceeds a certain critical value h_{crit} [10-13]. Therefore, its detailed study is a necessary step in searching for the optimal regimes that ensure preparation of compounds with given properties.

Numerical study of convective mass transport in the melt is carried out based on solution of a system of two-dimensional Boussinesq equations of concentration convection. It includes Navier–Stokes equations that are written in variables of the current function ψ, vortex ω, and two equations for determination of the functions X_A and X_C, which are concentrations expressed in atomic fractions of components dissolved in the liquid phase.

The solution of the problem lies in the rectangular region $G = [0, l] \times [0, h]$ with boundary $\Gamma = \Gamma_1 \cup \Gamma_2$. The Γ_1 value denotes that part of the calculation region where substrate of semiconducting material is placed in the actual experiment. The Γ_2 value corresponds to the graphite cell walls (Fig. 3b). It is assumed that both components of vector velocity revert to zero at the boundary Γ.

The concentration boundary conditions on Γ_1 are formulated assuming that the concentration of dissolved components at the crystallization front through the whole process remains in equilibrium, i.e., it satisfies the liquidus equation [14]

$$X_C = X_C^0(T) \frac{K(T)}{K(T) + X_A}, \tag{1}$$

where $X_C^0(T)$ is the solubility of component C in solvent B and $K(T)$ is the ratio of equilibrium constants for the formation reactions of the binary compounds AC and BC. These values are functions of the instantaneous temperature T of the epitaxial process. In the simplest case $T = T_0 - \alpha t$, where t is the time and α is the rate of temperature drop.

The second boundary condition is a result of the material balance equation at the melt–crystal interface

$$D_A (X_C^s - pX_C)\frac{\partial X_A}{\partial n}\bigg|_{\Gamma_1} = D_C (X_A^s - pX_A)\frac{\partial X_C}{\partial n}\bigg|_{\Gamma_1}. \qquad (2)$$

Here D_A and D_C are diffusion coefficients of components A and C in solvent B;

$$p = \frac{1}{2}\frac{\rho_B}{\rho_{BC}}\frac{M_{BC}}{A_B} ;$$

ρ is the density; A and M are atomic and molecular weights of the respective substances; X_A^s and X_C^s are the content of components A and C in the solid; and $X_C^s = 0.5$ and X_A^s is related to the concentration at the melt–crystal interface by the solidus equation

$$X_A^s = \frac{1}{2}\frac{X_A}{K(T) + X_A}. \qquad (3)$$

Modeling the preparation of binary compounds renders condition (2) superfluous whereas condition (1) takes the form

$$X_C = X_C^0(T).$$

The conditions $\partial X_A/\partial n\big|_{\Gamma_2} = \partial X_C/\partial n\big|_{\Gamma_2} = 0$ are fixed at the Γ_2 portion of the boundary. This implies the absence of currents of dissolved substances at the graphite cell walls.

The concentrations X_A and X_C are fixed in order to provide initial data. These correspond at a given temperature T_0 to the starting substrate composition $\bar{X}_A^s$, i.e., at $t = 0$

$$X_A = -\frac{2\bar{X}_A^s K(T_0)}{1 - 2\bar{X}_A^s} ,$$

$$X_C = X_C^0(T_0)(1 - 2\bar{X}_A^s).$$

Due to the insignificant thickness of the epitaxial layer (1-5% of the gap value), a change of the shape of the region related to its growth is not considered in the calculations. The melt temperature is also assumed to be uniform throughout the space and the laws of its change with time coincide with the temperature program at the boundary.

The growth rate is calculated from the flow value of any of the dissolved components onto the substrate. For example,

$$W_{\Gamma_1}(t, x) = pD_C \frac{\partial X_C}{\partial n}\bigg/(X_C^s - pX_C).$$

Then the thickness of the epitaxial layer at time $\bar{t}$

$$d(t, x) = \int_0^t W_{\Gamma_1}(t, x)\, dt.$$

The solid phase composition at the interface with the melt is determined from (3) at each time point. Knowing the layer thickness as a function of time, the distribution of component A in the solid can easily be determined.

With respect to numerical evaluation, the formulated problem has a number of features which must be considered in selecting an algorithm for its solution. First, this problem is nonsteady-state and its solution must be determined at sufficiently large time intervals $\bar{t}$. For several technological regimes, $\bar{t} \sim 2\cdot 10^3 t_{cr}$, where

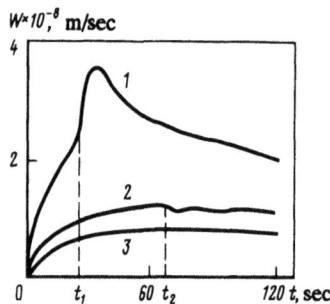

Fig. 4. Growth rate of an epitaxial layer: $l = 20$ mm, $h = 2$ mm. 1-3) $\overline{X}^s_{Al} = 0$, 0.1, and 0.2, respectively.

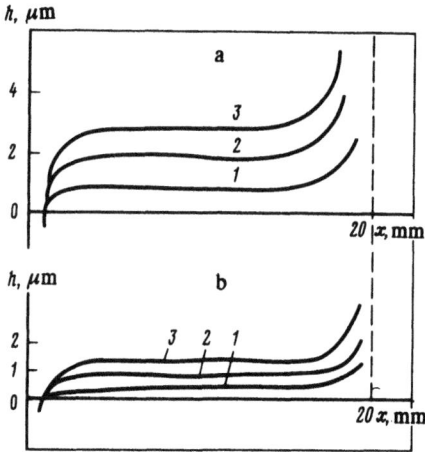

Fig. 5. Shape of surface epitaxial layer at various time points: $\overline{X}^s_{Al} = 0.1$ (a) and 0.4 (b). 1-3) $t = 33$, 66, and 99 min, respectively. $E = 2 \cdot 10^{-3}$ V/cm, $h = 1$ mm, $l = 20$ mm, and cooling rate $= 0.1$ deg/min.

$t_{cr} = h^2/\nu$ and ν is the kinematic viscosity of the melt B. Methods which permit calculations to be carried out in large time steps must be used in order to obtain a solution in a reasonable expenditure of machine time.

The matrix algorithm developed in [5] is used for numerical study of the convective mass transport process in the melt for solution of the Navier–Stokes equations. Experience in its use indicates that it is not encumbered by rigid limitations on the time step value. Also, a solution to the problem can be found over the whole range of parameters of practical interest. An implicit difference scheme, in which the current function and vortex at the internal grid points and at the interface are taken from the upper time layer, is the basis of the method.

The Newton method is used for solution of the nonlinear system of difference equations. The linear system of equations, which arises at each Newtonian iteration, is solved relative to the vector (ω, ψ) using an iteration process [16]. This is a generalization for the second-order matrix given in [17].

Additional difficulties connected with numerical evaluation of the nonlinear boundary conditions (1) and (2) appear in mathematical modeling of the growth of ternary compounds. In [18, 19], a matrix algorithm analogous to that used for solution of the Navier–Stokes equations was shown to be most effective for solution of equations for concentration determination. This evaluation was based on an analysis of the various solution methods. It is significant here that both desired functions X_A and X_C at the boundary conditions are taken

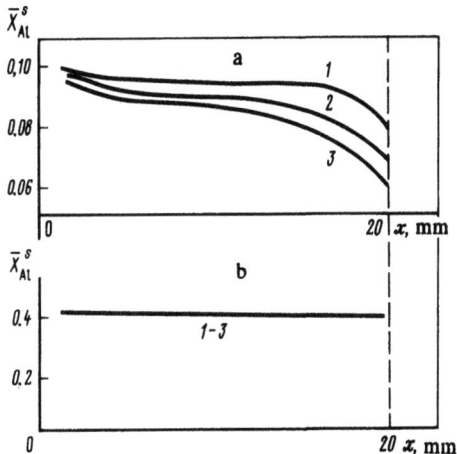

Fig. 6. Distribution of Al on a growing surface layer as a function of time. $\overline{X}^s_{Al} = 0.1$ (a) and 0.4 (b). 1-3) $t = 33, 66,$ and 99 min, respectively. $E = 2 \cdot 10^{-3}$ V/cm, $h = 1$ mm, $l = 20$ mm, and cooling rate = 0.1 deg/min.

from the upper time layer. Moreover, the nonlinear nature of the starting differential problem is retained upon changing to the difference approximation.

Numerical study of the convective mass transport process in the melt during preparation of epitaxial structures is carried out according to the problem as formulated above. In [20-22], special attention was paid to the study of mass exchange during growth of binary compounds. Here, we will present results which relate to preparation of the ternary compounds $Al_xGa_{1-x}As$.

Analysis of the calculations shows that the intensity of convective motion in the melt falls with an increase of aluminum content. Convection in the melt is practically absent and transfer of dissolved components to the crystallization front occurs predominantly by diffusion for an initial Al content in the melt of $\overline{X}^s_{Al} \geq 0.2$.

An important feature of epitaxial growth of layers from melts that contain aluminum is the decrease of growth rate as the aluminum concentration in the melt is increased. Figure 4 shows curves of the epitaxial layer growth rate as a function of time in the central part of the substrate for $X^s_{Al} = 0, 0.1,$ and 0.2. The sharp increase of gallium arsenide growth rate ($X^s_{Al} = 0$) at time point $[0, t_1]$ involves the significant contribution of convective motion to transfer to the substrate with a gap value $h = 2$ mm [23]. At present, electroepitaxy is being used more widely [24]. Without dwelling on the insignificant modifications that must be made in the statement of the problem for modeling this process, we will present some calculations. Figures 5 and 6 show the shape of the surface and composition of layers prepared by electroepitaxy when the voltage E of the applied electric field is directed along the substrate.

3. PREPARATION OF SOLID BINARY SOLUTIONS BY ZONE REFINING IN AMPULS

Zone refining in ampuls is used for production of solid binary solutions of semiconducting materials. Preparation of solid solutions of a given composition is a problem in this method. Use of mathematical modeling and computational experimentation can explain the features of the concentration profile in the crystallizing ingot, determine the parameters for control of the process, and help in designing controls that guarantee the required composition.

We will briefly describe the technological process. Let the liquid component A (at $0 \leq x \leq l_0$) and the solid component B (at $l_0 \leq x \leq L$) come into contact at the initial moment in a thin cylindrical ampul of length L. The ampul is placed in a furnace. A certain temperature profile (Fig. 7) is imposed on the inner cylindrical

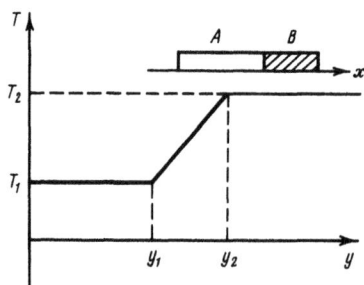

Fig. 7. Temperature profile at the inner furnace wall.

surface of the furnace. The ampul is kept motionless for a certain time. Dissolution of the solid component B and saturation of the liquid solution by it begins. Attainment of a certain concentration of component B in the liquid solution renders it unstable and crystallization begins in the colder end of the ampul at $x = 0$. At this time, the ampul begins to shift into the colder region at a certain rate u_0. The values of temperature and concentration of liquid and solid solutions at the interface obey relations determined by the phase diagram during dissolution and crystallization.

Mathematical modeling of this process is carried out based on numerical solution of the one-dimensional nonstationary problem on crystallization and melting of a binary system [25].

A nonlinear system of differential equations for temperature T and concentration C averaged over the cross section x includes the following equations:

1) the equation of thermal conductivity at the internal points of the region

$$\kappa\rho \frac{\partial T}{\partial t} = \frac{\partial}{\partial x}\left(k \frac{\partial T}{\partial x}\right) - q\,(T^4 - T_0^4\,(x, t)),$$

where κ, ρ, k, and q are certain constants and $T_0(x, t)$ is the temperature in the irradiation stream falling on the side surface of the ampul;

2) the equation of diffusion at the internal points of the region

$$\frac{\partial C}{\partial t} = \frac{\partial}{\partial x}\left(D \frac{\partial C}{\partial x}\right);$$

3) boundary conditions

$$k \frac{\partial T}{\partial x} = \sigma(T^4 - T_1^4\,(t)) \quad \text{at} \quad x = 0,$$

$$k \frac{\partial T}{\partial x} = -\sigma(T^4 - T_2(t)) \quad \text{at} \quad x = L,$$

$$\frac{\partial C}{\partial x} = 0 \quad \text{at} \quad x = 0, L,$$

where σ is a constant and $T_1(t)$ and $T_2(t)$ are fixed functions (temperature in the irradiation stream falling on the end of the ampul);

4) systems at the interfaces

$$k \frac{\partial T}{\partial x}\,(\xi_i + 0) - k \frac{\partial T}{\partial x}\,(\xi_i - 0) = (2i - 3)\,\rho\lambda \frac{\partial \xi_i}{\partial t},$$

$$D \frac{\partial C}{\partial x}\,(\xi_i + 0) - D \frac{\partial C}{\partial x}\,(\xi_i - 0) = [C(\xi_i - 0) - C(\xi_i + 0)] \frac{d\xi_i}{dt},$$

$$T\,(\xi_i, t) = (2 - i)\,\psi\,(C(\xi_i - 0, t)) + (i - 1)\,\varphi\,(C(\xi_i - 0, t)),$$

$$T\,(\xi_i, t) = (2 - i)\,\varphi\,(C(\xi_t + 0, t)) + (i - 1)\,\psi\,(C(\xi_i + 0, t)),$$

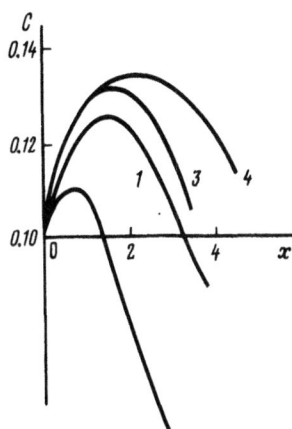

Fig. 8. Concentration distribution in a recrystallized ingot. 1) $l_0 = 2.5$, $D_L = 6 \cdot 10^{-2}$, $G = 10$; 2) 2.5, $3 \cdot 10^{-2}$, 20; 3) 2.5, $6 \cdot 10^{-2}$, 20; 4) 2.5, $3.6 \cdot 10^{-2}$, 20.

where λ is a constant, $\xi_1(t)$ is the crystallization front coordinate, $\xi_2(t)$ is the melting front coordinate, and φ and ψ are functions which are determined by the phase diagram.

A grid with a fixed number of points in each phase is introduced in the region $0 \leq x \leq L$. The melting and crystallization fronts are points on the spatial grid for each time layer. Therefore, the grid is reconstructed not only for each time layer but also for each iteration during solution of the difference equations for the layer.

Conserved difference schemes are built on the grid with changing time intervals. Schemes which are implicit in T, C, and ξ_i are used. A special iteration process is used for solution of the systems of nonlinear difference equations for the layer. The numerical method is described in detail in [26].

Several results of calculations are given in Fig. 8, which shows graphs of concentration $C(x)$ in an ingot which has been recrystallized at various values of the parameters l_0, D_L (diffusion coefficient in the liquid solution), and G (the gradient T in the range $y_1 \leq y \leq y_2$).

4. ENGINEERING COMPUTATIONAL EXPERIMENTATION
ON PROBLEMS OF SEMICONDUCTOR TECHNOLOGY

Expansion of the realm of applicability of computational experimentation requires an expansion of the circle of active users. The NEPTUN program set which was developed under the direction of the author at the M. V. Keldysh Institute of Applied Mathematics serves this purpose [28, 29]. This set is designated for numerical modeling of convective–diffusional heat and mass transfer taking into account varied physicochemical effects in various apparatuses. The mathematical model employed by it describes nonsteady-state two-dimensional and quasi-three-dimensional flows of a viscous noncompressed liquid (gas) in planar and cylindrical (with three velocity components) geometries, including convective–diffusional heat and mass transfer (accounting for thermal diffusion, the kinetics of chemical reactions, and phase transitions) in areas of complex shape, the multicomponent nature, the porosity and dispersivity of media, radiant heat exchange (including transfer of laser radiation), and the k–ε model of turbulence.

The NEPTUN program set not only covers an adequate range of physicomathematical models but can also be modified and adopted to a wide circle of users who are not specialists in computational mathematics. This is achieved through its modular nature, standardized documentation, and the simplicity and reliability of the numerical methods contained in it. At present, valuable recommendations for the improvement of actual processes and apparatuses have been received and much experience in its use in various organizations has been accumulated. In particular, this experience shows that an ordinary engineer without any experience in mathe-

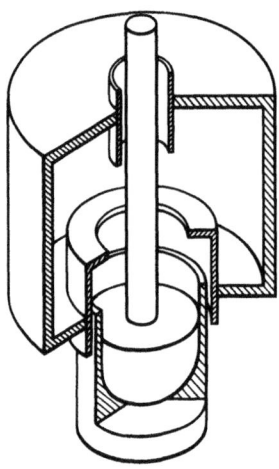

Fig. 9. Equipment for studying gas dynamics during the
Czochralski process.

matical modeling on a computer needs only 2-3 months for active assimilation of the program set. Also, the additional modules that are written by different organization are included in the set and become a contribution for the remaining users.

Figure 9 shows a general scheme of a Czochralski reactor as an example of the use of the NEPTUN set. The nature of mixing convection of a neutral gas in the reactor chamber was studied. Computational experimentation helped to explain the reasons for entry of silicon monoxide into the melt (the stagnant zone over the crucible) and measures for their removal (introduction of additional baffles).

REFERENCES

1.   L. D. Landau and E. M. Lifshits, *Hydrodynamics* [in Russian], Nauka, Moscow (1986).
2.   V. A. Smirnov, I. V. Starshinova, and I. V. Fryazinov, "Analysis of rate, temperature, and concentration distributions of dopants in the melt during growth of single crystals by the Czochralski method," in: *Growth of Crystals*, Vol. 14, Consultants Bureau, New York (1987).
3.   V. A. Smirnov, I. V. Starshinova, and I. V. Fryazinov, "Mathematical modeling of growth processes by the Czochralski method," in: *Mathematical Modeling. Preparation of Single Crystals and Semiconducting Structures* [in Russian], Nauka, Moscow (1986), pp. 40-59.
4.   W. E. Langlois and K. J. Lee, "Czochralski crystal growth in an axial magnetic field: effects of Joule heating," *J. Cryst. Growth*, **62**, 484-486 (1983).
5.   I. V. Starshinova and I. V. Fryazinov, "Statement and numerical study of problems on distribution of dopants in the Czochralski process," Preprint M. V. Keldysh Inst. Appl. Math. Acad. Sci. USSR, No. 94, Moscow (1982).
6.   I. A. Remizov, "Numerical modeling of concentration fields of dopants in a melt during growth of single crystals by the Czochralski method," *Fiz. Khim. Obrab. Mater.*, No. 2, 38-45 (1980).
7.   A. L. Goncharov and I. V. Fryazinov, "Difference schemes in nine-point cross models for solution of Navier–Stokes equations as applied to the Czochralski problem," Preprint M. V. Keldysh Inst. Appl. Math. Acad. Sci. USSR, No. 150, Moscow (1986).
8.   V. A. Smirnov, I. V. Starshinova, and I. V. Fryazinov, "Analysis of oxygen distribution in a silicon melt," *Kristallografiya*, **29**, No. 3, 560-565 (1984).
9.   V. I. Biberin, V. B. Osvenskii, V. A. Smirnov, et al., "Study of melt hydrodynamics during growth of gallium arsenide by the Czochralski method under a flux," *Kristallografiya*, **30**, No. 5, 980-985 (1985).
10. M. B. Small and J. Crossley, "The physical processes during liquid phase epitaxial growth," *J. Cryst. Growth*, **27**, 35-48 (1974).
11. W. R. Wilcox, "Influence of convection on growth of crystal from solution," *J. Cryst. Growth*, **65**, 133-142 (1983).
12. U. Konig and W. Keck, "Influence of cooling rate on convection during LPE growth of GaAs and (Ga, Al)As," *J. Cryst. Growth*, **65**, 588-595 (1983).
13. O. S. Mazhorova, Yu. P. Popov, V. A. Smirnov, and A. A. Shlenskii, "Numerical modeling of the growth process of single-crystalline layers of semiconducting materials by liquid-phase epitaxy," *Fiz. Khim. Obrab. Mater.*, No. 4, 81-90 (1983).
14. V. M. Andreev, L. M. Dolginov, and D. N. Tret'yakov, *Liquid-phase Epitaxy in Semiconducting Device Technology* [in Russian], Sov. Radio, Moscow (1975).

15. O. S. Mazhorova and Yu. P. Popov, "On numerical methods for solution of Navier–Stokes equations," *Zh. Vychisl. Mat. Mat. Fiz.*, 20, No. 4, 1005-1020 (1980).

16. O. S. Mazhorova, "Iterative method for solution of two-dimensional matrix equations," Preprint M. V. Keldysh Inst. Appl. Math. Acad. Sci. USSR, No. 48, Moscow (1979).

17. B. N. Chetverushkin, "On one iterative algorithm for solution of difference equations," *Zh. Vychisl. Mat. Mat. Fiz.*, 16, No. 2, 519-524 (1979).

18. O. S. Mazhorova, Yu. P. Popov, and V. I. Pokhilko, "On numerical solution of parabolic equations with nonlinear boundary conditions," Preprint M. V. Keldysh Inst. Appl. Math. Acad. Sci. USSR, No. 46, Moscow (1985).

19. O. S. Mazhorova and Yu. P. Popov, "Matrix algorithm for numerical solution of nonstationary problems of concentrational convection for multicomponent media," in: *Mathematical Modeling. Preparation of Single Crystals and Semiconducting Structures* [in Russian], Nauka, Moscow (1986), pp. 19-31.

20. L. A. Dmitrieva, O. S. Mazhorova, Yu. P. Popov, and É. A. Tvirova, "Mathematical modeling of liquid-phase epitaxy," Preprint M. V. Keldysh Inst. Appl. Math. Acad. Sci. USSR, No. 69, Moscow (1983).

21. L. A. Dmitrieva, O. S. Mazhorova, Yu. P. Popov, and É. A. Tvirova, "Study of mass transfer process during liquid-phase epitaxy in vertical systems," Preprint M. V. Keldysh Inst. Appl. Math. Acad. Sci. USSR, No. 118, Moscow (1984).

22. L. A. Dmitrieva, O. S. Mazhorova, Yu. P. Popov, É. A. Tvirova, and A. A. Shlenskii, "On numerical study of convective mass transfer during preparation of semiconductor material structures by liquid-phase epitaxy," in: *Mathematical Modeling. Preparation of Single Crystals and Semiconducting Structures* [in Russian], Nauka, Moscow (1986), pp. 83-101.

23. O. S. Mazhorova, Yu. P. Popov, and V. I. Pokhilko, "Mathematical modeling of mass transfer during liquid-phase epitaxy of $Al_xGa_{1-x}As$ layers," Preprint M. V. Keldysh Inst. Appl. Math. Acad. Sci. USSR, No. 194, Moscow (1986).

24. S. Yu. Karpov, M. G. Mil'vidskii, and S. A. Nikishkin, "Theory of current regulation of composition with respect to thickness of $Al_xGa_{1-x}As$ epitaxial layers," *Pis'ma Zh. Éksp. Teor. Fiz.*, 8, No. 14, 883-887 (1982).

25. L. I. Rubinstein, *The Stefan Problem*, American Mathematical Society, Providence, Rhode Island (1971).

26. O. I. Bakirova, "Numerical modeling of zone refining processes based on solution of the phase transfer problem in a binary system," in: *Mathematical Modeling. Preparation of Single Crystals and Semiconducting Structures* [in Russian], Nauka, Moscow (1986), pp. 142-158.

27. I. I. Voronov et al., "Study of the gas dynamics of the equipment chamber for growth of single crystal silicon by the Czochralski method," in: *Mathematical Modeling. Preparation of Single Crystals and Semiconducting Structures* [in Russian], Nauka, Moscow (1986), pp. 133-142.

28. B. P. Gerasimov et al., "Mathematical modeling of heat and mass transfer in an epitaxial reactor," in: *Mathematical Modeling. Preparation of Single Crystals and Semiconducting Structures* [in Russian], Nauka, Moscow (1986), pp. 112-132.

29. B. P. Gerasimov and A. V. Lesunovskii, "Effect of thermal diffusion with convective heat and mass transfer in a gas epitaxial reactor," Preprint M. V. Keldysh Inst. Appl. Math. Acad. Sci. USSR, No. 126, Moscow (1986), p. 28.

MATHEMATICAL MODELING OF MELT HYDRODYNAMICS
IN THE BRIDGMAN CRYSTALLIZATION PROCESS
WITH ACCELERATED CRUCIBLE ROTATION

V. D. Karpinskii and É. L. Lube

The necessity for melt mixing often arises during crystal growth from a melt by the Bridgman method. Rotation of the melt-containing crucible with alternating acceleration and deceleration, the so-called ACRT method [1-4], is effective for this purpose. The picture of flows in the crucible for this method was studied qualitatively by physical modeling [4]. The individual parameters were determined analytically [1, 2, 4].

Recently, crystallization processes have been widely studied using mathematical modeling. These methods examine in detail, with quantitative estimates, the heat and mass transfer during crystal growth [5]. The main efforts in computational experimentation on melt crystallization have been devoted to the Czochralski method. Schemes with only advancing movement of the crucible relative to the temperature zones are discussed in individual works on modeling of the Bridgman method.

The purpose of the present article is to formulate a mathematical model of the Bridgman method for detailed study of the melt hydrodynamics over a broad change of crystallization conditions including the rotation rate and dimensions of the crucible, physical properties of the melt, etc.

We introduce the following notation. The dimensionless variables are: θ, temperature; P, pressure; ν, kinematic viscosity; U, W, and V, the radial, axial, and circular components of the rate vector, respectively; R and Z, radial and axial coordinates, respectively; and t, time. The dimensional variables are: g, gravitational acceleration, $g = 981$ cm/sec^2; ρ, melt density in g/cm^3; β, coefficient of melt volume expansion in 1/K; ν, kinematic viscosity in cm^2/sec; a, coefficient of thermal conductivity of the melt in cm^2/sec; T_m and T_a, maximal and minimal melt temperature, respectively, in K; and H, melt height in cm (for remaining notation, see the text).

At present, a number of numerical methods which were developed for studying flows of nonuniform viscous liquids are known. The majority of these are based on solution of the Navier–Stokes equations in Helmholtz form [in variables of a vortex-function of current $(\omega - \psi)$] [6-8]. This limits the applicable area of these methods to two-dimensional and axially symmetric flows. The method of separation (MS) by physical factors has recently received wide attention [9-12]. This method is known to be applicable only up to Reynolds numbers Re $\leq 10^3$ [12]. Definite difficulties remain in approximation of boundary conditions although a step forward in this direction was recently taken [13].

Analysis of the computational algorithms showed a preference for MS since eventually models for cases which cannot be reduced to a planar or axially symmetric scheme can be developed. Melt flows with large

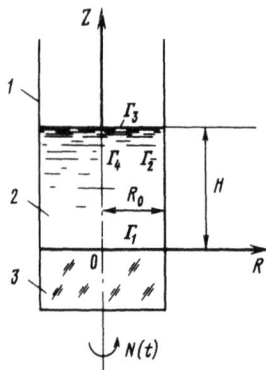

Fig. 1. The crucible–melt–crystal system.
1) Crucible; 2) melt; 3) crystal.

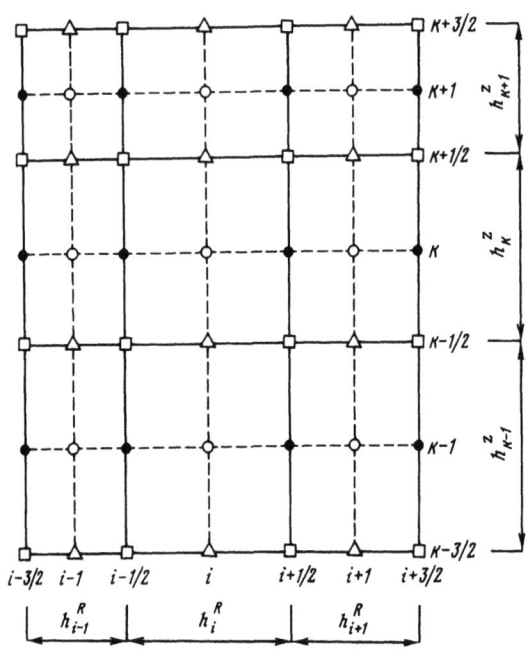

Fig. 2. Placement of functions on the grid. 1) ω, ψ; 2) W, $\widetilde{W}$, $\partial_z P$, and $\partial_z \delta P$; 3) U, $\widetilde{U}$, $\partial_R P$, and $\partial_R \delta P$; 4) V, $\widetilde{V}$, θ, ν, P, and δP.

Reynolds numbers are possible in Bridgman crystallization with accelerated crucible rotation. Accounting for these conditions, MS was developed by us for cases of flows with Re $\geq 10^4$. This was achieved by using an implicit function calculation in the determination of the intermediate rate range [12]. Thus, the convective terms are approximated by either the monotonic Samarskii scheme [14] or by a K–R scheme [15]. The convective terms in the heat transfer equation were approximated analogously.

Stratification of the melt according to viscosity is assumed in examination of the method. The crucible has a cylindrical (or simple) shape (Fig. 1). The force of gravity is directed vertically above. We will write the initial system of equations in the following dimensionless form:

$$U_{\tilde{t}} + R^{-1}\partial_R(RU^2) + \partial_z(UW) - R^{-1}V^2 = -\partial_R P + \text{Re}^{-1}\left[\partial_R(R^{-1}\partial_R(RU)) + \partial_{zz}U\right]\nu$$
$$+ 2\,\text{Re}^{-1}\,\nu_R U_R + \text{Re}^{-1}\,\nu_z\,(U_z + W_R).$$

$$(1)$$

$$W_{\bar{t}} + R^{-1} \partial_R(RUW) + \partial_z W^2 = -\partial_z P + \mathrm{Re}^{-1}[R^{-1}\partial_R(R\,\partial_R W) + \partial_{zz}W]\,\nu$$
$$+ 2\,\mathrm{Re}^{-1}\nu_z W_z + \mathrm{Re}^{-1}\nu_R(U_z + W_R) + \theta\,\mathrm{Fr}^{-1}, \tag{2}$$

$$V_{\bar{t}} + R^{-2}\partial_R(R^2 UV) + \partial_z(WV) = \mathrm{Re}^{-1}[\partial_R(R^{-1}\partial_R(RV)) + \partial_{zz}V]\,\nu$$
$$+ \mathrm{Re}^{-1}\nu_R R\,\partial_R(R^{-1}V) + \mathrm{Re}^{-1}\nu_z V_z, \tag{3}$$

$$\partial_R(RU) + \partial_z(RW) = 0, \tag{4}$$

$$\partial_{\bar{t}}\theta + R^{-1}\partial_R(RU\theta) + \partial_z(W\theta) = \mathrm{Pe}^{-1}[R^{-1}\partial_R(R\,\partial_R\theta) + \partial_{zz}\theta], \tag{5}$$

$$\nu = F(\theta). \tag{6}$$

The following are taken as scales: length, the crucible radius R_0; time, $(2\pi N_m)^{-1}$; rate, $\mathrm{Mv} = 2\pi N_m R_0$; temperature, $T_m - T_a$; viscosity, ν_0. Then, the dimensionless variables are

$$R = rR_0^{-1},\ Z = zR_0^{-1},\ \bar{t} = t\,2\pi N_m,\ U = u\mathrm{Mv}^{-1},\ W = w\mathrm{Mv}^{-1};$$
$$V = v\mathrm{Mv}^{-1},\ P = p(\rho\,\mathrm{Mv}^2)^{-1},\ \theta = (T-T_a)(T_m-T_a)^{-1},\ \nu = \nu\,\nu_0^{-1}. \tag{7}$$

The Reynolds, Peclet, and Froude numbers are

$$\mathrm{Re} = \mathrm{Mv}\,R_0\nu_0^{-1},\ \mathrm{Pe} = \mathrm{Re}\,\nu_0 a^{-1},\ \mathrm{Fr} = \mathrm{Mv}^2/(R_0 g\beta(T_m-T_a)). \tag{8}$$

The starting conditions are arbitrary. The boundary conditions are: at the solid surface $U = W = 0$ and V and θ are assumed to be unknown; at the free surface of the melt P and θ are known and $U = W = \partial_z V = 0$; on the axis $U = V = \partial_R W = \partial_R P = \partial_R \theta = 0$. The grid for our work is shown in Fig. 2. The functions V, P, θ, and ν are determined in the centers of cells (i, k), whereas the functions U and W are determined in the centers of cell faces at points $(i + 1/2, k)$ and $(i, k + 1/2)$, respectively. In this case, $h_j^x = X_{j+1/2} - X_{j-1/2}$; $\hbar_{j+1/2}^x = 0.5(h_j^x + h_{j+1}^x)$; $j = i, k$; $X = R, Z$. We define the difference derivatives as follows:

$$(\varphi_x)_{j+1/2} = (\varphi_{j+3/2} - \varphi_{j+1/2})/h_{j+1}^x,\ (\varphi_{\bar{x}})_{j+1/2} = (\varphi_{j+1/2} - \varphi_{j-1/2})/h_j^x;\ (\varphi_x^o)_{j+1/2} = 0.5(\varphi_x + \varphi_{\bar{x}})_{j+1/2},$$
$$(\varphi_{\hat{x}})_j = (\varphi_x)_{j-1/2},\ (\varphi_{\bar{x}x})_{j+1/2} = [(\varphi_x^o)_{j+1} - (\varphi_x^o)_j]/\hbar_{j+1/2}^x,\ \varphi = U, W;$$
$$(\varphi_x)_j = (\varphi_{j+1} - \varphi_j)/\hbar_{j+1/2}^x,\ (\varphi_{\bar{x}})_j = (\varphi_j - \varphi_{j-1})/\hbar_{j-1/2}^x,\ (\varphi_x^o)_j = 0.5(\varphi_x + \varphi_{\bar{x}})_j,$$
$$(\varphi_{\hat{x}}^o)_{j+1/2} = (\varphi_{\bar{x}})_j,\ (\varphi_{xx})_j = [(\varphi_x^o)_{j+1/2} - (\varphi_x^o)_{j-1/2}]/h_j^x,\ \varphi = V, \theta, P, \nu. \tag{9}$$

The scheme of the method of separation in an arbitrary coordinate system takes the form

Step I. $(\tilde{Y} - Y^n)\tau^{-1} + (Y^n\nabla)\tilde{Y} = \mathrm{Re}^{-1}\Delta\tilde{Y} - \nabla P^n + \hat{F}^n + \sigma^n,$ (10)

Step II. $\Delta(\delta P) = \nabla\tilde{V}\tau^{-1}$, where $\delta P = P^{n+1} - P^n,$ (11)

Step III. $(Y^{n+1} - \tilde{Y})\tau^{-1} = -\nabla(\delta P),$ (12)

Step IV. $(\theta^{n+1} - \theta^n)\tau^{-1} + (Y^{n+1}\nabla)\theta^{n+1} = \mathrm{Pe}^{-1}\Delta\theta^{n+1},$ (13)

Step V. $\nu^{n+1} = F(\theta^{n+1}),$ (14)

where $Y = (U, W, V)^T$; $\hat{F}$ are source terms in the Navier–Stokes equations; and σ are terms which contain viscosity gradients. Here, in contrast to [12], the implicit difference scheme (10) is applicable for determination of the intermediate rate range. Thus, calculations can be carried out with large time steps. Development of MS for flows with $\mathrm{Re} \geq 10^4$ was achieved by creating monotonicity by the Samarskii scheme [14] or by a K–R scheme [15] for Eqs. (10)-(13). In those cases when the values of the functions at the grid points which do not correspond to their positions in Fig. 2 must be defined, the arithmetic average should be used. For example,

$$W_{i,k} = 0.5(W_{i,k-1/2} + W_{i,k+1/2});\ W_{i+1/2,k} = (W1\,h_{i+1}^R + W2 h_i^R)/\hbar_{i+1/2}^R, \tag{15}$$

where

$$W1 = 0.5 (W_{i,k+1/2} + W_{i,k-1/2}), \quad W2 = 0.5 (W_{i+1,k+1/2} + W_{i+1,k-1/2}).$$

We deal analogously with the remaining functions. The quantities $\hat{F}$ and σ are approximated by central differences. Describing the axially symmetric case of the Step I equation by a five-point grid, we obtain three five-point equations for U, W, and V, respectively. Each of these is solved by the ICCG method using the Cholesky partial LU expansion for a ten-point grid (i, $i \pm 1$, $i \pm k_{max}$, $i \pm k_{max} \pm 1$) [16]. At Step II, we determine the pressure range from the solution of the Poisson equation [12]

$$R^{-1} \partial_R (R \, \partial_R \delta P) + \partial_{zz} \delta P = \tau^{-1} \Delta \tilde{Y}, \tag{16}$$

where

$$\nabla \tilde{Y} = R^{-1} \partial_R (R\tilde{U}) + \partial_z \tilde{W}; \quad \delta P = P^{n+1} - P^n. \tag{17}$$

Then, we determine the final rate range for the $(n + 1)$ time layer from the formulas

$$U^{n+1} = \tilde{U} - \tau \partial_R \delta P, \quad W^{n+1} = \tilde{W} - \tau \partial_z \delta P; \quad V^{n+1} = \hat{V}, \tag{18}$$

i.e., the rate range is corrected accounting for the pressure gradient. The equation of medium continuity $\nabla Y^{n+1} = 0$ is known to be fulfilled automatically [12]. Applying a divergence operator to the Eqs. (18), we obtain (16). We write Eqs. (16)-(18) in finite-difference form:

$$(R_i h_i^R)^{-1} [R_{i+1/2}(\delta P_{i+1,k} - \delta P_{i,k})/\hbar_{i+1/2}^R - R_{i-1/2}(\delta P_{i,k} - \delta P_{i-1,k})/\hbar_{i-1/2}^R]$$
$$+ (h_k^z)^{-1} [(\delta P_{i,k+1} - \delta P_{i,k})/\hbar_{k+1/2}^z - (\delta P_{i,k} - \delta P_{i,k-1})/\hbar_{k-1/2}^z] = \tau^{-1} \tilde{D}_{i,k}, \tag{19}$$

$$\tilde{D}_{i,k} = (R_i h_i^R)^{-1} [(R\tilde{U})_{i+1/2,k} - (R\tilde{U})_{i-1/2,k}] + (h_k^z)^{-1} (\tilde{W}_{i,k+1/2} - \tilde{W}_{i,k-1/2}) \tag{20}$$

$$U_{i+1/2,k}^{n+1} = \tilde{U}_{i+1/2,k} - (\tau \hbar_{i+1/2}^R)^{-1} (\delta P_{i+1,k} - \delta P_{i,k}), \tag{21}$$

$$W_{i,k+1/2}^{n+1} = \tilde{W}_{i,k+1/2} - (\tau \hbar_{k+1/2}^z)^{-1} (\delta P_{i,k+1} - \delta P_{i,k}), \tag{22}$$

$$V_{i,k}^{n+1} = \tilde{V}_{i,k}. \tag{23}$$

The pressure directly at the solid surface does not need to be calculated in the MS [12]. For example, using the continuity equation $D_{i,k}^{n+1} = 0$ and Eqs. (21) and (22), we obtain the uniform boundary condition for pressure at sites near the boundary with the crucible wall (Fig. 3)

$$[(R_i h_i^R \hbar_{i-1/2}^R)^{-1} R_{i-1/2} + (h_k^z \hbar_{k+1/2}^z)^{-1} + (h_k^z \hbar_{k-1/2}^z)^{-1}] \, \delta P_{i,k}$$

$$= (R_i h_i^R \hbar_{i-1/2}^R)^{-1} R_{i-1/2} \, \delta P_{i-1,k} + (h_k^z \hbar_{k-1/2}^z)^{-1} \delta P_{i,k-1} + (h_k^z \hbar_{k+1/2}^z) \delta P_{i,k+1} - \tau^{-1} \tilde{D}_{i,k}, \tag{24}$$

where

$$\tilde{D}_{i,k} = - (R_i h_i^R)^{-1} R_{i-1/2} \tilde{U}_{i-1/2,k} + (h_k^z)^{-1} (\tilde{W}_{i,k+1/2} - \tilde{W}_{i,k-1/2}). \tag{25}$$

Analogous conditions can be found on the axis and at the melt–crystal interface. At the free surface of the melt, $\partial P = P^{n+1} - P^n$ is assumed to be known. We solve (19) together with (24) by the ICCG method (see above). Calculation of the final rate range according to (21)-(23) is not difficult. Knowing the components of the rate range, the vortex range ω^{n+1} in cylindrical coordinates can easily be defined. At Step IV, we determine the temperature range. The convective terms are approximated by a monotonic Samarskii scheme [14] or by a K–R scheme [15], as was also done at Step I. The five-point equation obtained is also solved by the ICCG method [16] (see above). Knowing $\theta_{i,k}^{n+1}$, we determine from the formula

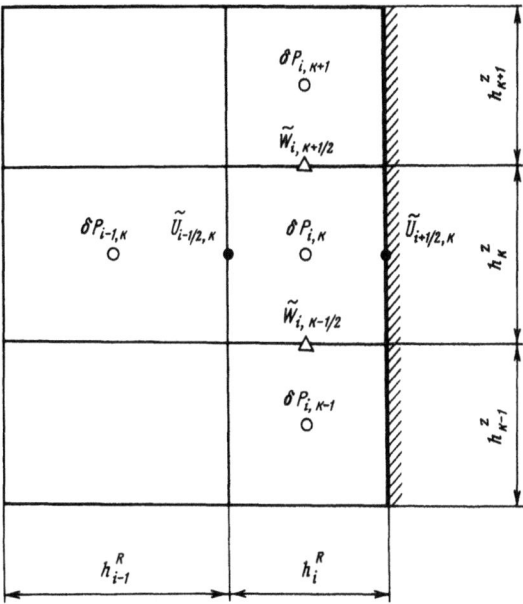

Fig. 3. Calculation grid for statement of boundary conditions
at the solid surface.

$$v_{i,k}^{n+1} = F(\theta_{i,k}^{n+1}), \tag{26}$$

after which we proceed to the calculation on the next time layer.

A calculation method that combines the merits of MS and methods that use variables of a vortex-function of current [5] is proposed for speeding up calculations in cases of planar and axially symmetric schemes. In this case, the equations of Steps II and III are transformed into the following:

$$\text{IIa.} \quad \omega^{n+1} = \tilde{\omega}, \quad B^\psi \Psi^{n+1} = \omega^{n+1}, \tag{27}$$

$$\text{IIIa.} \quad Y^{n+1} = A^\psi \Psi^{n+1} + A^Y \tilde{Y}, \quad \nabla P^{n+1} = \nabla P^n - \tau^{-1}(Y^{n+1} - \tilde{Y}), \tag{28}$$

where in the axially symmetric case

$$B^\Psi = \partial_R (R^{-1} \partial_R) + R^{-1} \partial_{zz}; \quad A^\Psi = R^{-1} (\partial_z, -\partial_R, 0)^T,$$

$$A^Y = \begin{pmatrix} 0 & 0 & 0 \\ 0 & 0 & 0 \\ 0 & 0 & 1 \end{pmatrix}. \tag{29}$$

The functions ω and Ψ are conveniently placed at sites of the type $(i + 1/2, k + 1/2)$ (Fig. 2). Then,

$$\omega_{i+1/2, k+1/2}^n = (U_{i+1/2,k+1}^n - U_{i+1/2,k}^n)/\hbar_{k+1/2}^z - (W_{i+1,k+1/2}^n - W_{i,k+1/2}^n)/\hbar_{i+1/2}^R;$$

$$U_{i+1/2,k}^n = R_{i+1/2}^{-1} (\Psi_{i+1/2,k+1/2}^n - \Psi_{i+1/2,k-1/2}^n)/h_k^z,$$

$$W_{i,k+1/2}^n = -R_i^{-1} (\Psi_{i+1/2,k+1/2}^n - \Psi_{i-1/2,k+1/2}^n)/h_i^R. \tag{30}$$

At Step IIa, we use the fact that the rate ranges $\tilde{Y}$ and Y^{n+1} have identical vortex characteristics at the internal points of the examined area. The value of $\tilde{\omega}$ is easily determined from (30). Then, solving the Poisson equation, we determine the current function range Ψ^{n+1} at the internal points of the area for the boundary condition $\Psi^{n+1}|_\Gamma = 0$. Here, we use the variable directions method with optimal selection of parameters as determined according to Vashpress [5]. Without a vector processor, the variable directions method is more effective than

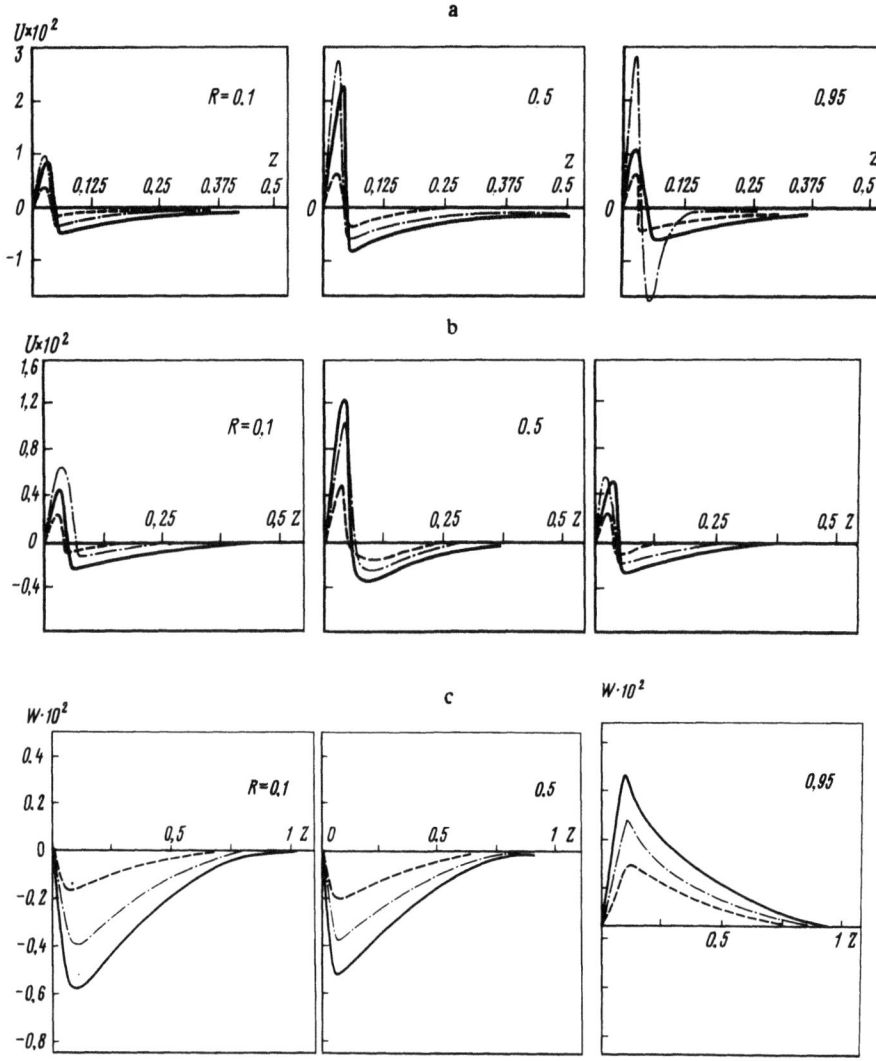

Fig. 4. Rates of radial [Al melt (a) and Al_2O_3 melt (b)] and axial [Al_2O_3 melt (c)] flows. 1-3) $t =$ 0.25, 0.5, and $0.75T_{sp}$, respectively.

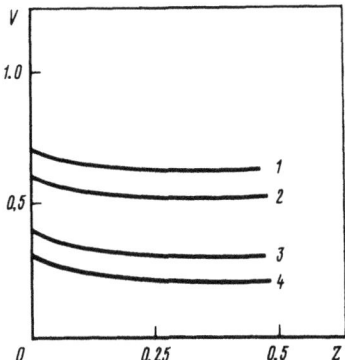

Fig. 5. Circular rate of melts. 1, 2) For Al_2O_3 and Al at $R = 0.95$; 3, 4) for Al_2O_3 and Al at $R = 0.5$. $t = 0.75T_{sp}$; for Al, Re = 1600, Pe = 80, and Fr = 8; for Al_2O_3, Re = 50, Pe = 1600, and Fr = 156.

ICCG. At Step IIIa, the rate and pressure gradient ranges, respectively, at the $(n + 1)$-th time layer can readily be determined by an implicit scheme. We note that the equations for Steps I, IV, and V remain unchanged.

The developed numerical methods comprise a program set written in Fortran-77 for an ES-1045 computer. Calculations were carried out for alloys of Al and Al_2O_3 at grid points 20×20, 30×40, and 40×60 which were condensed on the crucible wall and at the interface with the crystal. The crucible diameter and melt height were 3.2 cm. The spin time T_{sp}, which was determined from [1] for an instantaneous change of rotation rate, was 14 sec for Al and 2.5 sec for Al_2O_3. The crucible rotation rate was varied according to the law

$$N(t) = N_m t^{\tilde{\alpha}}/(a_0 + t^{\tilde{\alpha}}), \qquad (31)$$

where $N_m = 0.2$-0.5 rps, $a_0 = 0$-2, and $\tilde{\alpha} = 1$-3, which are similar to technical conditions. Figure 4 shows that the radial current is concentrated in a narrow layer that adjoins the interface with the crystal. The ascending flow along the axis is more evident at the crucible walls (Fig. 4b and c). Slippage of the cylindrical layers is illustrated in Fig. 5, from which the ratio of circular velocities of layers with different radii at the transitional portion where $t < T_{sp}$ is seen not to equal the ratio of the corresponding radii. The increased circular velocity at small Z values is explained by the effect of solid rotation adjoining the crystal surface. The difference in circular velocities at $Z = 0$ is related to the different crucible rotation rate at these time points. Comparison of the velocity pictures for Al and Al_2O_3 alloys indicates (Fig. 4) that the amplitude of radial and axial rates is substantially smaller and the transition to solid rotation is more rapid for the second melt with a high viscosity. The results obtained show that a high frequency of periodic acceleration—deceleration of crucible rotation is required in the case of a melt with high viscosity for melt mixing by the accelerated crucible rotation technique.

REFERENCES

1. E. O. Schulz-Dubois, "Accelerated crucible rotation: hydrodynamics and stirring effect," *J. Cryst. Growth*, **12**, No. 2, 81-87 (1972).
2. A. Horowitz, M. Goldstein, and Y. Horowitz, "Evaluation of the spin-up time during the accelerated crucible rotation technique," *J. Cryst. Growth*, **61**, No. 2, 317-322 (1983).
3. A. Horowitz, D. Gasit, J. Makovsky, and L. Ben-Dor, "Bridgman growth of Rb_2MnCl_4 via accelerated crucible rotation technique," *J. Cryst. Growth*, **61**, No. 2, 323-328 (1983).
4. P. Capper, J. J. G. Gosney, and C. L. Jones, "Application of the accelerated crucible rotation technique to the Bridgman growth of $Cd_xHg_{1-x}Te$: simulation and crystal growth," *J. Cryst. Growth*, **70**, No. 1/2, 356-364 (1984).
5. A. A. Samarskii, Yu. P. Popov, and O. S. Mazhorova, "Mathematical modeling," in: *Preparation of Single Crystals and Semiconducting Structures* [in Russian], Nauka, Moscow (1986).
6. B. G. Kuznetsov and G. G. Chernykh, "Numerical study of the behavior of a uniform spot in a liquid with ideally stratified density," Zh. Prikl. Mekh. Tekh. Fiz., No. 3, 120-126 (1973).
7. Yu. M. Lytkin and G. G. Chernykh, "On internal waves initiated by collapse of the mixing zone in a stratified liquid," in: *Mathematical Questions of Mechanics. Dynamics of Continuous Media* [in Russian], Nauka, Novosibirsk (1975), No. 22, pp. 116-132.
8. L. M. Simuni, "Numerical study of the blocking effect during circumvention of obstacles by a stratified liquid," *Izv. Akad. Nauk SSSR, Mekh. Zhidk. Gaza*, No. 4, 151-153 (1976).
9. O. M. Belotserkovskii, V. A. Gushchin, and V. V. Shchennikov, "Method of separation as applied to solution of problems on dynamics of a viscous noncompressed liquid," Zh. Vychisl. Mat. Mat. Fiz., **15**, No. 1, 197-207 (1975).
10. V. A. Gushchin, "Spatial circumvention of three-dimensional solids by a current of viscous liquid," Zh. Vychisl. Mat. Mat. Fiz., **16**, No. 2, 529-534 (1976).
11. S. O. Belotserkovskii and V. A. Gushchin, *Modeling Several Viscous Liquid Flows* [in Russian], Vychisl. Tsentr. Akad. Nauk SSSR, Moscow (1982).
12. O. M. Belotserkovskii, *Numerical Modeling in the Mechanics of Continuous Media* [in Russian], Nauka, Moscow (1984).
13. V. A. Gushchin and V. N. Kon'shin, *Numerical Modeling of Wavelike Liquid Motions* [in Russian], Vychisl. Tsentr. Akad. Nauk SSSR, Moscow (1985).
14. A. A. Samarskii, *Theory of Difference Schemes* [in Russian], Nauka, Moscow (1983).
15. S. G. Rubin and P. K. Khosla, "Navier–Stokes calculations by a strongly implicit method," *Comput. Fluids*, **9**, 163-180 (1981).
16. A. L. Goncharov, "Implementation of a method for partial LU-expansion of coupled gradients for solution of finite-difference equations in various patterns," Preprint M. V. Keldysh Inst. Appl. Math. Acad. Sci. USSR, No. 174, Moscow (1984).

ELECTROMAGNETIC METHODS FOR INFLUENCING
THE HYDRODYNAMICS AND HEAT AND MASS TRANSFER
DURING GROWTH OF LARGE SINGLE CRYSTALS
(REVIEW)

Yu. M. Gel'fgat, L. A. Gorbunov, and M. Z. Sorkin

Lately, electromagnetic methods for influencing a melt have found increasing application in the preparation of large single crystals. These methods are intended for active control of the hydrodynamics and heat and mass transfer both directly in the crucible and at the crystallization front [1]. These methods are based on a different kind of magnetohydrodynamic (MHD) effect which links the liquid with the electromagnetic field since the majority of semiconducting materials in the melt state have metallic conductivity $\delta \simeq 10^5\text{-}10^6 \ \Omega^{-1} \cdot \text{m}^{-1}$.

Applied magnetohydrodynamics possesses a broad spectrum of means for intensification of liquid movement and organization of currents of the required configuration in the melt and also for deceleration (suppression) of flows that exist under ordinary conditions, their increased stability, elimination of steady-state and nonsteady-state pulsations of rate and temperature, etc. [2]. Use of these MHD effects for directed control of heat and mass transfer during growth of single crystals can substantially raise their quality and increase the percent yield of available product [3-7].

The first separate studies on the effect of electromagnetic fields on preparation of single crystals were made rather long ago [8]. However, they were not extended into practice. Only recently has interest in the possibilities of the electromagnetic influence on a melt increased sharply due to the drastic increase in sizes and demands on the quality of single crystals and to the ever-broadening understanding of the role of hydrodynamic effects during growth of single crystals [3-7, 9]. However, in view of the fact that all of these studies were performed mainly by specialists in solid state physics, the MHD aspects of the problem have practically not been examined. Therefore, the possibilities and procedures for expedient application of the various MHD methods for affecting the preparation of single crystals have not been analyzed.

In the present work, a short review of efforts in this direction is given, criteria of the magnetic field effect on hydrodynamic and heat and mass transfer characteristics in the most widespread technologies for single crystal preparation (mainly applied to the Czochralski method) are analyzed, and the principal trouble spots in scientific—technical studies in this area are formulated.

Use of an electromagnetic influence is known to result in forced movement of liquid in a melt (influence of variable and traveling magnetic fields) or in weakening substantially the strength of steady-state and nonsteady-state convection currents occurring in it (influence of constant magnetic fields) and the heat and mass

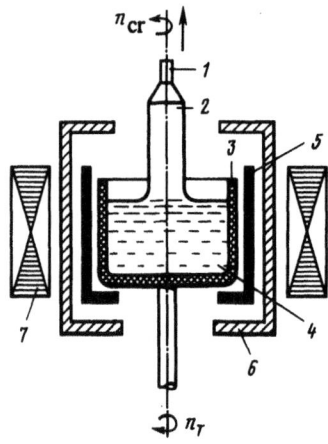

Fig. 1. Diagram of the Czochralski method
for growth of single crystals with electromagnetic influence on the melt. 1) Single-
crystalline seed; 2) grown single crystal; 3)
quartz crucible; 4) silicon melt; 5) graphite
heater; 6) hermetic growing chamber; 7)
electromagnetic influence inductor.

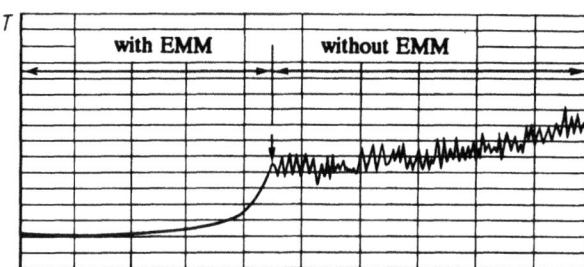

Fig. 2. Influence of electromagnetic mixing (EMM) on temperature oscillations in a melt [12].

transfer processes related to this movement [2]. One form or another of the influence on the melt (Fig. 1) can
be chosen depending on the actual technological problem on single crystal growth.

Thus, the effect of electromagnetic mixing (EMM) of a melt in the Czochralski method on the growth
conditions and quality of the ingot was studied in [10-12]. Azimuthal rotation of the melt in the crucible using
a rotating magnetic field inductor (analogously to the stator of an asynchronous electric motor) was found to
eliminate temperature oscillations in the melt and bands in the ingot related to them. It also leveled the distribution of dopant through the single crystal cross section. EMM of the melt in this case does not disturb the
single-crystallinity of the ingots produced by any growth procedures used in the experiments.

Parameters of the inductor and magnetic field were not introduced in these works. However, they do
indicate a dependence of the final results on the current to the inductor. In particular, small values of the
mixing effect (low melt rotation rate) do not affect the single crystal quality. Intensified azimuthal melt rotation
(to a maximum of ~100 rpm) simultaneous with practically complete elimination of temperature pulsations (Fig.
2) gives rise to ordering and disappearance of dopant bands in the ingot. The authors explain this effect by the
influence of forced melt mixing on the nonsteady-state natural convective currents up to their total suppression.

However, at the same time they emphasize that mixing in the subcrystal region (melt column) is apparently absent and is not reflected in the shape (slant) of the crystallization front.

The configurations of the mixing currents generated in the melt by the various modes and means of MHD influence can be exceedingly varied. This question is treated in most detail in [13-15]. Here it is important to emphasize that the nature of the mixing current trajectories with development of turbulent motion in the melt will apparently not substantially affect the final results of the single crystal growth process. This also occurs in studying metal purification by crystallization [16, 17]. Finally, electromagnetic criteria that determine the beginning of turbulent melt motion for each type of MHD influence and the relationships between the mixing currents and natural convective flows in the crucible are at present absent and require special investigations.

Conversely, the nature and shape of these motions in the bulk of the melt and especially in the subcrystal region for a stratified mixing regime will have a decisive role and should be created purposefully.

Many varied potential MHD mixing devices which in principle can be used in single crystal production techniques exist. Of these, the types which use electrical fields that flow through the heating elements of the equipment for creation of forced motion in the melt attract the most interest [13, 18-20]. These currents reach 1500-2000 Å in large aggregates, so that the combination in one package of melt heating and mixing is very alluring.

A similar type of mixing devices consisting of graphite resistance heaters divided into three sections and connected to commercial three-phase alternating current is studied in [1, 21]. Experiments were carried out on silicon. The rate of forced melt rotation around the crucible axis was 20-30 rpm. The effect of this forced liquid motion on the hydrodynamic melt structure was estimated indirectly from the axial and radial distribution of oxygen in the single crystal silicon which was grown.

A very substantial reduction of oxygen content in silicon ($c = (10-3) \cdot 10^{17}$ at./cm^3) with a sufficiently uniform distribution through the single crystal cross section was achieved in [21] at certain combinations of rates and directions of rotation of seed and crucible relative to the rate and direction of electromagnetic mixing. The explanation of the authors that this effect is due to the thick diffusional boundary layers formed at the crystallization front in certain EMM regimes and through which oxygen atoms pass with difficulty into the solid phase formation zone is clearly inadequate. Apparently, a hydrodynamic structure which, on the one hand, reduced washing of the crucible walls and consequently decreased entrainment of oxygen into the melt and, on the other, guaranteed feed to the crystallization front of silicon which was depleted in oxygen due to vaporization of the latter from the free surface, was formed in the melt in this situation.

Obviously, a similar hydrodynamic structure should depend strongly on the ratio of geometric sizes of the crystal and crucible, the melt volume, the magnitudes and directions of temperature gradients, etc. for otherwise equal conditions. It can be assumed that other values of the parameters indicated in [1] with approximately the same EMM conditions did not in this case allow such a significant reduction of oxygen in the single crystal, although the EMM influence on its content is also exhibited rather markedly here. The absence of data on the flow regime (stratified or turbulent), the structure of currents in the crucible, and the nature of melt motion in the subcrystal region prohibits making any final conclusions. Only the possible existence of a certain critical EMM regime can be noted by comparing results obtained during use of a sufficiently powerful external rotating field inductor [11, 12] and EMM using three-phase heaters [1, 21]. Temperature oscillations and nonsteady-state mass-transfer processes at the crystallization front which are related to them persist in the melt at mixing rates which are less than this critical value. The temperature fields in the melt are leveled, the nonsteady-state processes are greatly dissipated, and conditions of homogeneous solid phase growth in the subcrystal region are correspondingly improved when this hypothetical critical EMM regime is surpassed.

Apparently, the currents used today in the three-phase heater elements are insufficient for generation of the required EMM regimes, especially for large crucible sizes. We note that their increased effectiveness in the

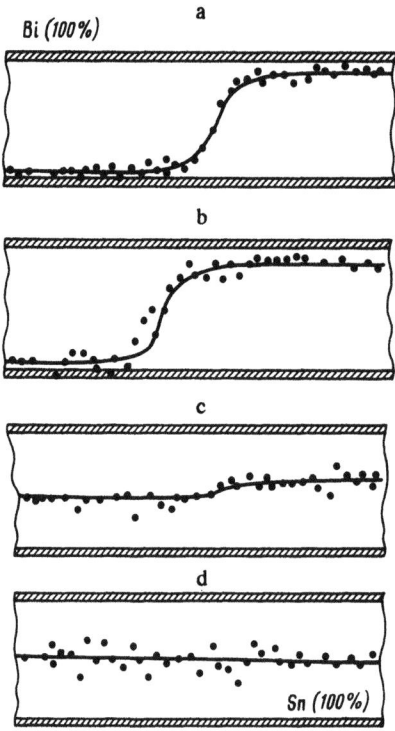

Fig. 3. Curves of the change in Bi–Sn system homogeneity along the ampul length under the influence of a rotating magnetic field on the melt [24]. Without the electromagnetic influence (EMI) at the initial time point ($t = 0$, $D = 0$) (a), without EMI ($t = 390$ min, $B = 0$) (b), with EMI ($t = 55$ min, $B = 0.3$ T) (c), and with EMI ($t = 130$ min, $B = 0.003$ T) (d).

case examined is possible by using special phase regulators which enable generation of the required heating regimes and the necessary electromagnetic field densities in the melt. Analogous results can be obtained by using feed sources with lowered or elevated working current frequency. In this regard, work [22] which analyzes the magnetic field of a three-phase heater as applied to Czochralski single crystal growth equipment is interesting.

With respect to external magnetic systems for EMM in the case examined, assorted literature on applied MHD can be used for their calculation and optimization. For example, the monograph [23] treats equipment with a traveling electromagnetic field, monograph [15] examines equipment with a pulsed electromagnetic field, etc. We also note that functions which determine the required intensity of electromagnetic influence on the melt for generation in it of a turbulent motion regime can be proposed for several MHD set-ups [24]. In particular, use of a rotating magnetic field inductor for determining the beginning of turbulent motion of a liquid in an ampul with a melt during growth of single crystals by directed crystallization is governed by the expression [24]

$$B_0 \gtrsim \left(\mathrm{Re}_{\mathrm{cr}}\, \nu^{3/2} / k R_a^3\, \omega_0 \right) \sqrt{\rho/\sigma}, \tag{1}$$

where B_0 is the amplitude of magnetic field induction; $\mathrm{Re}_{\mathrm{cr}}$ is the critical value of the Reynolds number; ν, ρ, and σ are the kinematic viscosity, density, and electric conductivity of the melt; R_a is the ampul radius; $\omega_0 = 2\pi f$ is the frequency of the current feeding the inductor; and $k = 4 \cdot 10^{-3}$ is an experimental constant.

We note that homogenization of multicomponent melts in the ampul occurs rather effectively upon attaining the turbulent regime indicated above. This is illustrated well by the experimental data for electromagnetic mixing of the Bi–Sn melt shown in Fig. 3.

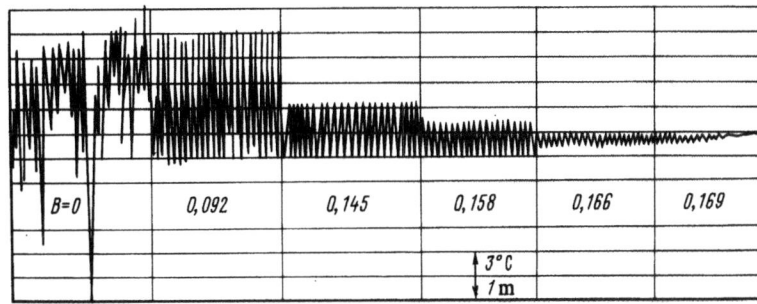

Fig. 4. Influence of a constant magnetic field on temperature oscillations in a melt
[34].

Concluding our review of efforts devoted to studying effects of alternating traveling magnetic fields on single crystal growth, it should be emphasized that all experiments using EMM have an exploratory nature. Analytical or phenomenological analysis of the mutual effect of intensity and direction of melt rotation and its hydrodynamic structure in the crucible, for example, on the action of traditional factors has not been attempted.

Works devoted to studying the effect of a constant magnetic field on growth of single crystals from a melt are somewhat more numerous. We will examine existing results as applied only to the Czochralski method [1, 31-44] without dwelling on similar studies on less widespread methods for single crystal preparation [25-30].

The majority of studies concern questions related to growth of single crystal silicon. Transverse [1, 31-34] and longitudinal [35-44] (relative to the axis of the single crystal) constant magnetic fields are used in the experiments, with identical (in principle) qualitative results on the influence of MHD on the growth process and ingot quality. In particular, practically complete suppression in the melt of temperature and rate oscillations, a substantially lower oxygen content with its uniform distribution through the ingot cross section, and a decrease of dislocation density and intensity of dopant bands in the structure of the prepared material were observed during preparation of single crystal silicon under the influence of a constant field.

These effects occurred in practically all cases even in spite of the fact that the specifics of the equipment for growth of single crystals cause formation of clearly asymmetric thermal and hydrodynamic fields in the case of the transverse field. They also give rise to an axially symmetric distribution of thermal and hydraulic characteristics in the melt for the longitudinal field.

Thus, it should be stressed that the absolute values of magnetic field induction (~0.15-0.2 T) at which the MHD influence is effective are approximately identical in both cases. Therefore, use of a constant longitudinal field is preferred, due to the ease of its use in actual magnetic systems, its smaller overall size, and the absence of additional electromagnetic forces that affect the heater elements.

These positive results are obviously attained based on known physical effects that arise in the conducting liquid under the influence of a constant magnetic field. These effects include deceleration of convective currents due to electromagnetic forces $\bar{f} = \bar{j} \times B$ that arise in the moving liquid, increase in flow stability, suppression of turbulent oscillations of rate and temperature, leveling of rate gradients along the field vector, etc. [2]. These effects also change heat and mass transfer conditions in the melt. Figure 4 shows a typical example [34] of the intensity of temperature oscillations near the crystallization front as a function of field. We note that a similar picture of very substantial suppression (up to complete disappearance) of temperature oscillations occurred in all the studies carried out. However, induction fields which strictly regulate the sinusoidal temperature oscillations at certain intermediate values were identified only in [34]. The reason for the appearance of this phenomenon requires further study.

We also note that the distribution of temperatures and the configurations of convective currents in the melt-containing crucible, as observed in [38], became symmetric under the influence of a longitudinal magnetic field. This effect is caused by a different interaction in the liquid current lines which are parallel and perpen-

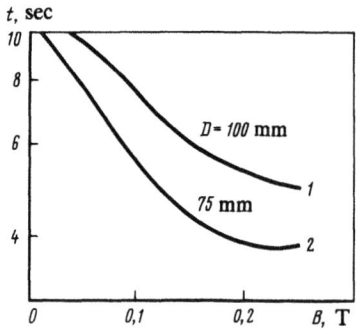

Fig. 5. Influence of a constant magnetic field on the change of oxygen concentration in silicon single crystals [1]. 1, 2) $D = 100$ and 75 mm, respectively.

dicular to it. As a result of this, the initially asymmetric convective currents, at a sufficiently large induction field strength and simultaneous with a reduction of their intensity, are arranged symmetrically relative to the crucible center in the form of a system of contours stretched out along the lines of the force field. In the end, the temperature distribution in the melt is determined only by its thermal conductivity, as would occur for a solid with comparable thermal conductivity. This result has great significance since growth of symmetric single crystals with crucible rotation or a seed in a longitudinal magnetic field can be proposed.

The reduction of oxygen content in the silicon single crystals ($C \approx (10\text{-}2) \cdot 10^{17}$ at./cm^3) observed in the experiments was shown to be caused by a change of mass transfer from the crucible walls. This change occurred due to a weakening of convective diffusion and a decrease, under the influence of the field, of the absolute values and gradients of the convective current rates. Generation of hydrodynamic structure in the melt occurs simultaneously. This structure results in currents in the crucible at the melt surface that are directed away from the walls and toward the growing single crystal. This condition ensures the possibility of vaporizing oxygen that enters the melt (in the form of the monoxide SiO) from its free surface. It also enables supplying the crystallization front with a melt having a substantially lower content of this deleterious impurity.

Figure 5 gives as an example the data of [1] on the influence of a constant field on the oxygen content in silicon single crystals of various diameters at otherwise equal process parameters. The path of the function indicates that the intensity of magnetic field influence has an asymptotic nature. The increase in the effect due to a reduction in oxygen content is insignificant above a certain field. The simultaneously different shapes of the curves in Fig. 5 confirms the conclusion, apparently first made for a process in a magnetic field in [33], that the oxygen content is a function of the ratio of the crystal and crucible diameters. This indirectly attests to the mentioned role of the direction and intensity of the resulting currents of forced and natural convection at the melt surface.

A substantially more uniform distribution of oxygen content through the cross section and length of the ingot is also achieved by the influence on the melt of a constant magnetic field at certain crystal and crucible rotation rates [1, 33, 35, 40]. This result is related to melt mixing conditions directly at the crystallization front and throughout the crucible volume. These conditions are ensured with suppression by the field of the undirected natural convection in the melt and by the combination, which is necessary for each actual case, of forced convection intensities from the crystal and crucible rotation.

We note that for the case of complete absence of crystal and crucible rotation, the reduction of oxygen content in the ingot was different through its cross section and length. A characteristic band structure, which was ascribed to asymmetry of the equipment thermal field, was found along the crystal axis [33, 34, 40]. Moreover, the reduction of oxygen content in the single crystal [less than $(3\text{-}4) \cdot 10^{17}$ at./cm^3] is characterized by the absence in it of specific defects independently of the quality of the oxygen distribution through the ingot cross section [1]. The authors of these studies did not even conduct qualitative estimates of the field influence on the hydrodynamics and heat and mass transfer in the melt during growth of single crystals. The observed effects were partially explained as being due to the suppression by the magnetic field of nonsteady-state convection currents due to elevation of the effective melt viscosity, taking into account the generation in it of a so-called additional magnetic viscosity. Attempts were made only to rectify the existence of elevated flow stability in the crucible according to the Rayleigh number for the cases $B = 0$ and $B \neq 0$, respectively [33, 34].

An analysis of the relationship between the influence of the magnetic field, the natural and forced convections, processes at the crystallization front, etc. is obviously an exceedingly complicated problem. Nevertheless, parametric evaluation of the field influence on several characteristic effects which determine the specifics of crystal growth can be attempted by separating the individual process steps.

We now estimate the degree of MHD influence on the hydrodynamic and heat and mass transfer processes in the melt during growth of silicon single crystals. This estimate is based on typical numerical data obtained using constant magnetic fields. To this end, we use the following thermophysical and geometric characteristics of the process: melt conductivity $\sigma = 1.28 \cdot 10^6 \ \Omega^{-1} \cdot m^{-1}$, density $\rho = 2.51 \cdot 10^3$ kg/m^3, coefficient of kinematic viscosity $\nu = 2.5 \cdot 10^{-7}$ m^2/sec, thermal conductivity of liquid silicon $\lambda = 50.24$ W/(m $\cdot$ deg), heat capacity $C_p = 912$ J/(kg $\cdot$ deg), coefficient of melt thermal conductivity $a = 2.2 \cdot 10^{-5}$ m^2/sec, coefficient of expansion $\beta = 2 \cdot 10^{-4}$ deg^{-1}, oxygen diffusion coefficient $D = 10^{-8}$ m^2/sec, crucible radius $R_c = 0.14$ m, melt height $h = 0.1$ m, single crystal radius $R_{cr} = 0.05$ m, temperature gradient $\Delta T \approx 40°C$, gravitational acceleration $g = 9.8$ m/sec^2, and external magnetic field induction $B = 0.2$ T.

Without a magnetic field, the natural convection is characterized by the Grashof number Gr $= \beta g \Delta T R_c^3 / \nu^2 = 4.3 \cdot 10^9$. Correspondingly, the Reynolds number for natural convection will be equal to Re$_n \approx$ Gr $= R_c V_n / \nu \approx 6.6 \cdot 10^4$ (V_n is the characteristic rate of natural convection). Forced convection is determined by the crystal rotation rate. For it, the Reynolds number is Re$_f = R_{cr} V_f / \nu$, where $V_f = V_{cr} = 2\pi R_{cr} n_{cr}$ (n_{cr} is the number of crystal revolutions per unit time). Characteristic values of Re$_f$ at n_{cr} = 5-30 rpm are $0.5 \cdot 10^4 \leq$ Re$_f \leq 3.24 \cdot 10^4$. We note that since the relative role of natural (thermal) and forced convection is determined by the parameter Gr/Re$_f^2$ [9], forced convection (Gr/Re$_f^2 \lesssim 1$) predominates in the melt in the absence of a magnetic field only for $n_{cr} > 30$ rpm. Natural convection currents in the melt prevail at smaller crystal rotation rates.

The ratio of convective heat transfer to molecular heat transfer at $B = 0$ is determined by the Peclet number Pe = PrRe. Here Pr $= \nu/a \approx 10^{-2}$ is the Prandtl number. For natural convection Pe$_n = 6.6 \cdot 10^2$, for forced convection at various n_{cr} the Pe$_f$ values are $50 \leq$ Pe$_f \leq 314$. The ratio of convective mass transfer to diffusional mass transfer is characterized by the corresponding diffusional Peclet number Pe$_d$ = ScRe, where Sc $= \nu/D \approx 25$ is the Schmidt number. For natural convection Pe$_{d,n} = 1.65 \cdot 10^6$, for forced convection it lies in the range $1.3 \cdot 10^5 \leq$ Pe$_{d,f} \leq 7.8 \cdot 10^5$.

The numerical estimates carried out show that heat transfer in the melt and mass transfer of oxygen in it in the case examined occur mainly due to a convective mechanism. Molecular processes are important only in narrow layers near the wall with a thickness on the order of $\delta_c/R_c \approx$ Pe$^{-1/2}$ and $\delta_d/R_c \approx$ Pe$_d^{-1/2}$. In this case, if the thermal boundary layer at the crucible wall δ_c is larger than the dynamic boundary layer $\delta/R_c \approx$ Re$^{-1/2}$, then the diffusional boundary layer δ_d is significantly less than the latter so that δ_d lies completely in the dynamic boundary layer. Thus, the convective currents in the melt at $B = 0$ completely determine the heat and mass transfer throughout the whole volume with the exception of the narrow zones near the crucible walls.

In agreement with [45], the characteristic rate of convection currents at Re $\gg$ 1 without a field is determined by the ratio of the Archimedian and electromagnetic forces as $u \approx (\beta g l \Delta T)^{1/2}$. With a field at Ha $\gg$ 1, it is determined by the ratio of Archimedian and electromagnetic forces as $u_f = \beta g \rho \Delta T / \sigma B^2$ (Ha $= B R_c (\sigma/\rho\nu)^{1/2}$

is the Hartmann number). Consequently, the reduction of intensity of convective currents under the field influence can be evaluated as

$$u/u_f = \text{Ha}^2/\text{Gr}^{1/2} = \text{Ha}^2/\text{Re} = \sigma B^2 R_c/2\rho\nu = N, \quad (2)$$

where N is a dimensionless parameter of MHD interaction. For a silicon melt with the parameters specified above, Ha = 1260. The MHD interaction parameter for natural convection $N = 24$ and has the values $50 \leq N \leq 304$ for forced convection depending on n_{cr}.

The change in the ratios of convective and molecular heat and mass transfer with a magnetic field will be determined by the parameters Pe/N and Pe$_d$/N. For natural convection, they are Pe$_n$/N = 27.5 and Pe$_{d,n}$/$N \approx 7 \cdot 10^4$. They depend on the n_{cr} value for forced convection.

Comparison of the parametric estimates obtained clearly indicates an exceedingly substantial reduction of convection current intensity and, consequently, of heat and mass transfer in the melt, under the influence of a field. Using the B, ΔT, and R_c values in the experiments, one can ensure practically complete absence of natural convection and control of melt hydrodynamics by purposefully organized forced convection in conjunction with electromagnetic influence. However, if realization of a molecular mechanism of heat transfer upon imposition on the melt of a magnetic field is in principle possible at $n_{cr} < 15$ rpm (Pe$_f$/$N \lesssim 1$), then the process is determined by convective transfer (Pe$_{d,f}$/$N \approx 430$) for oxygen mass transfer even at minimal values of $n_{cr} = 5$ rpm.

We now examine further the field effect on thermocapillary convection in the melt. Until very recently, these convective currents were not at all considered in the Czochralski process since they were thought to be negligibly small [46]. However, estimates carried out in [9] showed that the hydrodynamic effects which are related to the surface tension gradient on the free surface of the melt can in a number of cases be comparable to the examined convection mechanisms and lead both to a change of the flow structures in the volume and to appearance of oscillatory components. This question has also not been studied sufficiently for the case without magnetic field influence on the melt. Nevertheless, an evaluation of the field effect on the thermocapillary flows in the Czochralski process is absolutely of interest.

Following the example of [9], we will evaluate in this case the scale of the rate, as for flows like the boundary layer

$$U_\alpha = [(\partial\alpha/\partial T)^2 \Delta T^2 \nu/\rho^2 \nu^2 D]^{1/3}. \quad (3)$$

This expression for the scale of the relative rate transforms into

$$\text{Re}_\alpha = \frac{L}{\nu} U_\alpha = \frac{1}{A}\left(\frac{\text{Ma}}{\text{Pr}}\right)^{2/3}. \quad (4)$$

Here α is the coefficient of surface tension, ΔT is the characteristic temperature drop, L and D are two different size characteristics (along the vertical and horizontal axes), $A = D/L$, and Ma = $(\partial\alpha/\partial T)\Delta TD/\rho\nu a$ is the Marangoni number.

Thus, the characteristic thermal and diffusional Peclet number for thermocapillary convection are determined as

$$\text{Pe}_{T,\alpha} = \frac{1}{A}\left(\frac{\text{Ma}}{\text{Pr}}\right)^{2/3}\text{Pr} = \frac{1}{A}\text{Ma}^{2/3}\text{Pr}^{1/3}, \quad (5)$$

$$\text{Pe}_{d,\alpha} = \frac{1}{A}\left(\frac{\text{Ma}}{\text{Pr}}\right)^{2/3}\text{Sc}. \quad (6)$$

For evaluation of the magnetic field effect on thermocapillary convection, we set the influence of viscous tension in the surface layer of the moving liquid and the tension of surface forces equal to the influence of the electromagnetic forces. From the condition of their equality, we obtain for the scale of the rate

$$U_B = \left[\left(\frac{\partial \alpha}{\partial T}\right)\Delta T/\rho \nu L\right]\sqrt{\nu \rho/\sigma B^2}. \tag{7}$$

In this case

$$\mathrm{Re}_{f,\alpha} = \frac{L}{\nu} U_B = \frac{\mathrm{Ma}}{\mathrm{Pr}\,\mathrm{Ha}}\frac{1}{A^2}. \tag{8}$$

The following can be taken as a condition for the initiation of field influence on the thermocapillary convection process:

$$\mathrm{Re}_\alpha/\mathrm{Re}_{f,\alpha} = A(\mathrm{Ma}/\mathrm{Pr})^{-1/3}\mathrm{Ha} > 1. \tag{9}$$

When the electromagnetic forces exert the predominant effect, the Peclet number can be written as

$$\mathrm{Re}^B_{T,\alpha} = \frac{1}{A^2}\frac{\mathrm{Ma}}{\mathrm{Ha}}, \tag{10}$$

$$\mathrm{Pe}^B_{d,\alpha} = \frac{1}{A^2}\frac{\mathrm{Ma}}{\mathrm{Ha}}\frac{\mathrm{Sc}}{\mathrm{Pr}}. \tag{11}$$

Using the given physicochemical constants of a silicon melt and also the characteristic sizes and taking $\partial \alpha/\partial T = -0.51 \cdot 10^{-3}$ N/(m · deg), we obtain that $\mathrm{Re}_\alpha = 4.94 \cdot 10^4$. This corresponds to the scale of the rate of thermogravitational convection, i.e., the intensity of currents caused by the action of the thermocapillary effect is the same as for the currents that arise due to the action of bouyancy forces.

It is easy to determine from the condition $A(\mathrm{Ma}/\mathrm{Pr})^{-1/3}\mathrm{Ha} = 1$ that the influence of the field should be felt at $B = 0.04$ T. The ratio $\mathrm{Re}_\alpha/\mathrm{Re}_{B,\alpha} = 4.8$ at $B = 0.2$.

However, in this case $\mathrm{Pe}_{T,\alpha} \gg 1$ and $\mathrm{Pe}^B_{d,\alpha} \gg 1$, so that raising heat and mass transfer to the molecular level only is practically impossible with the use of a magnetic field and in the presence of thermocapillary convection.

The estimates obtained were carried out for a laminar flow regime although it is apparent that a turbulent regime is important in the melt at $\mathrm{Re} > 2 \cdot 10^3$. Calculation of the field effect in this case is naturally much more complicated. Therefore, we will limit ourselves here to only a few qualitative remarks on possible consequences of the field effect on turbulence.

In agreement with [47], a thin viscous sublayer, the thickness of which is approximately equal to $\delta/R_T \approx \alpha/\mathrm{Re}^{1/2}$ (where $\alpha \approx 10$), can be considered to exist on the crucible walls in the turbulent regime when $B = 0$. Practically complete melt mixing occurs at the core of the current due to convection. The presence of turbulent oscillations of the rate leads to turbulent Prandtl and Schmidt numbers which are approximately equal to unity so that a similar distribution of rates and temperature and concentration fields exists at the current core.

A feature of the field effect on the heat and mass transfer in this case is the complete or partial extinction of turbulent disturbances in the melt and a transition of flows to a laminar regime [2]. Such a regime is characterized by the absence of turbulent rate oscillations at $\mathrm{Re} > \mathrm{Re}_{cr}$ and the generation near the solid boundaries of characteristic layers near the walls. Thus, layers $\sim\mathrm{Ha}^{-1}$ form on the walls which are perpendicular to the field induction vector. Layers $\sim\mathrm{Ha}^{-1/2}$ form on the walls which are parallel to the vector. The beginning of complete current lamination can be defined roughly by the parameter $(\mathrm{Re}/\mathrm{Ha})_{cr} \lesssim 130$. This is done during flow of liquid metal in pipes [2]. Finally, such an estimate is rather arbitrary and determines only the order of magnitude of the magnetic field induction necessary for attainment of a laminar regime. However, it agrees

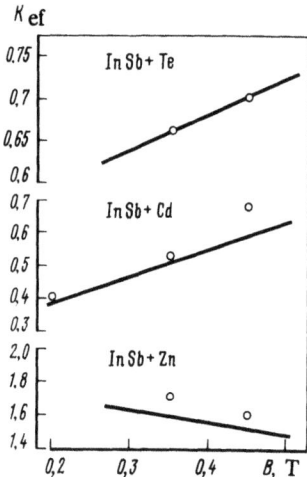

Fig. 6. Calculated and experimental data for the influence of a constant field on the effective tellurium, zinc, and cadmium distribution coefficients in single crystals of indium antimonide.

sufficiently well with the results obtained in [34, 35] if the field and flow parameters at which temperature oscillations in the currents disappear are taken as the beginning of the laminar regime.

The ratio of convective heat and mass transfer to molecular heat and mass transfer in the case of turbulent flow is determined by expressions of the form $Pe_T \approx 10^3$ and $Pe_d \approx 10^2$, i.e., it does not depend on the Re number of current. However, convective heat and mass transfer also prevails in this case, such that the given approximate estimates remain valid, changing only slightly due to suppression of the molar transfer mechanism in a magnetic field.

We now examine further the effect of constant magnetic field on the redistribution of dopants at the crystal–melt interface during growth of doped single crystals, i.e., on the value of the effective distribution coefficient K_{ef} [52]. The value K_{ef} is known to be determined by the Barton–Prim–Schlichter formula:

$$K_{ef} = k_0/[k_0 + (1 - k_0)e^{-V_L\delta\,d/D}],\tag{12}$$

where k_0 is the equilibrium distribution coefficient, V_L is the extrusion rate of the crystal, δ_d is the thickness of the boundary diffusion layer at the crystallization front, and D is the dopant diffusion coefficient in the melt. Since the magnetic field influences the hydrodynamic characteristics of the melt, it is reasonable to assume that the thickness of the diffusion layer at the interface also changes and, consequently, so does K_{ef}.

As a criterion for estimating the change of δ_d under the influence of a field, we assume that forced convection in the melt due to crystal rotation prevails over natural and thermocapillary convections and that the action of the electromagnetic forces on the flow is the determining factor ($Ha^2/Re > 1$). Using the following expression for the thickness of the boundary diffusion layer

$$\delta_d/L \sim 1/Re^{1/2}Sc^{1/3}\tag{13}$$

and assigning the subscript B to values which relate to the case of field influence with induction B, we have

$$\delta_{d,B}/L \sim 1/Re_B^{1/2}Sc^{1/3}.\tag{14}$$

Here L is the characteristic size and $Re = VL/\nu$ and $Re_B = V_BL/\nu$ are the Reynolds numbers at $B = 0$ and $B \neq 0$, respectively. Considering (2), for the ratio of rates V/V_B and (13) and (14), we have

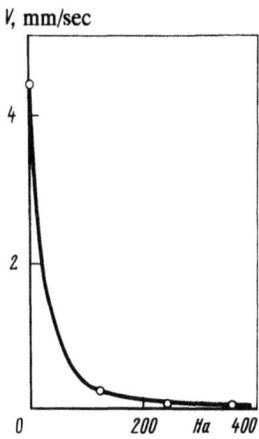

Fig. 7. Calculated change of horizontal rate component in a silicon melt in the Czochralski process with a longitudinal magnetic field.

$$\delta_{d,B} \approx \delta_d(\mathrm{Ha}/\mathrm{Re}^{1/2}).\tag{15}$$

Substituting (15) into (12), we obtain K_{ef} as a function of the magnetic field strength.

Figure 6 shows experimental points and calculated curves for $K_{ef} = f(B)$ for tellurium, zinc, and cadmiun in single crystals of indium antimonide prepared by the Czochralski method in a transverse magnetic field [52]. The agreement of the calculated and experimental data here is quite satisfactory. The path of the function $K_{ef} = f(B)$ can be explained by an increase of B in the melt which substantially reduces the intensity of the mixing currents. This leads to a growth of the δ-layer thickness, in agreement with (13). An increase of dopant enrichment at the crystallization front occurs with an increase of the δ-layer for $k_0 < 1$. For $k_0 > 1$, it is depleted in comparison to crystal growth without a field influence.

Results of experiments and calculations show that the K_{ef} value can be controlled using a magnetic field. This occurs without changing such process parameters as rate of rotation and crystal extrusion. The K_{ef} regulation range can be sufficiently broad.

Thus, the electromagnetic influence can control the intensity of convective heat and mass transfer in a melt over rather significant limits which are suitable for practical use. An *a priori* vague combination of magnetic field strength with crystal and crucible rotation procedures is required for attainment of uniform dopant distribution and elimination of micro- and macrodefects in the single crystals. Thus, reduction of the oxygen content in silicon single crystals, besides the indicated effects, must involve a resulting hydrodynamic structure such that the decreased entry of oxygen into the melt occurs simultaneously with its unhindered exit from the free surface. If it is assumed that entry and exit of oxygen from the surface occurs through a thin boundary layer near the walls and the fluxional diffusion layer on the melt free surface and that their complete mixing occurs in the central zone, then examination of the oxygen balance in the melt suggests that its concentration in the main crucible volume is determined only by the ratio of the crucible surface area which contacts the melt to the melt free surface. A decreased oxygen content in this case can be obtained due to supply to the growing single crystal of a melt surface layer that is depleted in oxygen with corresponding coordination of melt movement in the crucible and in the subcrystal region [48, 49].

Results of existing studies confirm that MHD methods for influencing a melt are effective for controlling hydrodynamics and heat and mass transfer directly both in the crucible and at the crystallization front of the growing single crystal. The initial task in solving this problem is establishment of the relationship between growth conditions and hydrodynamic and heat and mass transfer characteristics of the process, which are needed for the preparation of high-quality material, and knowing the parameters of the electromagnetic influence. Investigations should be directed in this direction.

Experiments on an industrial scale are encumbered by their complexity and high cost. Therefore, pilot experiments must be preceded by model studies on low-melting simulated alloys where the technique and method for conducting the studies are substantially simpler. Numerical experiments, which in a number of cases permit different variants of MHD influence to be carried out without costly expenditures, take on a meaning which is no less important.

In this scenario, recently developed numerical methods for calculating the growth of single crystals without an electromagnetic influence on the melt [50, 51] and the first efforts at taking into account the effect on the process of an axial constant magnetic field [36, 39] are attracting much interest. In [37, 39], a very strong field effect on the hydrodynamics and heat and mass transfer in a silicon melt (Fig. 7) is conclusively demonstrated. In [36], the nonexistence of a contribution to the energy balance of the Czochralski process of Joule dissipation from electrical currents induced by the field in the melt was proved rather convincingly. The numerical method developed for modeling the whole process with a field defines an expedient combination of crystal and crucible rotation procedures in order to attain the required results.

Studies of the MHD influence on a melt in the Czochralski process are not limited only to preparation of silicon single crystals. The first studies of effects that arise during growth of single crystals of gallium arsenide [41-43] and indium antimonide [3, 52] using pulsed magnetic fields [44] and others have appeared. Without going into the details of their analysis, we point out only that in this case completely unexpected, new effects occurred along with results similar to those described (suppression of temperature oscillations in the melt, elimination of striations in the structure, uniform distribution of dopant and background impurities, etc.). For example, it was observed in [42] that gallium arsenide single crystals grown in a transverse constant magnetic field quite distinctly changed their electric resistivity, going from an insulator with 10^8 $\Omega \cdot$ cm resistance to a semiconductor with 10 $\Omega \cdot$ cm resistance. This totally unexpected phenomenon was explained by a significant decrease of the concentration and level of defects in the single crystal grown in the field.

Electromagnetic methods for influencing melts during preparation of large single crystals have also been used as a means for modelling conditions typical of outer-space technology under weightless conditions [24, 54, 55]. The first data obtained in these experiments have already given encouraging results [24, 54].

Obviously, studies in this direction should be continued. Thus, considering the broad possibilities of MHD methods for controlling hydrodynamic and heat and mass transfer processes relative to their simple implementation in operating equipment and the ease of regulating the intensity of the MHD influence, one can expect that the use of these methods in preparing single crystals can solve one of the important problems of current materials science.

REFERENCES

1.   G. Fiegl, "Recent advances and future directions in CZ–silicon crystal growth technology," *Solid State Technol.*, **26**, No. 8, 121-131 (1983).

2.   Yu. M. Gel'fgat, O. A. Lielausis, and É. V. Shcherbinin, *Liquid Metal under the Influence of Electromagnetic Forces* [in Russian], Zinatne, Riga (1976); Express-inform., Tsentr. Nauchno-Issled. Inst. Tsvetn. Met. (1983), Scr. V, No. 16, 16.1; (1982), Ser. V, No. 48, V-48.1; (1981), Ser. V, No. 4, V-4.1, V-10.3; (1980), Ser. V, No. 46, V-46.1.

4.   "Growth of silicon crystals in a magnetic field," Byull. Inostr. Nauchno-Tekh. Inform., TASS (1984), No. 20 (2161); *Chem. Eng.*, **91**, No. 5, 11-12 (1984).

5.   C. Cohen, "Magnetic field breeds skylab-like semiconductors," *Electronics*, July 3, 83-84 (1983).

6.   C. Cohen, "Magnetic field breeds defectless crystals," *Élektronika*, No. 15, 18-19 (1980).

7.   R. B. Swaroop, "Advances in silicon technology for semiconductor industry," *Solid State Technol.*, No. 6, 111-114 (1983).

8.   R. Laudise and R. Parker, *Growth of Single Crystals*, Prentice-Hall, Englewood Cliffs, New Jersey (1970).

9.   S. Ostrach, "Fluid mechanics in crystal growth. The 1982 Freeman scholar lecture," *Teor. Osn. Inzh. Rasch.*, **105**, No. 1, 89-107 (1983).

10. K. F. Hulme and J. B. Mullin, "Advances in crystal growth for the semiconductor industry," *Solid State Electron.*, 5, 211-247 (1962).

11. Yu. M. Shashkov, N. Ya. Shushlebina, and B. M. Kanevskii, "Influence of magnetic mixing on dopant behavior during silicon single crystal growth by the Czochralski method," *Nauchn. Tr. Gos. Nauchno-Issled. Proektn. Inst. Redkomet. Promst.*, 33, 10-12 (1971).

12. Yu. M. Shashkov and N. Ya. Shushlebina, "On the influence of electromagnetic melt mixing on silicon single crystal growth," *Fiz. Khim. Obrab. Mater.*, No. 1, 34-36 (1972).

13. V. N. Vigdorovich and A. S. Zherebovich, *Methods and Equipment for Melt Mixing in Purification of Metals and Semiconducting Materials by Crystallization* [in Russian], Tsvetmetinformatsiya, Moscow (1969).

14. V. N. Vigdorovich and F. V. Markov, "Use of electromagnetic mixing for intensification of purification by crystallization and growth of single crystals," *Nauchn. Tr. Probl. Mikroelektron., Ser. Khim.*, No. 19, 172-182 (1974).

15. L. L. Tir and M. Ya. Stolov, *Electromagnetic Equipment for Control of a Circulating Melt in Electric Furnaces* [in Russian], Metallurgiya, Moscow (1975).

16. V. N. Vigdorovich, T. I. Darvoid, N. A. Nordanskaya, and O. Yu. Mamaev, "Study of silver dopant distribution during thallium purification by crystallization," *Zh. Prikl. Khim.*, 35, No. 10, 2165-2170 (1962).

17. R. V. Ivanova and A. A. Bel'skii, "Selection of the conditions for electromagnetic stirring of a melt during zone refining of metals," *Izv. Akad. Nauk SSSR, Metally*, No. 5, 93-95 (1968).

18. K. M. Rozin, Inventor's Certificate No. 167305 (USSR), *Byull. Izobret.*, No. 18 (1965).

19. K. M. Rozin, Inventor's Certificate No. 165311 (USSR), *Byull. Izobret.*, No. 18 (1964).

20. Patent No. 49-49307 (Japan) (1974).

21. K. Hoshikawa, H. Kohda, H. Hirata, and H. Nakanishi, "Low oxygen content Czochralski silicon crystal growth," *Jpn. J. Appl. Phys.*, 19, No. 1, L33-L36 (1980).

22. V. F. Keskyula and G. N. Petrov, "Magnetic field of a three-phase heater," *Tr. Tallin. Politekh. Inst.*, No. 539, 37-46 (1982).

23. Yu. K. Krumin', *Principles of the Theory and Calculation of Equipment with a Traveling Magnetic Field* [in Russian], Zinatne, Riga (1983).

24. Yu. M. Gel'fat and L. A. Gorbunov, "On homogenization of melts of multicomponent systems under weightless conditions," *Izv. Akad. Nauk SSSR, Ser. Fiz.*, 49, No. 4, 667-672 (1985).

25. G. F. Vaiderov, "Equipment for growth of semiconducting single crystals in a magnetic field," *Prib. Tekh. Éksp.*, No. 2, 221-223 (1972).

26. G. F. Vaiderov, V. V. Krygin, and G. M. Mokrousov, "Elimination of dopant bands in Ge and GaAs crystals by a magnetic field," *Izv. Akad. Nauk SSSR, Neorg. Mater.*, 10, No. 2, 231-233 (1974).

27. S. Sen, R. A. Lefever, and W. R. Wilcox, "Influence of magnetic field on vertical Bridgman-Stockbarger growth of $In_xGa_{1-x}Sb$," *J. Cryst. Growth*, 43, No. 4, 526-530 (1978).

28. A. S. Krylov and V. N. Romanenko, "Control of the composition of crystals grown in electric and magnetic fields," *Izv. Akad. Nauk SSSR, Neorg. Mater.*, 14, No. 1, 9-12 (1978).

29. N. De Leon, I. Guldberg, and I. Salling, "Growth of homogeneous high resistivity FZ silicon crystals under magnetic field bias," *J. Cryst. Growth*, 55, No. 3, 406-408 (1981).

30. H. Kimura, M. F. Harvey, D. I. O'Connor, et al., "Magnetic field effects on float-zone Si crystal growth," *J. Cryst. Growth*, 62, No. 4, 523-531 (1983).

31. A. E. Witt, C. J. Herman, and H. C. Gatos, "Czochralski-type crystal growth in transverse magnetic fields," *J. Mater. Sci.*, No. 5, 822-824 (1970).

32. Patent No. 56-45889 (Japan). "Method for growth of semiconductor single crystals," Patent No. 56-104791 (Japan) (1981).

33. T. Suzuki, N. Isawa, Y. Okubo, and K. Hoshi, "CZ-silicon crystals grown in a transverse magnetic field," in: Semiconductor Silicon, Proc. of the IVth Int. Symp. on Silicon Mater. Sci. and Technol., Tokyo (1981), pp. 90-100.

34. K. M. Kim, "Suppression of thermal convection by transverse magnetic fields," *J. Electrochem. Soc.*, 129, No. 2, 427-429 (1982).

35. K. Hoshiawa, "Czochralski silicon crystal growth in the vertical magnetic field," *Jpn. J. Appl. Phys.*, 21, No. 9, L545-L547 (1982).

36. W. S. Laglois and K.-J. Lee, "Czochralski crystal growth in an axial magnetic field: effects of Joule heating," *J. Cryst. Growth*, 62, No. 4, 481-486 (1983).

37. W. E. Langlois and I. S. Walker, "Czochralski crystal growth in an axial magnetic field," in: Proc. IInd Int. Conf. Bound. and Inter. Layers – Comput. and Asympt. Methods, Trinity College, Dublin (1982), pp. 299-304.

38. H. Hirate and N. Inoue, "Study of thermal symmetry in Czochralski silicon melt under a vertical magnetic field," *Jpn. J. Appl. Phys.*, 23, No. 8, L527-L530 (1984).

39. Yu. M. Gel'fgat, V. A. Smirnov, I. V. Starshinova, and I. V. Fryazinov, "Numerical modeling of melt hydrodynamics during growth of single crystals by the Czochralski method in a magnetic field," in: XIth Riga Conf. on Magnetohydrodynamics, Vol. 11 [in Russian], (1984), pp. 139-142.

40. K. Hoshikawa, H. Konda, and H. Hirata, "Homogeneous dopant distribution of silicon crystal growth by vertical magnetic field applied Czochralski method," *Jpn. J. Appl. Phys.*, 23, No. 1, L37-L39 (1984).

41. K. Terashima and T. Fukuda, "A new magnetic field applied pulling apparatus for LEC GaAs single crystal growth," *J. Cryst. Growth*, 63, No. 3, 423-425 (1983).

42. K. Tetrashima, T. Katsumata, F. Orito, T. Kukuta, and T. Fukuda, "Electrical resistivity of undoped GaAs single crystals grown by magnetic field applied LEC technique," *Jpn. J. Appl. Phys.*, 22, No. 6, L325-L327 (1983).

43. J. Osaka, H. Konda, T. Kobayashi, and K. Hoshikawa, "Homogeneity of vertical magnetic field applied LEC GaAs crystal," *Jpn. J. Appl. Phys.*, 23, No. 4, L195-L197 (1984).

44. R. Miller, "Qualitative effects of oscillating magnetic fields on crystal melts," *J. Cryst. Growth*, 20, No. 2, 310-312 (1973).

45. P. S. Lykoudis, "Natural convection of an electrically conducting fluid in the presence of a magnetic field," *Int. J. Heat Mass Transfer*, 5, 23-25 (1962).

46. W. E. Langlois, "Convection of Czochralski growth melts," *Physicochem. Hydrodyn.*, 2, No. 4, 245-261 (1981).

47. L. G. Loitsyanskii, *Mechanics of Liquid and Gas* [in Russian], Moscow (1970).

48. W. Zulehner, "Czochralski growth of silicon," *J. Cryst. Growth*, 65, No. 2, 189-216 (1983).

49. K. I. Lee, W. E. Langlois, and K. M. Kim, "Digital simulation of oxygen transfer and oxygen segregation in magnetic Czochralski growth of silicon," *Physicochem. Hydrodyn.*, 5, No. 2, 135-141 (1984).

50. V. I. Polezhaev and A. I. Prostomolotov, "Study of hydrodynamic and heat and mass transfer processes during growth of crystals by the Czochralski method," *Izv. Akad. Nauk SSSR, Mekh. Zhidk. Gaza*, No. 1, 55-56 (1981).

51. V. A. Smirnov, I. V. Starshinova, and I. V. Fryazinov, "Analysis of rate, temperature, and concentration distributions of dopants in the melt during growth of single crystals by the Czochralski method," in: *Growth of Crystals*, Vol. 14, Consultants Bureau, New York (1987).

52. V. S. Zemskov, M. R. Raukhman, D. P. Mgaloblishvili, Yu. M. Gel'fgat, and M. Z. Sorkin, "Impurity distribution coefficients in the growth of indium antimonide single crystals under the action of a magnetic field on the melt," *Fiz. Khim. Obrab. Mater.*, No. 2, 64-67 (1986).

53. V. S. Zemskov, M. R. Raukhman, and D. P. Mgaloblishvili, "Effect of a magnetic field on impurity inhomogeneities in indium antimonide single crystals," *Fiz. Khim. Obrab. Mater.*, No. 5, 50-56 (1985).

54. H. C. Gatos, J. Lagowski, L. M. Pawlowcz, et al., "Crystal growth of GaAs in space," in: Proc. XIIIth Int. Conf. Def. Semiconductors, Colorado (1984), pp. 1-5.

55. J. Ochial, N. Kuwahara, M. Morioka, et al., Experimental Study on Marangoni Convection, COSPAR, Tokyo (1984), No. 25.

56. V. S. Zemskov, M. R. Raukhman, Yu. M. Gel'fgat, M. Z. Sorkin, and D. P. Mgaloblishvili, "Influence of a magnetic field on temperature oscillations in a melt and layered inhomogeneity in indium antimonide single crystals," *Fiz. Khim. Obrab. Mater.*, No. 3, 27-33 (1986).

GPSR Compliance

The European Union's (EU) General Product Safety Regulation (GPSR) is a set of rules that requires consumer products to be safe and our obligations to ensure this.

If you have any concerns about our products, you can contact us on ProductSafety@springernature.com

In case Publisher is established outside the EU, the EU authorized representative is:

Springer Nature Customer Service Center GmbH
Europaplatz 3
69115 Heidelberg, Germany

Batch number. 09625694

Printed by Printforce, the Netherlands